METAL-LIGAND INTERACTIONS
IN ORGANIC CHEMISTRY AND BIOCHEMISTRY
PART 1

THE JERUSALEM SYMPOSIA ON
QUANTUM CHEMISTRY AND BIOCHEMISTRY

Published by the Israel Academy of Sciences and Humanities,
distributed by Academic Press (N.Y.)

1st JERUSALEM SYMPOSIUM: *The Physicochemical Aspects of Carcinogenesis*
(October 1968)
2nd JERUSALEM SYMPOSIUM: *Quantum Aspects of Heterocyclic Compounds in*
Chemistry and Biochemistry (April 1969)
3rd JERUSALEM SYMPOSIUM: *Aromaticity, Pseudo-Aromaticity, Antiaromaticity*
(April 1970)
4th JERUSALEM SYMPOSIUM: *The Purines: Theory and Experiment*
(April 1971)
5th JERUSALEM SYMPOSIUM: *The Conformation of Biological Molecules and Polymers*
(April 1972)

Published by the Israel Academy of Sciences and Humanities,
distributed by D. Reidel Publishing Company (Dordrecht and Boston)

6th JERUSALEM SYMPOSIUM: *Chemical and Biochemical Reactivity*
(April 1973)

Published and distributed by D. Reidel Publishing Company (Dordrecht and Boston)

7th JERUSALEM SYMPOSIUM: *Molecular and Quantum Pharmacology*
(March/April 1974)
8th JERUSALEM SYMPOSIUM: *Environmental Effects on Molecular Structure and*
Properties (April 1975)

VOLUME 9

METAL-LIGAND INTERACTIONS IN ORGANIC CHEMISTRY AND BIOCHEMISTRY

PART 1

PROCEEDINGS OF THE NINTH JERUSALEM SYMPOSIUM ON
QUANTUM CHEMISTRY AND BIOCHEMISTRY HELD IN JERUSALEM,
MARCH 29TH–APRIL 2ND, 1976

Edited by

BERNARD PULLMAN

*Institut de Biologie Physico-Chimique
(Fondation Edmond de Rothschild) Paris, France*

and

NATAN GOLDBLUM

The Hebrew University, Hadassah Medical School, Jerusalem, Israel

D. REIDEL PUBLISHING COMPANY

DORDRECHT-HOLLAND/BOSTON-U.S.A.

Library of Congress Cataloging in Publication Data

Jerusalem Symposium on Quantum Chemistry and Biochemistry, 9th, 1976.
    Metal-ligand interactions in organic chemistry and biochemistry.

    (The Jerusalem symposia on quantitative chemistry and biochemistry; v. 9)
    Bibliography: p.
    Includes index.
        1. Ligands—Congresses. 2. Chemistry, Organic—Congresses.
    3. Biological chemistry—Congresses. I. Pullman, Bernard, 1919–
    II. Goldblum, Natan. III. Title. IV. Series.
    QD474.J47 1976   547'.1'2242   76–44910

The set of two parts: ISBN 90–277–0771–5

ISBN-13: 978-94-010-1172-3   e-ISBN-13: 978-94-010-1170-9
DOI: 10. 1007/978-94-010-1170-9

Published by D. Reidel Publishing Company,
P.O. Box 17, Dordrecht, Holland

Sold and distributed in the U.S.A., Canada, and Mexico
by D. Reidel Publishing Company, Inc.
Lincoln Building, 160 Old Derby Street, Hingham,
Mass. 02043, U.S.A.

TABLE OF CONTENTS

TABLE OF CONTENTS OF PART 2

PREFACE

The 9th Jerusalem Symposium was dedicated to the memory of Professor
Ernst David Bergmann. An imposing and deeply moving memorial session,
chaired by Professor Ephraim Katzir, the President of the State of Israel and a close friend of Professor Bergmann preceded the Symposium
itself. During this session, Professor Bergmann's personality, scientific achievements and contributions to the development of his country
were described and praised, besides President Katzir, by Professor A.
Dvoretzky, President of the Israel Academy of Sciences and Humanities,
Professor D. Ginsburg, Dean of the Israel Institute of Technology in
Haifa and the author of these lines.
     May I just quote short extracts from these speeches.

President Katzir: "As we open this ninth in the series of symposia
initiated in 1967, it is difficult for me as, I am sure, for many of
Ernst Bergmann's friends, co-workers and students, to be here without
him. He was not only a great scientist and a beloved teacher, he was
one of the most important founders of science in this country. To him
we owe many institutes and the establishment here of many branches of
science."

Professor Dvoretzky: "Ernst Bergmann's greatness did not stem from one
component overshadowing all the others. It was a multifaceted greatness consisting of the harmonious coalescing of seemingly contrasting
entities into a wonderful unity ... A deep universality dwelt in him
jointly and in full harmony with an ardent love for his people and nation. A deep respect for tradition and the heritage of the past, with
an admiration for progress and innovation. A faultless memory, with
originality. Bold initiative, with prudent sagacity. Great were his
achievements and no less great the achievements which he inspired others to make. He headed great projects, but he was equally ready to
lend active support to any project, no matter how small, which he considered deserving. His approach to public affairs was always imperson-
al and guided by sober analysis; yet he displayed a wonderful warmth
in his personal contacts. He was most demanding of himself and understanding towards his fellow men."

Professor Ginsburg: "It is not possible to encompass in a short session or even to attempt to do justice to the accomplishments of the
phenomenon named Ernst David Bergmann. Those of you who did not know
him should know, after all has been said and done, that the whole was
more than the sum of the descriptions."

May I also quote from my own speech: "I do not know any other man like Ernst. None, whose knowledge and wisdom, simplicity and depth, devotion to science and love of his country, were comparable to his. He was a unique combination of all these elements fused together by an exceptional human warmth, which irresistibly conquered all those who had the priviledge of approaching him and knowing him... I believe that during the last years of his life, the preparation, organization and yearly holding of the Jerusalem Symposia were for Ernst one of his preferred tasks. I also believe that this task was a joyful and pleasant one, although a hard one, for him. Just as he always did, in all his scientific works, he paid the utmost attention and attached the utmost importance to the perfection of the organization. He looked after all the details himself, always searching to improve, caring for everybody's satisfaction and pleasure. I know also, that the success and the growing renown of these Symposia were a great joy to him."

I wish once more to express my deep thanks to all those who in one way or another contribute to the possibility and success of the Jerusalem Symposia. Our deepest thanks are due to the Baron Edmond de Rothschild whose understanding and generosity make these Symposia possible and to the Israel Academy of Sciences and Humanities, and the Hebrew University of Jerusalem for their support, and collaboration.

                                                              BERNARD PULLMAN

SEARCH FOR GENERAL FEATURES IN CATION-LIGAND BINDING. *AB INITIO* SCF STUDIES OF THE INTERACTION OF ALKALI AND ALKALINE EARTH IONS WITH WATER

ALBERTE PULLMAN
Institut de Biologie Physico-Chimique, Laboratoire de Biochimie Théorique associé au C.N.R.S., 13, rue P. et M. Curie - 75005 Paris

We have investigated recently the possibility of utilization of SCF *ab initio* computation for the study of the interactions of alkali and alkaline-Earth cations with various ligands and explored the reliability in this respect of the possible combinations of different basis sets on the cation and the ligand. The technical aspect and conclusions of this study are presented elsewhere [1]. Broadly speaking it was shown that the poor results obtained with a standard minimal STO 3G basis set were due partly to the poor intrinsic representation it gives of the ligand and to the simultaneous overavailability of empty orbitals on the cation, leading to a spurious charge-transfer stabilization of the complex. Successive improvements of the representation of the cation and of the ligand have shown that reasonable binding characteristics may be obtained using a good atomic minimal basis.

In the afore-mentioned study, we have utilized the fact that the binding energy, ΔE, obtained in an *ab initio* SCF computation may be decomposed into its Coulomb — or pure electrostatic — , (E_C), exchange (E_{EX}), and delocalization components (E_{DEL}) as:

$$\Delta E = E_C + E_{EX} + E_{DEL}$$

where the last term comprises both the polarization and the charge-transfer effects [2, 3]. An examination of the values of the various components has shown the existence of interesting regularities in their evolution and of interrelations between the cations in their binding to a given ligand (which, in our model computations, was water). We would like to present here this aspect of the results which may be of direct interest to a number of participants in this Symposium.

Figure 1 gives the evolution of the Coulomb energy upon the approach to water of Na^+, K^+, Mg^{++} and Ca^{++}, compared to the same evolution computed for an approaching proton. It is observed that in all cases, E_C starts for quite large distances where it is the sole component of the binding energy. For the monocations, E_C remains identical to the corresponding value for the proton, until a distance at which the exchange energy starts to be non-zero. This corresponds to the moment where the electron clouds of the ion and of the water molecule begin to overlap appreciably. Below this distance, $|E_C|$ increases more rapidly than it does for a point charge, due to the so-called penetration effect, E_{PEN}

B. Pullman and N. Goldblum (eds.), Metal-Ligand Interactions in Organic Chemistry and Biochemistry, part 1 , 1 - 6. All Rights Reserved.
Copyright 1977 D. Reidel Publishing Company, Dordrecht-Holland.

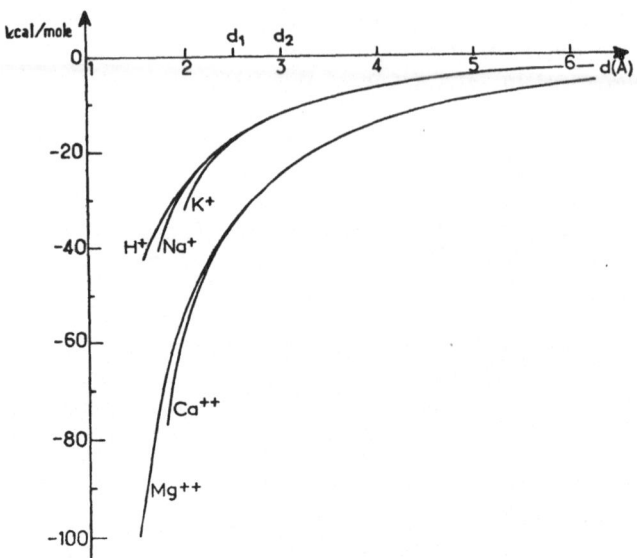

Fig. 1. Evolution of the *Coulomb* component of the interaction energy for Na$^+$, K$^+$, Mg^{++} and Ca^{++}. d_1 and d_2 indicate the distances where the exchange energy starts for Na$^+$ and K$^+$, respectively.

[3, 4]. For the dications, the value of E_C is exactly twice that for the monocations (or H$^+$), again until overlap starts and the penetration effect lowers the curves.

Then, for a given distance, the penetration term is in the order (in absolute value)

$$E_{PEN}\ (K^+) > E_{PEN}\ (Ca^{++}) > E_{PEN}\ (Na^+) > E_{PEN}\ (Mg^{++}) \tag{1}$$

The corresponding evolution of the exchange energy is shown in Figure 2.

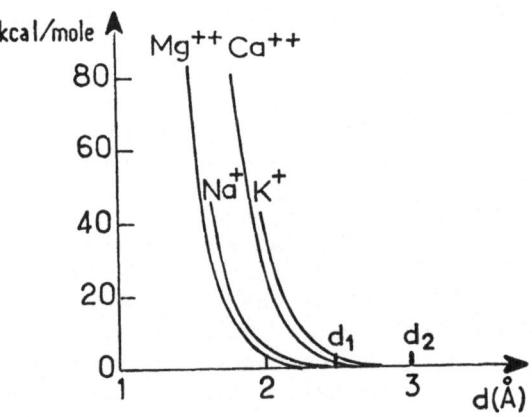

Fig. 2. Same as Figure 1 for the *exchange* component of the binding energy.

One observes that it starts first for K^+ (d_2), then for Ca^{++}, very close after, then later for Na^+ (d_1), closely followed by Mg^{++}, but the order of E_{EX} at a given distance is always:

$$E_{EX} (K^+) > E_{EX} (Ca^{++}) > E_{EX} (Na^+) > E_{EX} (Mg^{++}), \qquad (2)$$

the same as the order of decreasing $|E_{PEN}|$. Both E_{EX} and E_{PEN} are a direct function of the overlap between the electron cloud of water and that of the approaching cation but E_{EX} increases much more rapidly with approach than E_{PEN} [3] so that the global (exchange + penetration) — contribution, ϵ, is repulsive and in the order:

$$\epsilon (K^+) > \epsilon (Ca^{++}) > \epsilon (Na^+) > \epsilon (Mg^{++}) \qquad (3)$$

for a given distance. This order reflects the decrease in the extension of the electron cloud around the cations as shown by the increasing exponents of the occupied 2sp or 3sp shell [1]. The numerical values and the distance where exchange and penetration begin differ according to the basis set used in the computations but the qualitative facts of the evolution remain the same.

The remaining constituent of the binding energy is the delocalization term. As indicated by the curves of Figure 3, this (attractive) term starts relatively early, due to the polarization effect, as observed in the approach of a proton [5] and upon formation of a hydrogen bond [3]. At large distances ($d>d_2$) it is roughly twice (or somewhat more than twice as large for the divalent ions than for the monovalent ones. Its detailed evolution upon closer approach becomes more difficult to analyze on accoun of the parasit charge transfer brought about by the basis set extension error [1]. But the best of our computations and those using more extended basis [6, 7, 8] indicate that it remains a small part of the total interaction.

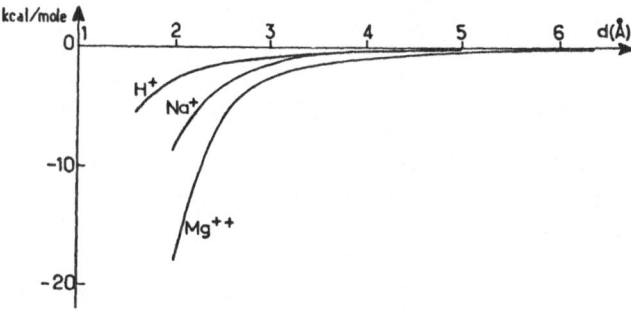

Fig. 3. Same as Figure 1 for the *delocalization* component of the binding energy.

This set of observations allows a rationalization of the order to be expected of the binding energies for the various cations. In the first place, inside a given series, say Li^+ to K^+, the attractive part of the energy being the same at large distances, and the repulsion starting in the order $K^+ > Na^+ > Li^+$, the global binding energies $|\Delta E|$ will be in decreasing order:

$$Li^+ > Na^+ > K^+ \tag{4}$$

with the converse order for the equilibrium distances. A similar relationship exists inside the series Be^{++} to Ca^{++}. Moreover the interrelations seen in the attractive components for the pairs Ca^{++}/K^+ and Mg^{++}/Na^+, and the fact that the repulsion for the monocation of the pair is greater than that for the dication, imposes for $|\Delta E|$:

$$Ca^{++} > 2K^+, \quad \text{and} \quad Mg^{++} > 2Na^+ \tag{5}$$

Thus, *a fortiori*:

$$Ca^{++} > K^+, \quad \text{and} \quad Mg^{++} > Na^+ \tag{6}$$

These interrelations are compatible either with the order:

$$Mg^{++} > Ca^{++} > Na^+ > K^+ \tag{7}$$

or

$$Mg^{++} > Na^+ > Ca^{++} > K^+ \tag{8}$$

for $|\Delta E|$.

But it is easily seen that the attractive component of the energy for Na^+ does not increase rapidly enough upon decreasing distance to allow the corresponding binding energy to trespass that of Ca^{++} *, so that it is the order (7) which occurs.

For the same reasons Li^+ should still remain above Ca^{++} and the order of $|\Delta E|$ for the whole series should be:

$$Be^{++} > Mg^{++} > Ca^{++} > Li^+ > Na^+ > K^+ \tag{9}$$

(For larger and larger dications this may not be true anymore and Li^+ may pass below).

The order found is independent of the nature of the ligand approaching the cation (provided the site is well exposed [10]). It explains the fact that the binding energies to carbonyl [11] and phosphate groups

---

* If d is the equilibrium distance for Ca^{++} and d' that for Na^+, the attraction for Na^+ at d' (roughly $1/d'$) should be greater than that for Ca^{++} at d (roughly $2/d$) for an inversion to occur. This would require $d' < d/2$ which obviously cannot occur if one considers the relative location of d_1 and d_2 where the repulsion starts for the mono- and dications respectively.

[12, 13, 14] have been found in the same order as the binding energies to water for the series considered.

REFERENCES

 1. Pullman, A., Berthod, H., and Gresh, N.: *Intern. J. Quant. Chem. Symp.* 10, 1976, in press.
 2. Dreyfus, M., and Pullman, A.: *C.R. Acad. Sci. Paris* 271, 457 (1970).
 3. Dreyfus, M., and Pullman, A.: *Theor. Chim. Acta* 19, 20 (1970).
 4. Howard, B.B.: *Chem. Phys.* 39, 2524 (1963).
 5. Pullman, A.: *Chem. Phys. Letters* 20, 29 (1973).
 6. Ruessegger, P., and Schuster, P.: *Chem. Phys. Letters* 19, 254 (1973).
 7. Schuster, P., Marius, A., Pullman, A., and Berthod, H.: *Theor. Chim. Acta* 40, 323 (1975).
 8. Diercksen, G.H.F., Kraemer, W.P., and Roos, B.O.: *Theor. Chim. Acta* 36, 249 (197).
 9. Beyer, A., Lischka, H., and Schuster, P.: *Theor. Chim. Acta,* in press.
10. Perahia, D., Pullman, B., and Pullman, A.: in preparation.
11. Perricaudet, M., and Pullman, A.: *FEBS Letters* 34, 222 (1973).
12. Pullman, B., Gresh, N., and Berthod, H.: *Theor. Chim. Acta* 40, 71 (1975).
13. Pullman, A., and Pullman, B.: these Proceedings.
14. Marynick, D.S., and Schaeffer III, H.F.: *Proc. Natl. Acad. Sci. U.S.* 10, 3794 (1975).

DISCUSSION

Veillard: A proposal which has been advocated in order to remedy to the error associated with the use of small basis sets (for instance for the calculation of hydrogen bond energy or conformational barrier) is the use of a counterpoise basis set. Namely the energy of each constituent of a supermolecule is computed separately with the *supermolecule basis set*. Then the binding energy is obtained as the difference of the energies of the supermolecule and of the fragments computed with this same basis set. Do you know if this procedure has been used for the calculation of hydration energies with small basis sets and what are the results?

A. Pullman: The counterpoise method could be used in this field in order to try to improve the poor results of the STO 3G basis. Aside from the fact that the procedure generally gives too high a correction, it has the inconvenience to increase the computation time because it increases the number of basis functions. Thus we did not try to implement this kind of improvement but went rather in the direction I indicated.

Sue Hanlon: Would you anticipate that the same order of binding energies $(Be^{++} > Mg^{++} > Cu^{++} > Li^+ > Na^+ > K^+)$ appropriate for interaction with H_2O would also obtain for interactions with organic phosphates?

A. Pullman: Yes, with the restriction mentioned in my talk, concerning the definition of the site. In fact I have the values computed for DMP^-

(dimethylphosphate anion taken as a model of the phosphodiester linkage) in the gauche-gauche, (gg) gauche-trans (gt) and trans-trans (tt) conformations, for binding of Na^+, K^+, Mg^{++}, Ca^{++} and the values obtained do indead show the order indicated. Here is the set of values of $-\Delta E$(kcal mole)

	gg	gt	tt
Mg^{++}	340.1	335.2	330.0
Ca^{++}	285.0	280.3	275.2
Na^+	150.3	148.3	146.0
K^+	128.3	126.3	124.0

But it must be mentioned again that the binding site is the same in all cases, namely the most stable position found for the cation on the bissectrix of the O_1PO_3 angle (O_1, O_3 being the anionic oxygens) (Pullman, B., Pullman, A., Gresh, N., and Berthod, H.: *Theor. Chim. Acta*, in press.

Rode: The influence of correlation is not important in the case of Li^+/H_2O as you have mentioned. Do you think that we can extrapolate this statement to other ligands, which may contain also π-electron systems, where possibly a considerable electron rearrangement will take place upon ion bonding?

A. Pullman: For the time being the correlation effect has been computed in the case of Li^+ binding to water only. Very recently a similar result has been found for NH_3-Li^+ (W. Goddard III *et al.*, in press). One may hope that this may be generalized, but there is no certitude. Correlation effects are very difficult to extrapolate from one case to another. More computations of the kind that Dr. Diercksen has performed for Li^+H_2O should be performed on typical systems.

CHELATE METAL COMPLEXES OF PURINES, PYRIMIDINES AND THEIR
NUCLEOSIDES:
METAL-LIGAND AND LIGAND-LIGAND INTERACTIONS

THOMAS J. KISTENMACHER and LUIGI G. MARZILLI
*Department of Chemistry, The Johns Hopkins
University, Baltimore, Md. 21218, U.S.A.*

ABSTRACT. Structural, synthetic and spectroscopic studies on chelate
metal complexes of purines, pyrimidines and their nucleosides are
discussed. All these studies suggest that chelate metal complexes can,
via interligand interactions, react selectively with nucleic acid
components.

1. INTRODUCTION

Considerable attention has recently been focused on the binding of
metal ions (Eichhorn, 1973) and metal complexes (Marzilli, 1977) to
purines, pyrimidines, nucleosides and nucleotides and nucleic acids.
The motivations for these investigations are manifold:
      (a) Metal ions mediate the biological function of nucleotides and
nucleic acids;
      (b) Heavy-metal complexes can be used for isomorphic replacement
in X-ray crystallographic studies of tRNA and for sequencing investigations by electron microscopy;
      (c) Advantageous properties of metal centers (paramagnetism, high
specific density, affinity for donor ligands and solid supports) can be
exploited in numerous studies into (i) the mechanism of enzyme reactions, (ii) the structure of nucleic acids, and (iii) the purification
and isolation of nucleic acids.
      (d) Some complexes, wich are effective anti-tumor agents, probably
interact with the nucleic acids in tumor cells.
      Selective reactions of a metal-containing species with the common
bases, Figure 1, of the nucleic acids is essential in (a), (b) and (c)
above. Nonetheless, most studies have relied on the naturally greater
affinity of metal species such as Hg^{2+} for one of the given heterocyclic
bases (thymidine, in the case of Hg^{2+}).
      The electrostatic potentials at the various nitrogen binding sites
do not markedly differ (Bonaccorsi *et al.*, 1972), and it is clear that
the affinity of metal ions and metal complexes for the base binding
sites cannot solely be a function of the nucleophilicity of the coordination site (Eichhorn, 1973). Our approach to solving the selectivity

*B. Pullman and N. Goldblum (eds.), Metal-Ligand Interactions in Organic
Chemistry and Biochemistry, part* 1, *7 - 40. All Rights Reserved.
Copyright © 1977 by D. Reidel Publishing Company, Dordrecht-Holland.*

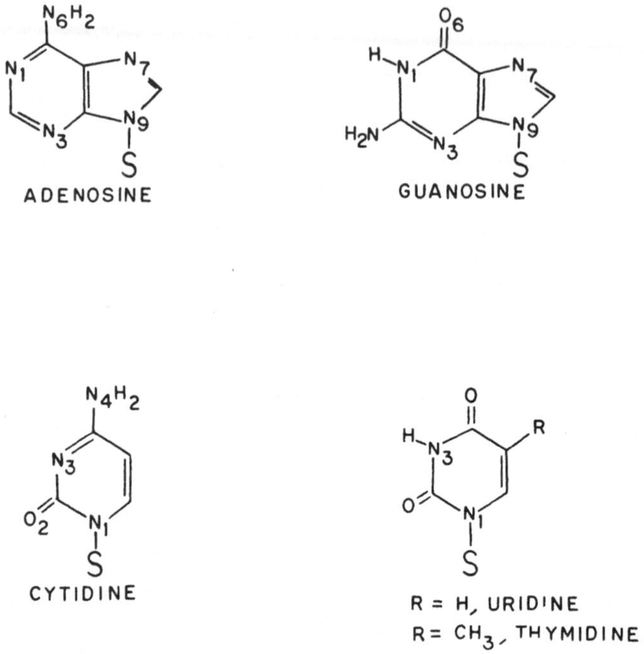

Fig. 1.   The five common purine and pyrimidine nucleosides.
In any case, S = the ribosyl functional moiety.

problem is to utilize the interactions between chelate ligands and the
exocyclic functional groups which border potential nitrogen base binding sites on the nucleosides. These interactions can be favorable (interligand hydrogen bonding) or unfavorable (nonbonding repulsions).
The bases themselves display a variety of potential nitrogen binding sites, Figure 1. Adenosine is the most complex in this respect with
potential binding sites at N(1), N(3) and N(7). It is probable, however, that the presence of the ribose group at N(9) will sterically
hinder complexation at N(3). Thus, N(1) and N(7) are the most likely
binding sites on adenosine. Guanosine, like adenosine, presents similar
possibilities with N(7) as the most likely binding site, but with N(1)
as a possibility if deprotonation accompanies complexation. Cytosine
has only one available heterocyclic nitrogen, N(3), and it is likely
that this site will be used for metal binding. Thymidine(uridine) is
unique among the four bases in that there are no available nitrogen
base binding sites for metal complexation, unless deprotonation accompanies complexation, Figure 1.
We will discuss in this paper the results of our crystallographic
investigations which serve to delineate the types of interactions which
can and do occur between chelate ligands and the exocyclic functional
groups on the bases, and our solution studies which we feel clearly
indicate that such interactions can lead to selectivity in base binding.

2. METAL CENTERS

We will confine our discussion to two transition metal centers, copper (II) and cobalt(III). There are major differences in the conformational, chemical and physical properties of complexes containing these two transition metals, as outlined in (a) and (b) below.

(a) *Copper(II)*

Copper in its +2 formal oxidation state is a $3d^9$ system. Generally, the coordination geometry about Cu(II) is described as square planar, but significant coordination affinity exist along the open axial positions (Freeman, 1967; Hathaway, 1973) leading to the so-called (4+1) and (4+1+1) coordination geometries. The metal-ligand distances in the axial positions are usually 0.2—0.9 Å longer than the metal-ligand distances in the equatorial coordination positions (Freeman, 1967; Hathaway, 1973). Cu(II) complexes are normally paramagnetic and substitution labile.

(b) *Cobalt(III)*

Cobalt in its formal +3 oxidation state is a $3d^6$ system. The normal coordination geometry is octahedral with all coordination sites of approximately equivalent affinity and metal-ligand bond distance. Co(III) complexes are normally diamagnetic and substitution inert.

3. CHELATE LIGANDS

Chelate complexes offer a wide variety of attractive features: (1) normally chelate complexes are more stable to ligand substitution reactions than complexes containing unidentate ligands of comparable ligand field strength; (2) chelate ligands offer the possibility of restricting the number of available coordination position about the metal center; (3) chelate ligands have in many cases a wide flexibility in terms of modification of the ligand near or far from the metal center; (4) suitable chelate ligands exist or can be constructed which offer a wide variety of potential interactions with the heterocyclic functional groups on the nucleic acid constituents.

   The ligands we have employed in our solution and structural studies are presented in Figure 2. The chelate ligands shown have a variety of interesting features. They range from bidentate ligands (en, acac, dmg), through tridentates (dien, glygly, Mesal, Mebenzo) to quadradentate (trien). The arrangement of the ligands in Figure 2 follows essentially along the following lines. The first three ligands are all polyamine ligands which have N-H groups near the metal center which are capable of acting as donors in an interligand hydrogen bond. The second three ligands, the glycylglycine dianion and the two Schiff base systems, offer to an incoming ligand both a functional group which is a potential hydrogen bond donor, the primary or secondary amine groups, and a functional group which is a potential hydrogen bond acceptor, the coordinated carboxylate oxygen on glygly or the coordinated salicylidene

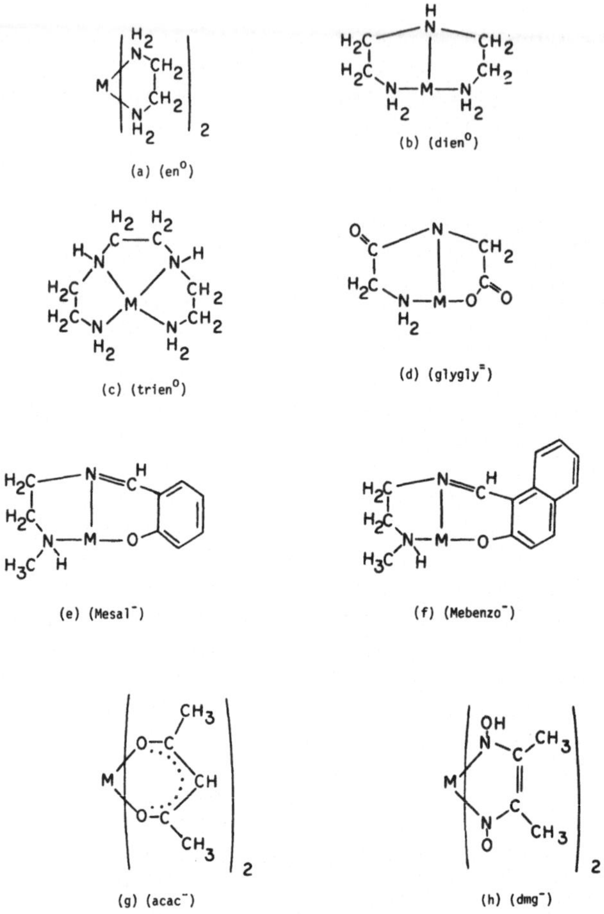

Fig. 2.   Chelate ligands employed in these studies.

oxygen of the Schiff base chelates. Finally, the last two ligands offer solely hydrogen bond acceptor groups for interaction with incoming ligands.

    We have investigated the question of whether or not interligand interactions can induce selectivity. We will first survey the range of interligand interactions as established by X-ray crystallography.

4. CRYSTALLOGRAPHIC STUDIES

4.1. Binding Interactions in Adenine Derivatives

As we have noted above, adenine derivatives have the largest number of unprotonated, heterocyclic donor-nitrogen atoms [N(1), N(3) and N(7)] available for metal coordination. The masking of N(9), by a methyl group as in 9-methyladenine or a ribose moiety as in adenosine, appears to sterically hinder metal coordination at the pyrimidine site N(3). In N(9) substituted molecules, then, the likely metal binding sites are N(1) and N(7). P.m.r. line broadening studies (Berger and Eichhorn, 1971; Eichhorn *et al.*, 1966) and logitudinal relaxation data (Marzilli *et al.*, 1975) clearly indicate that these sites are employed by Cu(II) complexes. Similar conclusions have been drawn for the binding of Pd(II)- and Pt(II)-chelate complexes to adenosine (Marzilli, 1977). There are also a variety of crystallographic studies which support the contention that (N(1) and N(7) are likely binding sites for metal ions in N(9)-substituted adenine compounds (Hodgson, 1977).

We will briefly describe the molecular conformations and inter-ligand interactions we have observed in several chelate complexes of adenine derivatives.

(a) [(Glycylglycinato)(Aquo)(9-Methyladenine)Copper(II)]

This complex exists in the crystal in a slightly distorted square-pyramidal geometry, Figure 3 (Kistenmacher *et al.*, 1976). The four equatorial sites are occupied by the tridentate glycylglycine dianion,

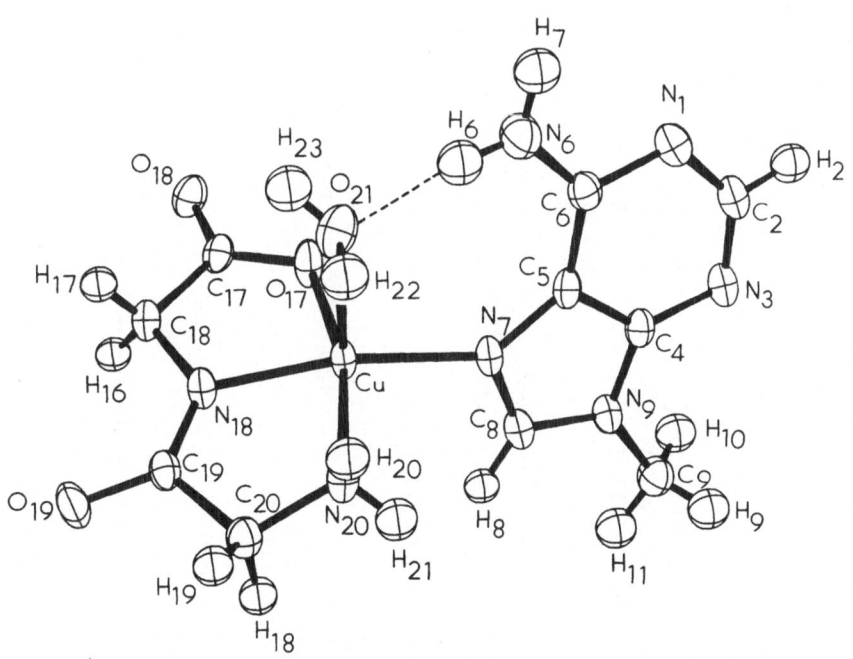

Fig. 3. The molecular structure of [(glycylglycinato)(aquo) (9-methyladenine) copper(II)].

ligand (d) in Figure 2, and N(7) of the 9-methyladenine ligand, while one of the axial positions is occupied by a water molecule. As we have

TABLE I

Parameters in the Interligand Hydrogen Bonds in Chelate Metal Complexes of Purines and Pyrimidines

Complex	D	A	D...A	H...A	∠D-H...A
(a) *Adenine Residues*					
[(glygly)(H$_2$O)(9-MeAd)Cu(II)][1]	N(6)H[1]	O(H$_2$O)	2.820A	1.96A	168°
[(Mesal)(H$_2$O)(9-MeAd)Cu(II)][2]	N(6)H[2]	O(Mesal) AX	2.870	1.93	161
[(acac)(NO$_2$)(dAdo)Co(III)][3]	N(6)H[3]	O(acac) EQ	2.84	2.13	139[AVE]
	N(6)H	O(acac)'	3.04	2.38	134[AVE]
(b) *Theophylline Anion Residues*					
trans[(Cl)(Theo)(en)$_2$Co(III)][+]	NH(en)[4]	O(6)	2.824	2.04	161
	NH(en)[5]	O(6)	2.976	2.19	150
cis[(Cl)(Theo)(en)$_2$Co(III)][+]	NH(en)[5]	O(6)	2.942	2.14	153
	NH(en)	O(6)	2.814	1.99	158
[(Mesal)(Theo)Cu(II)]	NH(Mesal)[6]	O(6)	2.839	2.05	150
[(Mebenzo)(Theo)]	NH(Mebenzo)[7]	O(6)	3.103	2.29	152
[(Theo)$_2$(dien)Cu(II)]	OH(water)	O(6)	2.868	1.88	159
	NH(dien)[8] AX	O(6) EQ	2.996	2.31	136
(c) *Cytosine Residues*					
[(Mesal)(Cy)Cu(II)][+]	NH(Mesal)[9]	O(2)	3.065	2.30	134

[1] Figure 3
[2] Figure 4
[3] Figure 6
[4] Figure 9
[5] Figure 8
[6] Figure 10
[7] Figure 11
[8] Figure 12
[9] Figure 17

noted, solution studies clearly indicate that N(1) and N(7) are likely binding sites — with N(7) perhaps being slightly favored. The binding mode, Cu(II)-N(7), found in the present structure should not be construed as being indicative of any dramatic difference in the coordination affinities of N(1) and N(7) for N(9)-substituted adenine residues. The more likely explanation for the observance of the N(7) isomer may be associated with the axial water ligand. In the observed conformation of the complex, the exocyclic amine, N(6)H$_2$, on the 9-methyladenine ligand forms an interligand hydrogen bond with the coordinated water molecule, the interligand hydrogen bond being denoted by a dashed line in Figure 3. The parameters in this hydrogen bond are in accord with a strong hydrogen bond, Table I.

(b) [N-Salicylidene-N'-Methylethylenediamine) (Aquo) (9-Methyladenine) Copper(II)]$^+$

This complex cation shows approximately square-pyramidal coordination geometry in the solid, Figure 4 (Szalda et al., 1975). The four equatorial positions are occupied by the tridentate Schiff base chelate, ligand (e) Figure 2, and N(7) of the 9-methyladenine. The primary coordination

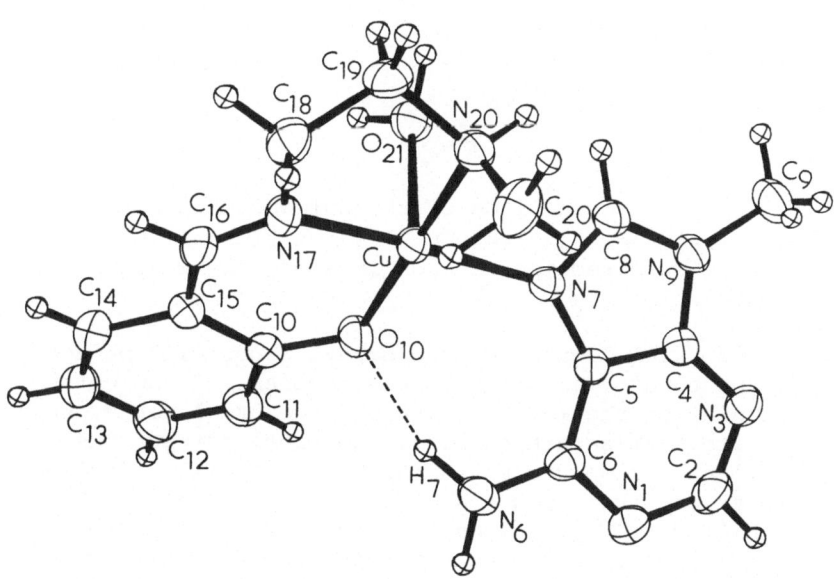

Fig. 4.   The molecular structure of the [(N-salicylidene-N'-methylethylenediamine) (aquo) (9-methyladenine) copper(II)]$^+$ cation.

sphere is completed by the occupation of one of the axial positions by a water molecule. The most interesting aspect of the structure is that an interligand hydrogen bond is formed between the exocyclic amine,

N(6)H$_2$, of the 9-methyladenine and the equatorial oxygen atom, O(10),
of the salicylidene ring. The parameters in this interligand hydrogen-
bond system, Table I, are again consistent with a strong hydrogen bond.
      It is interesting to compare the interligand hydrogen bonding found
in this Schiff base complex, involving the equatorial salicylidene
oxygen, and that found above in the glycylglycinatocopper (II) complex,
where the axial water acts as the acceptor, see above. The reason for
the change in the hydrogen bond acceptor site in the Schiff base complex
is probably related to the presence of the methyl substituent on the
terminal amine nitrogen of the Schiff base chelate. The role of this
substituent in determining the molecular conformation and the interligand hydrogen bonding site appears to be twofold: (1) As can be seen
in Figure 5, which compares the glygly and Mesal complexes, the methyl
group in the Schiff base complex appears to strongly hinder 'on axis'
coordination at one of the axial positions. It is likely then that a

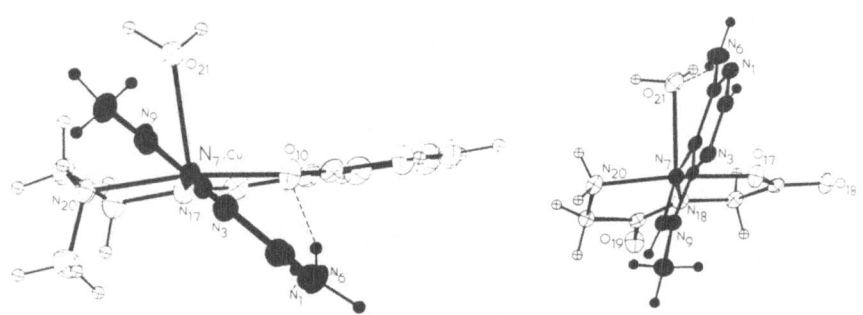

Fig. 5.   A comparison of the molecular structures of
[(glygly)(H$_2$O)(9-MeAd)Cu(II)] and [(mesal)(H$_2$O)(9-MeAd)Cu(II)]$^+$
The view direction in each is along the N(7)–Cu(II) vector.

strongly bound axial ligand and the methyl group will be anti with
respect to the equatorial plane of the complex. (2) Accepting this
anti-positioning of the methyl group and the axial water, space filling
models indicate that severe nonbonded contacts between the
N-methylethylenediamine group and the 9-methyladenine framework would
occur in an attempt to produce a molecular conformation which would
lead to an N(6)-H...OH$_2$ (axial water) hydrogen bond system. The equatorial
salicylidene oxygen provides an alternate acceptor site which is then
used in the formation of the interligand hydrogen bond. The net result
is an energetically favorable situation in which both an axial ligand
is present in the complex as well as the interligand hydrogen bond.

(c) [Bis(acetylacetonato)(Nitro)(Deoxyadenosine)Cobalt(III)]

The molecular conformation of this cobalt(III) complex is illustrated
in Figure 6 (Sorrell et al., 1976). The coordination sphere is approximately octahedral with the equatorial plane defined by the two acac

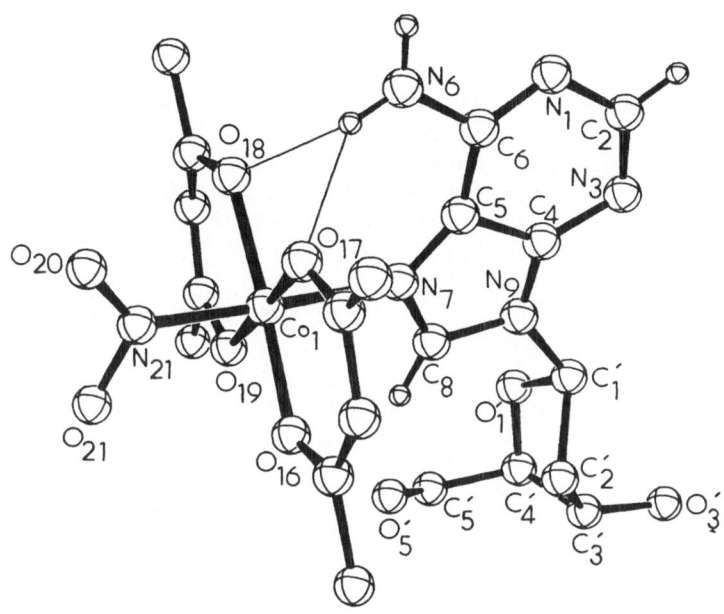

Fig. 6.   The molecular structure of [bis(acetylacetonato)
(nitro)(deoxyadenosine)cobalt(III)].

ligands, ligand (g) Figure 2, in *trans* positions and the two axial
positions occupied by the nitro and the deoxyadenosine ligand. The
nucleoside is coordinated to the metal through N(7), and this complex
represents the first structural study of a nucleoside coordinated to
Co(III). The conformation of the complex is such that the plane of the
adenosine ligand approximately bisects the O–Co–O bonds of two of the
acac oxygen atoms; in the adopted conformation the exocyclic amine
N(6)H$_2$ forms a bifurcated hydrogen bond system to two of the acac
oxygens, Table I and Figure 6.

Summary

The important molecular features of the above three substituted adenine
systems are that in each case the metal binding site is N(7) and in
each instance the exocyclic amine group on the adenine framework is
involved in an interligand hydrogen bond with an acceptor group on
another ligand in the coordination sphere. We have not isolated any
N(1) complexes thus far, but it is probable that metal binding at this
site could also be accompanied by interligand hydrogen bonding *via* the
amino group. The main feature then that distinguishes adenine-like
purines is the presence of the amine function near the base binding
sites.

4.2. *Binding Interactions in Guanine Derivatives*

As we have discussed above, N(7) is the most likely metal binding site
in guanine derivatives, but N(1) is a possibility if deprotonation
accompanies complexation. In all of our studies to date we have employed
the monoanion of the substituted purine theophylline as a model for
metal binding to the guanine class of purines, i.e., those purines
with a carbonyl group at position 6 of the pyrimidine ring. Figure 7
illustrates the molecular structures of both theophylline and guanine.

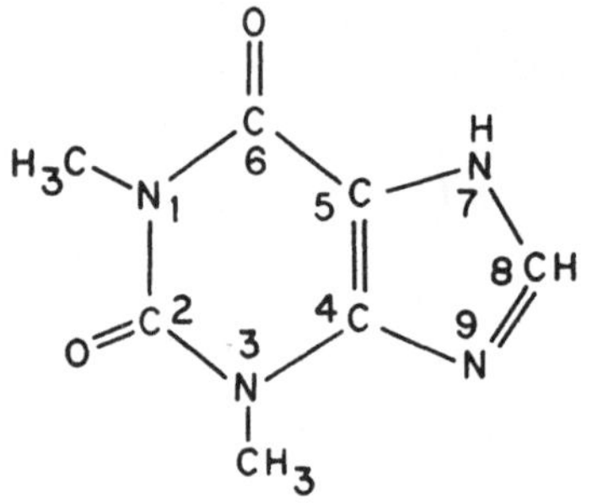

Fig. 7. A comparison of the molecular structures of
theophylline and guanine.

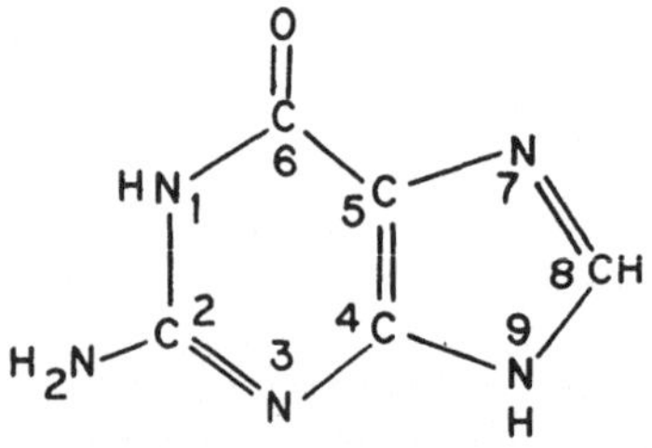

The important feature to note is that when a metal coordinates to the
theophylline monoanion through N(7), the steric and interligand hydrogen bonding potential near the metal-binding site of theophylline and
guanine are very similar. There have now been a variety of crystallographic studies on metal binding to guanine derivatives, and in all
known instances, where N(9) is blocked, the metal binding site is N(7)
(Hodgson, 1977). It should be noted that in 6-thiopurines evidence has
been presented that the thio function can also bind to the metal center
forming a five-membered chelate ring (Hodgson, 1977).

(a) *Cis* and *Trans*[(Chloro)(Theophyllinato)bis(Ethylenediamine)Cobalt
     (III)]$^+$

We have isolated and determined the molecular structure of both the
cis (Marzilli *et al.*, 1974; Kistenmacher and Szalda, 1975) and the *trans*

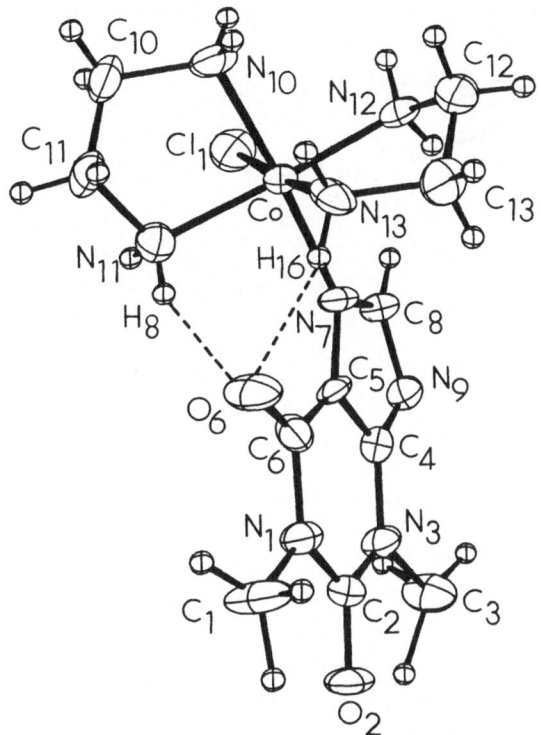

Fig. 8.   The molecular structure of the *cis*[(chloro)
(theophyllinato) bis (ethylenediamine)cobalt(III)]$^+$ cation.

Marzilli *et al.*, 1973; Kistenmacher, 1975) isomers of the theophylline
monoanion complex of [(en)$_2$(Cl)Co(III)]$^{2+}$, where en is ligand (a) in
Figure 2. The molecular conformations and bonding characteristics of
the *cis* isomer are presented in Figure 8, while those of the *trans*
isomer are presented in Figure 9. In each isomer the metal binding site
is N(7) on the theophylline monoanion, and there are two interligand
hydrogen bonds from different ethylenediamine ligands to O(6) on the
substituted purine, Figures 8 and 9 and Table I.

(b) [(N-Salicylidene-N'-Methylethylenediamine)(Theophyllinato)Copper(II)]

A perspective view of the complex [(Mesal)(theophyllinato)copper(II)],
where Mesal is ligand (e) in Figure 2, is presented in Figure 10. The
copper is approximately square planar with the tridentate Schiff base
ligand Mesal and N(7) of the theophyllinato moiety occupying the four
coordination sites (Kistenmacher *et al.*, 1975a). An interligand hydrogen
bond exists in this complex, Figure 10, between the carbonyl oxygen
O(6) and the secondary amine group at the N-methylethylenediamine
terminus of the Schiff base. The parameters in this interligand hydrogen
bond, Table I, are consistent with a strong hydrogen bond.

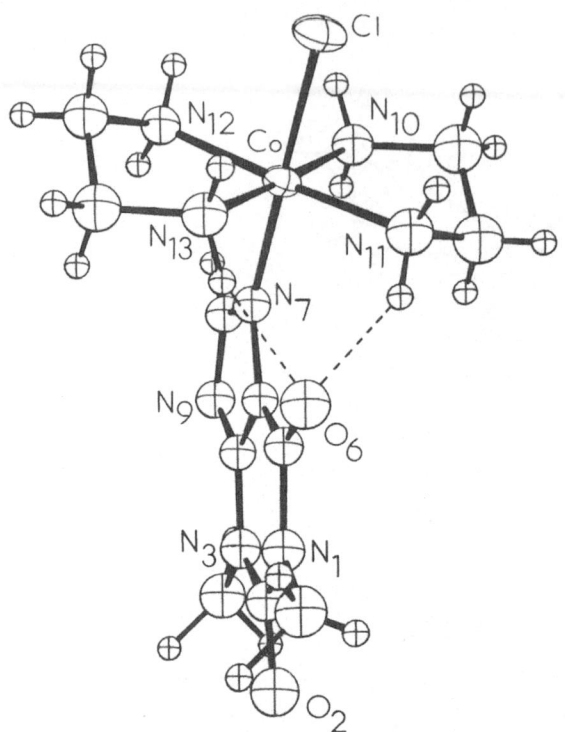

Fig. 9. The molecular structure of the *trans*[(chloro)
(theophyllinato) bis (ethylenediamine)cobalt(III)]⁺ cation.

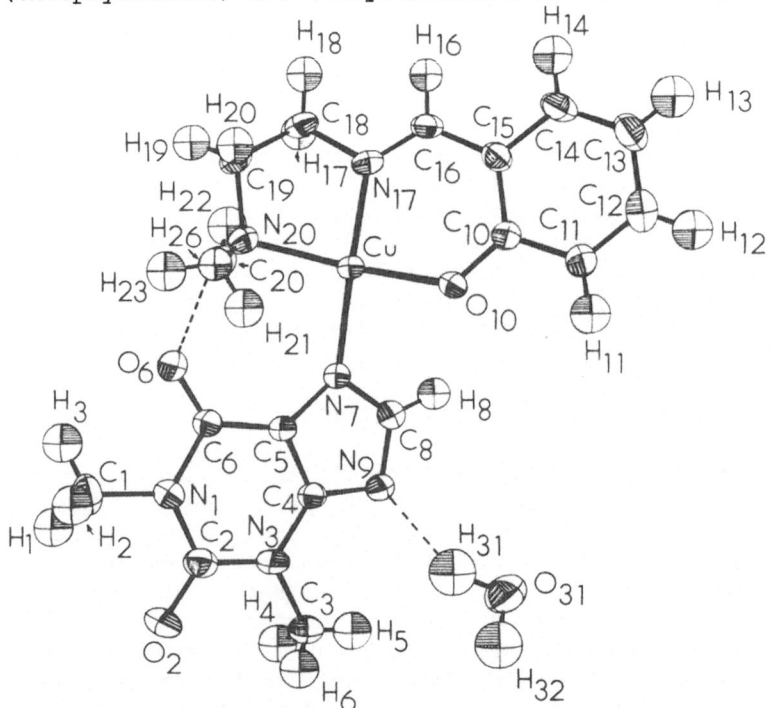

Fig. 10. The molecular structure of [(N-salicylidene-
N'-methylethylenediamine)(theophyllinato)copper(II)].

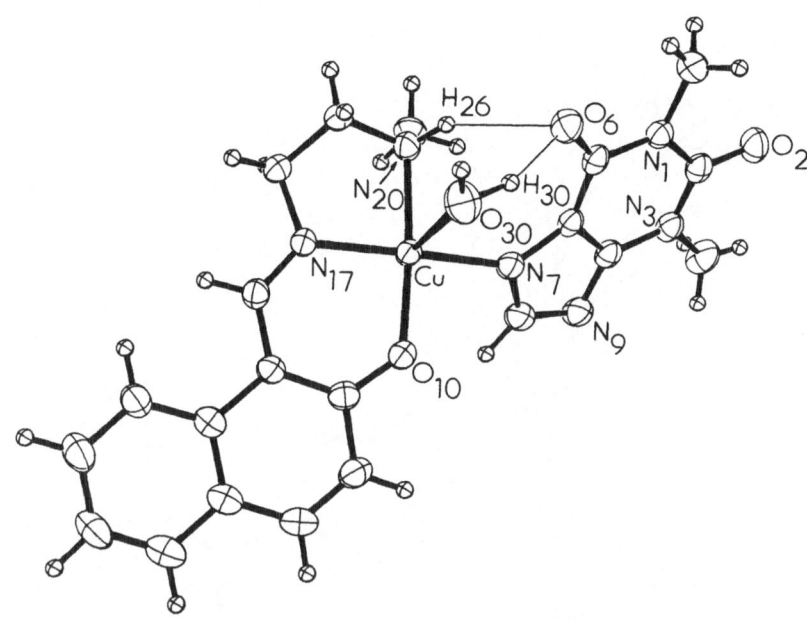

Fig. 11.   The molecular structure of [(N-3,4-benzosalicylidene-
N'-methylethylenediamine)(theophyllinato)(aquo)copper(II)].

(c) [(N-3,4-Benzosalicylidene-N'-Methylethylenediamine)(Theophyllinato)
     (Aquo)Copper(II)]

The molecular structure of this Schiff base copper(II) complex of the
theophylline monoanion is presented in Figure 11. The copper (II) is
in a square-pyramidal environment with the tridentate Schiff base
chelate Mebenzo, ligand (f) Figure 2, and N(7) of the theophylline
anion in equatorial positions and a water molecule in one of the axial
positions (Szalda *et al.*, 1976a). There are two interligand hydrogen
bonds to O(6) of the purine ring, one from the secondary amine of the
ethylenediamine terminus of the Schiff base chelate and one from the
axially bound water molecule, Table I.

(d) [Bis (Theophyllinato)(Diethylenetriamine)Copper(II)]

The molecular conformation of the bis(theophyllinato)(diethylenetriamine)
copper(II) complex is illustrated in Figure 12. The primary coordination
sphere about the copper (II) center is approximately square pyramidal
(Sorrell *et al.*, 1976). The tridentate chelate diethylenetriamine,
ligand (b) in Figure 2, with its terminal amine nitrogen atoms in the
commonly observed *trans* positioning, occupies three of the four equatorial coordination sites. The coordination sphere is completed by a
strongly bound equatorial theophylline monoanion, Cu(II)-N(7) bond
length = 2.007(3)A, and a weaker bound axial theophylline anion, Cu-N(7)
bond length = 2.397(3)A. The binding of both of the purine anions
through the imidazole ring nitrogen N(7) is of particular interest.

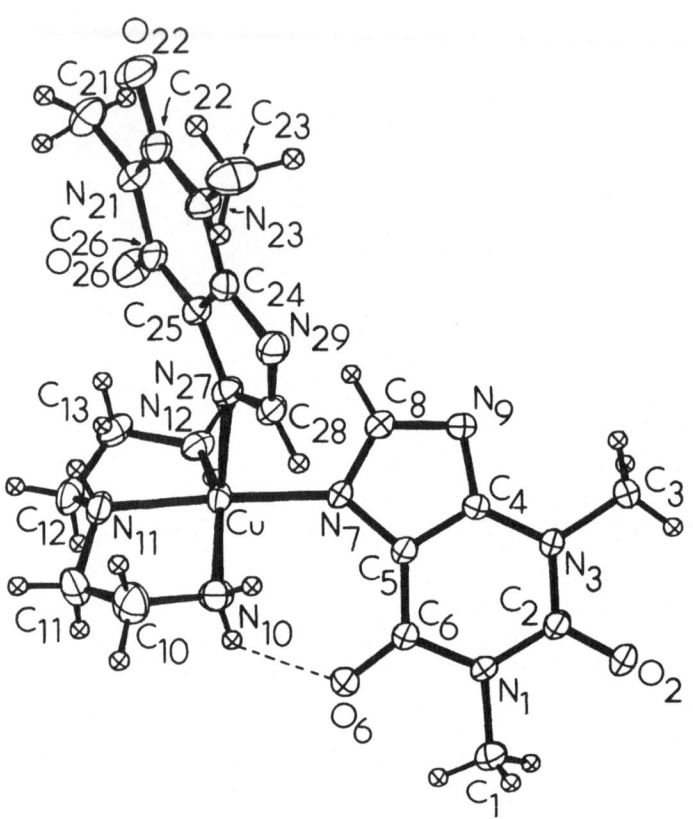

Fig. 12. The molecular structure of [bis(theophyllinato)
(diethylenetriamine)copper(II)].

There are no particular steric interactions between the two purine
ligands. In fact, the relative orientation of the axial theophylline
anion seems to be primarily dictated by interactions with the dien
chelate. As in the above complexes the carbonyl oxygen O(6) of the
equatorial purine anion is hydrogen bonded to one of the primary amine
groups on the chelate ligand, Figure 12 and Table I.

(e) [(N-3,4-Benzosalicylidene-N',N'-Dimethylethylenediamine)
    (Theophyllinato)Copper(II)]

A perspective view of this complex is presented in Figure 13. In this
complex the Schiff base chelate shown in Figure 2(f) has been modified
by the addition of a second methyl group to the N-methylethylenediamine
terminus; thus this chelate has no hydrogen bond donor near the metal
binding site. The theophylline anion complex of this system has square
pyramidal geometry about the Cu(II) with the Schiff base chelate occupying three of the four equatorial sites and N(7) of the purine anion
in the fourth equatorial site (Szalda, *et al.*, 1976b). The axial position

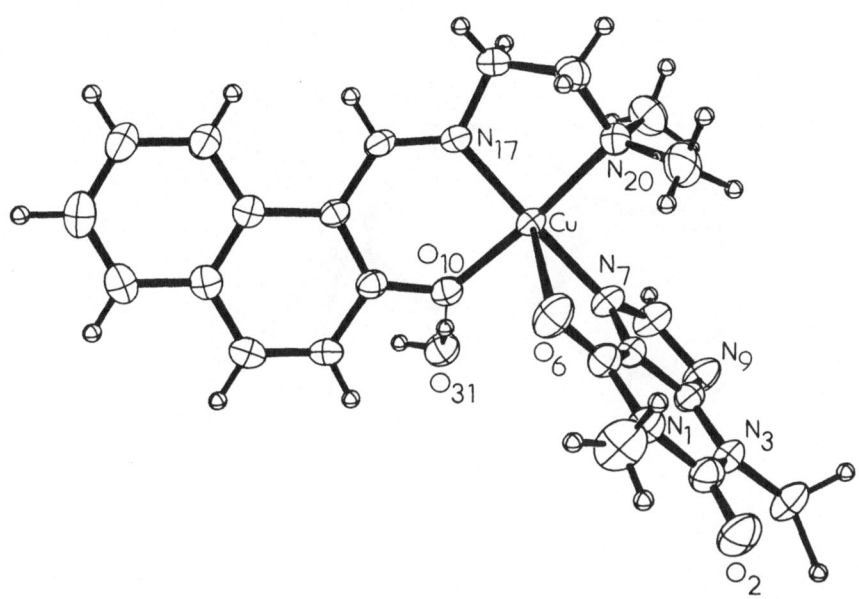

Fig. 13.   The molecular structure of [(N-3,4-benzosalicylidene-N',N'-dimethylethylenediamine)(theophyllinato)copper(II)].

on one side of the equatorial plane is occupied by O(6) of the coordinated theophylline monoanion, Figure 13. The Cu(II)-O(6) bond length at 2.919(3)A is by far the shortest we or other workers have observed (Hodgson, 1977). It is clear from this study that in the absence of interligand hydrogen bonding and with the steric crowding of the axial positions by the two methyl substituents, that O(6) will also bind to the metal center making the theophylline a bidentate ligand.

(f) [(N-Salicylidene-N',N'-Dimethylethylenediamine)(Theophyllinato)
(Aquo)Copper(II)]

This complex represents the second system we have studied in which the chelate ligand has been modified such that no hydrogen bond donor is present near the metal center. The theophyllinato complex of this chelate system is depicted in Figure 14. The coordination sphere about the copper(II) center is (4+2) with the equatorial positions occupied by the tridentate chelate and N(7) of the theophylline anion (Szalda *et al.*, 1976c) and the two axial positions occupied by weakly bound ligands, a water molecule and O(6) of the purine, Figure 14. The weaker interaction of O(6) in this complex is probably related to both steric factors and the presence of the axial water ligand.

Summary

Three important features emerge from the above studies: (1) in all of

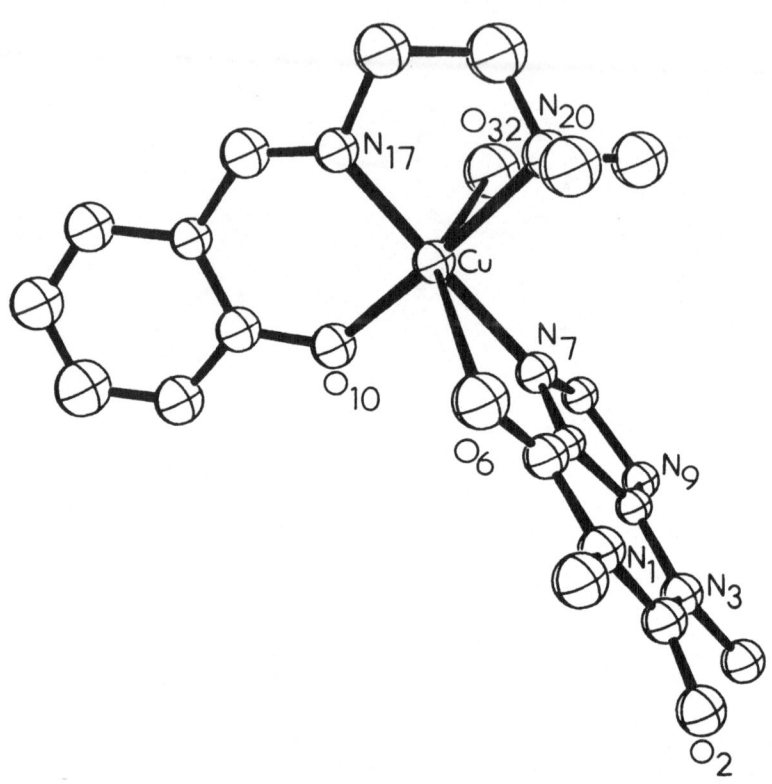

Fig. 14. The molecular structure of [(N-salicylidene-N',
N'-dimethylethylenediamine)(theophyllinato)(aquo)copper(II)].

the complexes N(7) is the metal binding site; (2) in those complexes
where a hydrogen bond donor group exists on the chelate ligand, an
interligand hydrogen bond is formed to the O(6) acceptor group on the
coordinated purine; (3) in those complexes where no hydrogen bond
donor group exists on the chelate ligand, a significant — but weak —
Cu-O(6) axial interaction is indicated from the crystallographic studies.
     We have not obtained any N(1) bonded complexes of guanine deriva-
tives in our studies, but it is possible that if metal binding took
place at N(1) that again O(6) could be used as an acceptor in an interligand hydrogen bond with potential donor groups on a chelate ligand.
     The principal feature that distinguishes guanine-like purines is
the presence of the carbonyl oxygen near the metal binding site and
the employment of this carbonyl oxygen as an acceptor in interligand
hydrogen bonds.

4.3. *Structural Parameters in Chelate Metal Complexes of Purines*

Some of the important structural parameters observed in our studies on
chelate metal complexes of purine bases are collected in Tables II and
III. The metal-N(7) bond lengths show some variation, 1.96 Å to 2.04 Å,

TABLE II
Some Important Structural Parameters in Chelate Metal Complexes of Purines

Complex	Coordination Geometry	M-N(7) Distance	∠M-N(7)-C(5)	∠M-N(7)-C(8)	∠C(5)-N(7)-C(8)	Interligand Hydrogen Bonding
(a) Adenine Residues						
[(glygly)(H$_2$O)(9-MeAd)Cu(II)]	(4+1)[1]	2.021(4) A	137.7(2)°	116.2(2)°	104.9(4)°	N(6)-H...O(H$_2$O)
[(Mesal)(H$_2$O)(9-MeAd)Cu(II)]$^+$	(4+1)[2]	2.037(2)	136.2(2)	119.6(2)	104.2(3)	N(6)-H...O(Mesal)
[(acac)$_2$(NO$_2$)(dAdo)Co(III)]	(6)[3]	2.02(2)	133(2)	116(2)	110(3)	N(6)-H...O(acac)[2]
		1.97(2)	135(2)	120(2)	104(3)	
(b) Theophylline Anion Residues						
trans[(Cl)(Theo)(en)$_2$Co(III)]$^+$	(6)[4]	1.956(2)	132.1(2)	124.9(2)	102.8(3)	O(6)...H-N(en)[2]
cis[(Cl)(Theo)(en)$_2$Co(III)]$^+$	(6)[5]	1.984(8)	131.8(5)	123.3(5)	104.4(8)	O(6)...H-N(en)[2]
[(Mesal)(Theo)Cu(II)]	(4)[6]	1.986(1)	138.0(1)	118.9(1)	103.0(3)	O(6)...H-N(Mesal)
[(Mebenzo)(Theo)(H$_2$O)Cu(II)]	(4+1)[7]	2.000(3)	136.1(2)	117.0(2)	103.1(3)	O(6)...H-N(Mebenzo)
[(Theo)$_2$(dien)Cu(II)]	(4+1)[8]	2.007(3) EQ	135.9(3)	120.2(3)	102.9(3)	O(6)...H-O(H$_2$O)
		2.397(3) AX	142.1(3)	115.7(3)	101.9(3)	O(6)...H-N(dien)
[(Me$_2$benzo)(Theo)Cu(II)]	(4+1)[9]	1.955(7)	118.6(6)	138.9(7)	102.2(7)	None
[(Me$_2$sal)(Theo)(H$_2$O)Cu(II)]	(4+2)[10]	1.99	126	126	108	None

[1]Figure 3   [3]Figure 6   [5]Figure 8   [7]Figure 11   [9]Figure 13

[2]Figure 4   [4]Figure 9   [6]Figure 10   [8]Figure 12   [10]Figure 14

TABLE III
Magnetic Moments for Some Chelated Copper(II) Complexes of Nucleic Acid
Bases

Complex		Bond Lengths(A) in Primary Coordination Sphere		Shortest Cu...Cu Distance(A)	μ(B.M.)
(a) *Adenine Residues*					
[(glygly)(H$_2$O)(9-MeAd)Cu(II)]	(4+1)[1]	O(17)	1.963(4)	6.237(1)	1.9
		N(18)	1.917(4)		
		N(20)	2.023(4)		
		N(7)	2.021(4)		
		O(21)	2.347(4)	(axial water)	
[(Mesal)(H$_2$O)(9-MeAd)Cu(II)]$^+$	(4+1)[2]	O(10)	1.908(2)	7.633(1)	1.9
		N(17)	1.959(3)		
		N(20)	2.045(2)		
		N(7)	2.037(2)		
		O(21)	2.353(2)	(axial water)	
(b) *Theophylline Anion Residues*					
[(Mesal)(Theo)Cu(II)]	(4)[3]	O(10)	1.902(1)	4.810(1)	2.0
		N(17)	1.947(1)		
		N(20)	2.020(1)		
		N(7)	1.986(1)		
[(Mebenzo)(Theo)(H$_2$O)Cu(II)]	(4+1)[4]	O(10)	1.902(2)	6.913(1)	1.9
		N(17)	1.935(3)		
		N(20)	2.017(3)		
		N(7)	2.000(3)		
		O(30)	2.740(3)	(axial water)	
[(Theo)$_2$(dien)Cu(II)]	(4+1)[5]	N(10)	2.047(3)	8.022(1)	2.1
		N(11)	2.020(3)		
		N(12)	2.040(3)		
		N(7)	2.007(3)	(equatorial Theo anion)	
		N(27)	2.397(3)	(axial Theo anion)	
[(Me$_2$benzo)(Theo)Cu(II)]	(4+1)[6]	O(10)	1.930(2)	6.925(1)	1.9
		N(17)	1.939(3)		
		N(20)	2.068(3)		
		N(7)	1.956(3)		
		O(6)	2.919(3)		
[(Me$_2$sal)(Theo)(H$_2$O)Cu(II)]	(4+2)[7]	O(10)	1.921(9)	6.986(2)	2.0
		N(17)	1.957(11)		
		N(20)	2.114(11)		
		N(7)	1.993(10)		
		O(6)	3.343(10)		
		O(32)	3.282(11)		

TABLE III *(continued)*

Complex		Bond Lenghts (A) in Primary Coordination Sphere		Shortest Cu...Cu Distance (A)	μ(B.M.)
(c) *Cytosine Residues*					
[(glygly)(Cyt)Cu(II)]	(4+1)[8]	O(7)	1.99(2)	4.716(1)	1.8
		N(8)	1.87(2)		
		N(10)	2.02(2)		
		N(3)	2.01(2)		
		O(2)	2.74(2)	(intramolecular)	
[(glygly)(Cy)Cu(II)]	(4+2)[9]	O(7)	1.983(3)	4.333(1)	2.0
		N(8)	1.892(3)		
		N(10)	2.011(3)		
		N(3)	1.979(3)		
		O(2)	2.819(3)	(intramolecular)	
		O(2)	2.713(3)	(intermolecular)	
[(Mesal)(Cy)Cu(II)][+]	(4+2)[10]	O(10)	1.922(1)	7.453(1)	1.9
		N(17)	1.938(1)		
		N(20)	2.048(1)		
		N(3)	2.008(1)		
		O(2)	2.772(1)	(intramolecular)	
		O(7)	2.806(1)	(NO_3^- anion)	
(d) *Thymine residues*					
[(Thy)(H₂O)(dien)Cu(II)][+]	(4+1)[11]	N(4)	2.002(3)	6.341(1)	1.9
		N(5)	2.009(3)		
		N(6)	2.040(3)		
		N(1)	1.989(3)		
		O(7)	2.465(2)	(axial water)	

* Magnetic moment in Bohr Magnetons corrected for diamagnetic effects.

[1] Figure 3   [7] Figure 14

[2] Figure 4   [8] Figure 15. Parameters averaged over the two inde-
                          pendent complexes in the unit cell.
[3] Figure 10

[4] Figure 11   [9] Figure 16

[5] Figure 12   [10] Figure 17

[6] Figure 13   [11] Figure 18

and this trend has been attributed to various factors: (1) electronic
effects, (2) steric factors, (3) the strength of the axial interactions
in the copper(II) systems.

One principal feature we have noted in all of the interligand hydrogen bonded species is that the exocyclic bond angles at N(7) are quite dissymmetric, Table II; the dissymmetry is such that in all of these complexes the bond angle M–N(7)–C(5) is usually about $20°$ larger than the bond angle M–N(7)–C(8). We have attributed this dissymmetry to the requirements for the formation of the interligand hydrogen bond.

In the complexes where no interligand hydrogen is formed, the exocyclic bond angles at N(7) are either very close to being the same, as in [(Me$_2$sal)(theo)(H$_2$O)Cu(II)], or the dissymmetry in the exocyclic bond angles at N(7) is almost exactly reversed from the hydrogen–bonded conplexes, [(Me$_2$benzo)(theo)Cu(II)] see Table II. In this latter complex the dissymmetry at N(7) is such as to bring O(6) near to the copper(II) axial position.

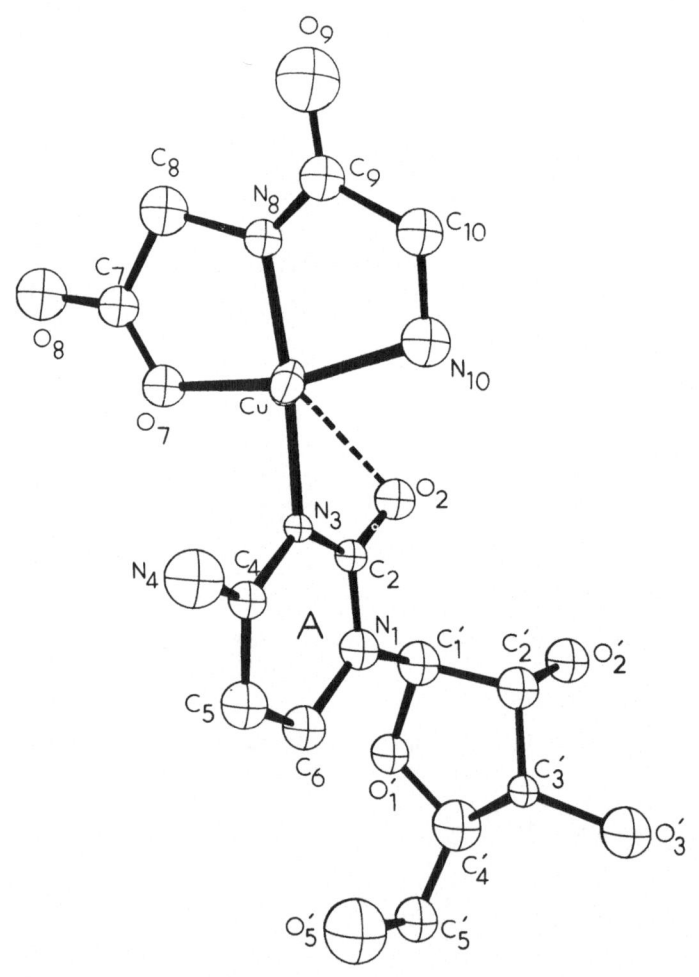

Fig. 15. The molecular structure of [(glycylglycinato)
(cytidine)copper(II)].

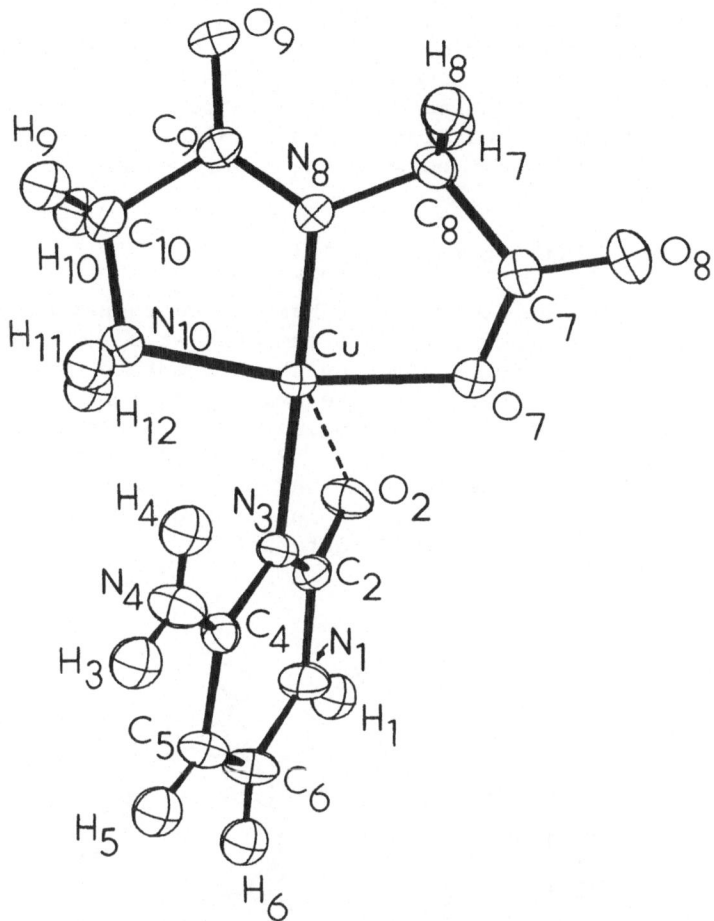

Fig. 16.   The molecular structure of [(glycylglycinato)(cytosine)copper(II)].

4.4 *Binding Interactions in Cytosine and Cytidine*

As we have indicated above, N(3) is the most likely binding site for transition metal complexes of cytidine. To date, we have studied three complexes of cytosine or cytidine, and in each case we have observed the metal binding site to indeed be N(3).

(a) [(Glycylglycinato)(Cytidine)Copper(II)]

The molecular structure of the cytidine complex of glycylglycinato-copper(II) is illustrated in Figure 15. The coordination geometry about the copper(II) center is approximately square planar with the glycyl-

glycine dianion and N(3) of cytidine occupying the four equatorial coordination sites. Furthermore, the exocyclic carbonyl oxygen, O(2), on
the cytidine ring approximately occupies one of the axial positions,
Cu-O(2) bond length = 2.74(1)A [averaged over the two independent molecules in the unit cell], extending the coordination geometry to square
pyramidal (Szalda, Marzilli and Kistenmacher, 1975a). A feature to note
is that there is no interaction between the sugar residue and the Cu(II)
atom.

(b) [(Glycylglycinato)(Cytosine)Copper(II)]

[(Glycylglycinato)(cytosine)(copper(II)] has the coordination geometry
(4+2) in the solid state, Figure 16 (Kistenmacher et al., 1975b). The
four equatorial positions are occupied by the tridentate glygly dianion
and N(3) of the coordinated cytosine. The axial positions are occupied
by Cu-O(2) interactions, one intramolecular Cu-O(2) distance = 2.819(3)A,
and one intermolecular, Cu-O(2) distance = 2.713(3)A. In essence, then,
the molecular structures of the cytidine and the cytosine complexes of
[(glygly)Cu(II)] are virtually identical in terms of the metal binding
site and the intramolecular axial O(2) interaction.

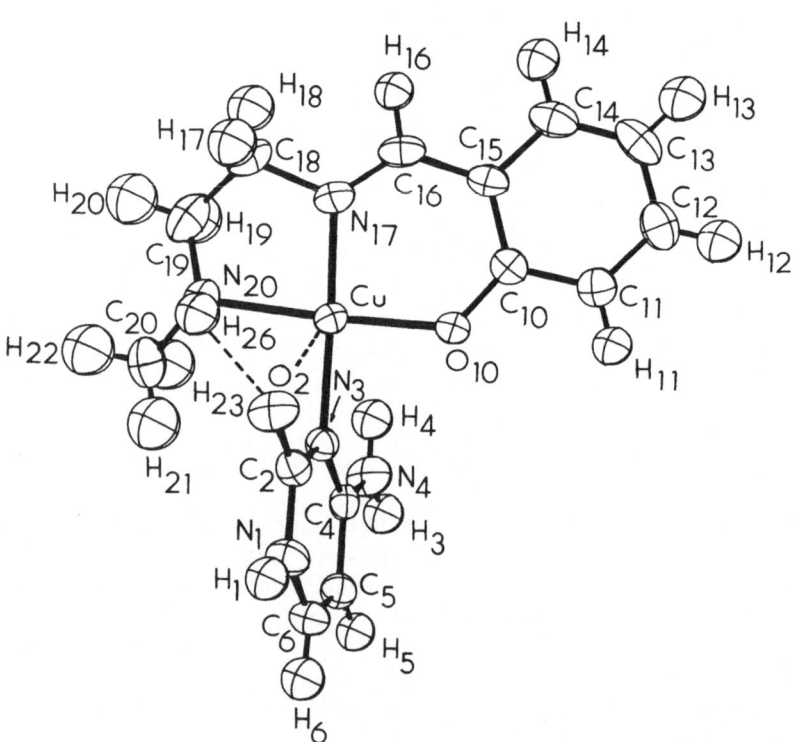

Fig. 17.   The molecular structure of the [(N-salicylidene-
N'-methylethylenediamine)(cytosine)copper(II)]$^+$ cation.

(c) [(N-Salicylidene-N'-Methylethylenediamine)(Cytosine)Copper(II)]$^+$

This cytosine complex cation fits the general pattern shown by the above two systems. The binding of the copper has taken place at N(3), and there is a significant interaction between O(2) on the cytosine ring and the Cu(II) through one of the axial positions, Cu-O(2) distance = 2.772(1)A (Figure 17). The coordination sphere is completed *via* an axial interaction with an oxygen atom of the nitrate counter ion, Cu-ONO$_2^-$ distance = 2.806(1)A. Furthermore, in this complex there is an interligand hydrogen bond formed between O(2) and the hydrogen atom at the N-methylethylenediamine terminus of the Schiff base chelate, Table I (Szalda *et al.*, 1975b).

Summary

The principal features which carry over among all the known cytosine or cytidine complexes of copper(II) are: (a) binding of the pyrimidine or pyrimidine nucleoside through N(3); (b) formation of an intramolecular, axial, Cu-O(2) interaction. The essentially constant geometric features exhibited in these Cu-O(2) interactions, Table IV, suggests that the Cu-O(2) axial interaction is important in Cu-cytosine(cytidine) complexes. Furthermore, there would seem to be some theoretical justification for this semichelation of cytidine to copper(II). A molecular electrostatic potential calculation (Bonaccorsi *et al.*, 1972) for cytosine shows that the presence of the carbonyl group C(2)-O(2) adjacent to the pyrimidine nitrogen N(3) establishes a wide attractive region for electrophilic agents. The calculation shows two deep minima, one in the direction of the lone pair at N(3) and one at an angle of 55° to the C(2)-O(2) bond. The minima at N(3) and O(2) are significantly deeper than at similar sites in adenine or thymine, respectively. Thus the simultaneous binding of copper(II) to N(3) and O(2) may be thought of as a natural consequence of the tendency of copper(II) to extend its coordination sphere *via* the open axial positions and the electrostatic potential distribution inherent to the cytosine(cytidine) molecular framework. In fact, the presence of the Cu-O(2) interaction in all of the known Cu-cytidine (cytosine) complexes has prompted us to suggest that this interaction may be important in the recognition of cytidine residues in polynucleotides by copper(II) (Szalda *et al.*, 1975a).

One further point is worth mentioning. With metal binding at N(3) it is also possible that the exocyclic amino group at C(4) could involve itself in hydrogen bonds with suitable acceptors on other ligands in the coordination sphere. We do not at this point, however, have any crystallographic evidence for such interactions.

4.5. *Binding Interactions in Thymine and Thymidine*

As we indicated earlier, thymidine (thymine) is unique among the nucleic acid constituents in that there are no free lone pairs available for metal complexation. In fact, solution and preparative studies indicate that except for Hg^{2+} and RHg$^+$, the interaction between transition metal complexes and thymidine is weak (Marzilli, 1977). We have investigated

TABLE IV
Some Important Structural Parameters in Chelate Metal Complexes of Pyrimidines

Complex	Coordination Geometry	Cu-N Dist.	Cu-O(2) Dist.	Cu-C(2) Dist.	∠Cu-O(2)-C(2)	Interligand Hydrogen Bonding
(a) *Cytosine Residues*						
[(glygly)(Cyt)Cu(II)][N(3)]	(4+1)[1]	1.97(2)A	2.80(2)A	2.76(3)A	76(2)°	None
		2.04(2)	2.68(2)	2.73(3)	78(2)	
[(glygly)(Cy)Cu(II)][N(3)]	(4+2)[2]	1.979(3)	2.819(3)	2.775(3)	75.5(2)	None
[(Mesal)(Cy)Cu(II)]+[N(3)]	(4+2)[3]	2.008(1)	2.772(1)	2.762(1)	76.7(3)	O(2)...H-N(mesal)
[(Cl)₂(Cy)₂Cu(II)][N(3)]	(4+2)[4]	1.97(1)	2.74(1)	2.71(1)	75.7(6)	None
		1.95(1)	2.88(1)	2.81(1)	74.6(6)	
(b) *Thymine Anion Residue*						
[(Thy)(dien)Cu(II)]+[N(1)]	(4+1)[5]	1.989(3)	3.159(3)	2.935(3)	68.3(2)	None

[1] Figure 15
[2] Figure 16
[3] Figure 17
[4] Sundaralingam and Carrabine (1971)
[5] Figure 18

one copper(II) complex of thymine; in this case the metal binding site is N(1) and the base is in its monoanionic form.

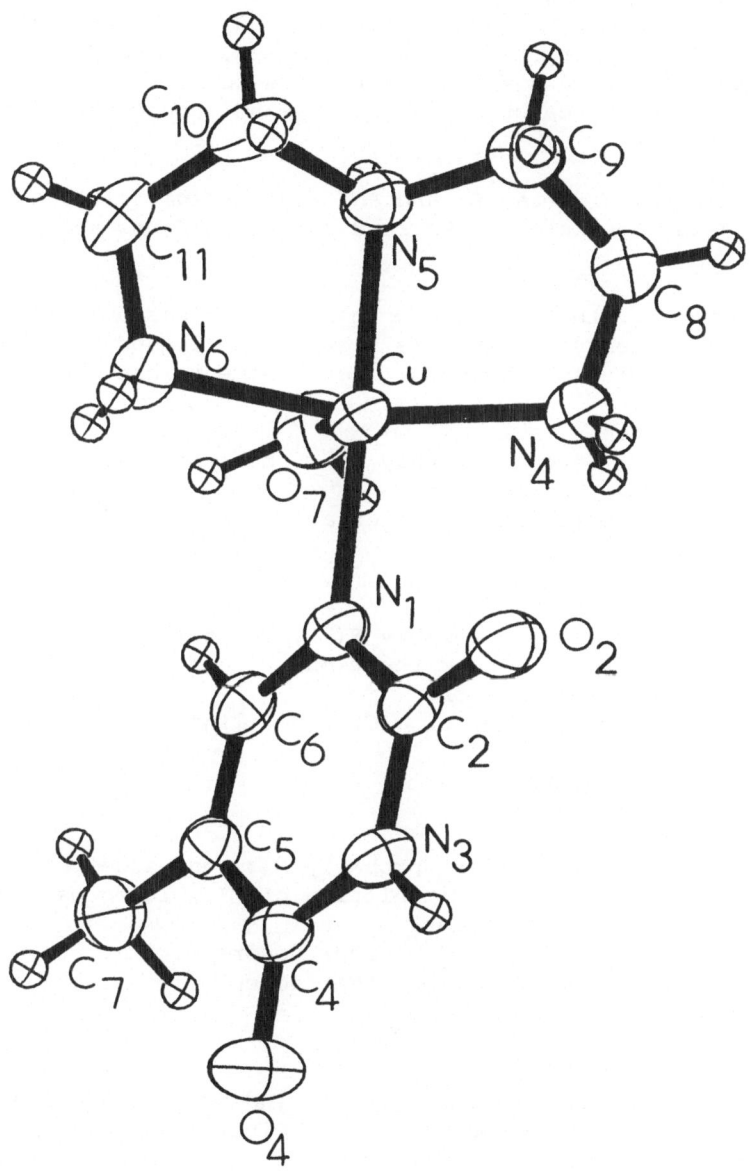

Fig. 18.   The molecular structure of the [(thyminato)(aquo) (diethylenetriamine)copper(II)]⁺ cation.

(a) [(Thyminato)(Aquo)(Diethylenetriamine)Copper(II)]⁺

The molecular conformation of this complex is illustrated in Figure 18.

The primary coordination sphere is approximately square pyramidal with the tridentate diethylenetriamine ligand and N(1) of the thymine monoanion occupying the four equatorial coordination sites. One of the axial positions is occupied by a water molecule (Kistenmacher et al., 1975). A point of some interest is that we find no evidence for a significant Cu-O(2) interaction in this complex, Cu...O(2) distance = 3.159(3)A. The lack of such an interaction is consistent with the molecular electrostatic calculation for thymine (Bonaccorsi, et al., 1972).

4.6. General Summary of Metal-Ligand and Ligand-Ligand Interactions in Chelate Metal Complexes of Purines and Pyrimidines and Their Nucleosides

The crystallographic studies which we have summarizes above lead to several general comments we can make about metal-ligand and ligand-ligand interactions in such systems. These comments are given below:

(a) *Adenosine Residues*
    (1) *Metal Binding Site* — Probably N(7), but possibly N(1)
    (2) *Ligand-Ligand Interactions*
        (i)   N(6)H$_2$...O(ligand) or other acceptors
        (ii) Steric factors

(b) *Guanosine Residues*
    (1) *Metal Binding Site* — Probably N(7), but possibly N(1) with
                             deprotonation
    (2) *Ligand-Ligand Interactions*
        (i)   O(6)...H-N(ligand) or O(6)...H-O(ligand)
           or other donors
        (ii) Steric factors

(c) *Cytidine Residues*
    (1) *Metal Binding Site* — Very probably N(3) and aided by
                             semichelation of O(2) in copper(II) systems
    (2) *Ligand-Ligand Interactions*
        (i)   O(2)...H-N(ligand) or other donors
        (ii) N(4)H$_2$...O(ligand) or other acceptors
        (iii) Steric factors

(d) *Thymidine Residues*
Least likely residue for transition metal binding as there are no heterocyclic nitrogens with lone pairs. Possible metal binding at N(3) accompanied by deprotonation.

5. BINDING SPECIFICITY THROUGH METAL-LIGAND AND LIGAND-LIGAND INTER-ACTIONS: SOLUTION STUDIES

We have performes a variety of solution studies which we feel indicate that ligand-ligand interaction in chelate metal complexes can induce

selectivity in the binding of such complexes to nucleic acid constituents.
    It is well known that steric interactions are of paramount impor-
tance in determining the nature and type of complexes which can be
formed with cobalt(III). We have recently made a study of the reaction
of $[cis$-β-$Co(III)(trien)Cl_2]^+$, where trien is ligand (c) in Figure 2,
with ^{14}C-labeled deoxythymidine, deoxycytidine, deoxyguanosine and
deoxyadenosine (Marzilli *et al.*, 1974). A representation of the possible
cis interactions between the NH or NH_2 groups of coordinated trien and
the deoxynucleoside bases is shown in Figure 19. If thymidine coordinates
at N(3)(deprotonated), the two contiguous exocyclic carbonyl oxygens
can hydrogen bond to the NH_2 or the NH groups of trien, Figure 19.

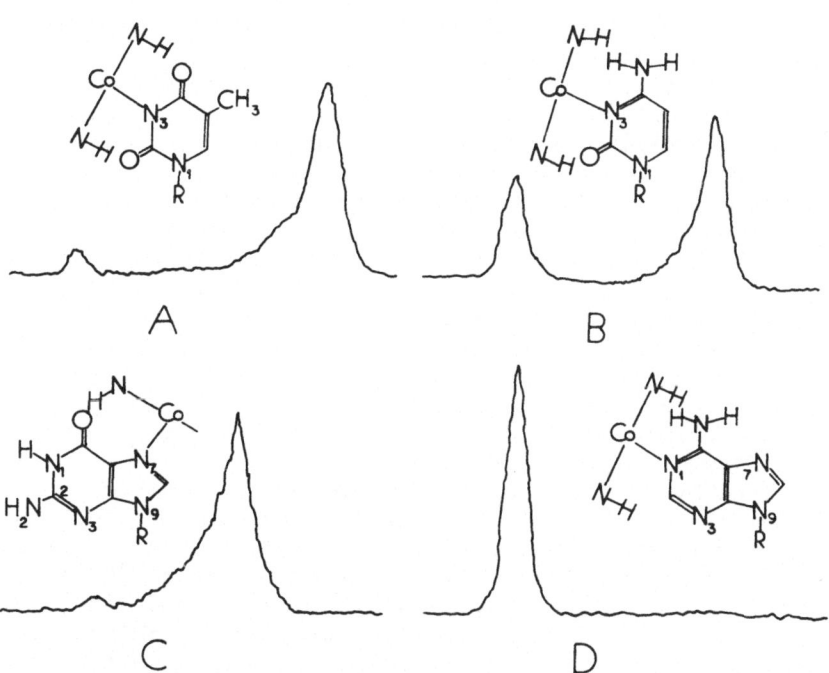

Fig. 19. Scan of radioactivity of electrophoretic charts of
the reaction mixtures of $[Co(trien)Cl_2]Cl$ and the nucleosides,
(A) dT, (B) dC, (C) dG, and (D) dA.

Coordination of cytidine *via* N(3) could produce a favorable hydrogen
bond to trien *via* O(2) but a repulsive interaction would occur *via* the
amino group at C(4). Coordination of the purine nucleosides through
N(7) leads to only favorable interligand hydrogen bonding with guanosine,
Figure 19, and to only repulsive interactions with adenosine, Figure 19.
Coordination of adenosine through N(1) would also lead to only repulsive
interactions.
    Detailed results of an electrophoretic study of the reaction of

[cis-β-Co(III)(trien)Cl$_2$]$^+$ with the above four deoxynucleosides is
presented in Table V. It can be seen that dT, with its capability of
forming two interligand hydrogen bonds, reacts most rapidly, and dG,
with its capability of forming one hydrogen bond, reacts second most
rapidly. Given sufficient time at [Co] = 0.1M, both these nucleosides
react almost completely. Cytidine, with its favorable hydrogen bonding
capability via O(2) and its repulsive steric interaction through N(4)H$_2$,
ranks third in order of reactivity. Adenosine, which cannot form interligand hydrogen bonds in this system, does not react at all. The relative
affinity for binding of deoxynucleosides to the trien complex of cobalt
(III) is thus: dT > dG > dC >> dA. This ordering is consistent with our
expectations based on the interligand interactions observed in the
crystallographic studies and reiterated above. Unfortunately, although
the results illustrate the importance of interligand interactions
between nucleoside ligands and other coordinated ligands, the nucleoside
complexes are relatively unstable.

   More recently we have studied the reaction of the complex anion
[Co(III)(acac)$_2$(NO$_2$)$_2$]$^-$, where acac is ligand (g) in Figure 2, with the
four commonly occurring nucleosides on a preparative scale in aqueous
solution (Sorrell, Kistenmacher, Marzilli and Epps, 1976). Adenosine
and deoxyadenosine rapidly formed neutral complexes with the formulation
[(adenosine)Co(III)(acac)$_2$NO$_2$] and [(deoxyadenosine)Co(III)(acac)$_2$NO$_2$].
We have presented in Figure 6 the results of our crystallographic
investigation of the deoxyadenosine complex. In the solid, it is evident
that interligand hydrogen bonds are formed between the exocyclic amino
group at C(6) and two of the acac equatorial oxygens, Figure 6. We
observed no formation of a solid product with any of the other
nucleosides. P.m.r. evidence supports this selectivity. Although Co(III)
complexes are generally inert, a fairly rapid reaction takes place (in
dimethylsulfoxide) in which adenosine (or deoxyadenosine) is displaced
to some extent by DMSO (~50% in 0.05M solution). Addition of free
adenine nucleosides can force the reaction towards the nucleoside
complex. However, the addition of guanosine, cytidine or uridine is
relatively ineffective in displacing the DMSO ligand. These results
can be explained in terms of the possible interligand interactions
between the nucleoside ligands and the acac ligands in much the same
way as for the Co(III)trien system described above. In this case, however
it is clear that adenosine has a far greater affinity for binding to
[Co(III)(acac)$_2$(NO$_2$)] than any of the other nucleosides.

   Somewhat similar results were obtained in studies of the reaction
of purine and pyrimidine bases with the trans[(nBu)$_3$PCo(III)(dmg)$_2$]$^+$
cation (where dmg is ligand (h) in Figure 2) (Marzilli et al., 1975).
The steric requirements of the trans[(nBu)$_3$PCo(III)(dmg)$_2$]$^+$ moiety are
greater than those for the [Co(III)(acac)$_2$NO$_2$] system. Only completely
unhindered purine derivatives will react with this complex and, thus,
it was not possible to prepare a complex with the theophylline anion,
normally an excellent ligand. Adenine did form a complex, but coordinatio
was probably via N(9). The complexes trans[(nBu)$_3$PCo(III)(dmg)$_2$
(xanthinato)] and trans[(nBu)$_3$PCo(III)(dmg)$_2$(hypoxanthinato)] were
isolated, and the structure of the former complex established bonding
at the sterically unhindered site N(9). Some of these trans[(nBu)$_3$PCo(III

TABLE V
Percent Reaction of Deoxyribonucleosides with $[Co(trien)Cl_2]Cl$

Nucleoside	1 Day	3 Days
	Co = 0.025 M	
dT	65	65
dG	45	50
dC	10	10
dA	0	0
	Co = 0.1 M	
dT	85 — 90	90 — 95
dG	65	>95
dC	10	65 — 70
dA	0	0

$(dmg)_2(purinato)]$ complexes proved to be useful intermediates for the synthesis of N(7) alkylated purines.

The conclusion that we draw from these solution and preparative studies, consistent with the crystallographic results, is that *inter-ligand interactions can be effective in introducing selectivity into the reaction of chelate metal complexes with nucleic acid components.*

6. MAGNETIC PROPERTIES OF COPPER(II)-CHELATE COMPLEXES OF NUCLEIC ACID CONSTITUENTS

Finally, we would like to very briefly mention our work on the solid state magnetic properties of copper(II)-chelate complexes of nucleic acid constituents. Table III lists the room temperature magnetic moments for the eleven copper(II) complexes we have studied. We also include in Table III the shortest Cu(II)...Cu(II) distances found in the crystallographic investigations. Two features emerge from these limited studies: (1) the range of magnetic moments is fairly small, with the smallest, 1.8BM, for [(glygly)(Cyt)Cu(II)] and the largest, 2.1BM, for [(Theo)$_2$(dien)Cu(II)]; (2) the complexes should be considered to be magnetically dilute with the shortest intermolecular Cu...Cu distance of 4.333(1)A in the complex [(glygly)(Cy)Cu(II)], see Table III.

ACKNOWLEDGMENTS

The authors would like to thank the National Institutes of Health and the Petroleum Research Fund, administered by the American Chemical Society, for support of this research. We would also like to thank our co-workers Professor Michael Beer, Professor Donald Hollis, David Szalda, Chien-Hsing Chang, Theophilus Sorrell, Robert Stewart, William Trogler, Brian Hanson and others who have contributed to this work.

REFERENCES

Berger, N.A. and Eichhorn, G.L;: 1971, *Biochem.*, 10, 1847
Bonaccorsi, R., Pullman, A., Scrocco, E., and Tomasi, J.: 1972,
  Theoret. Chim. Acta(Berlin), 24, 51
Eichhorn, G.L.: 1973, G.L. Eichhorn, (ed.), *Inorganic Biochemistry*
  Elsevier Press, Amsterdam.
Eichhorn, G.L., Clark, P., and Becker, E.D.: 1966, *Biochem.*, 5, 245.
Freeman, H.C.: 1967, *Advanc. Protein Chem.*, 22, 235.
Hathaway, B.J.: 1973, *Struct. Bond.*, 14, 49.
Hodgson, D.J.: 1977, S.J. Lippard, (ed.), *Progress in Inorganic*
  Chemistry, John Wiley, New York.
Kistenmacher, T.J.: 1975, *Acta Cryst.*, B31, 85.
Kistenmacher, T.J., Marzilli, L.G., and Szalda, D.J.: 1976, *Acta Cryst.*,
  B32, 186.
Kistenmacher, T.J., Sorrell, T., and Marzilli, L.G.: 1975, *Inorg.*
  Chem., 14, 2479.
Kistenmacher, T.J. and Szalda, D.J.: 1975, *Acta Cryst.*, B31, 90.
Kistenmacher, T.J., Szalda, D.J., and Marzilli, L.G.: 1975a, *Inorg.*
  Chem., 14, 1686.
Kistenmacher, T.J., Szalda, D.J., and Marzilli, L.G.: 1975b, *Acta Cryst.*,
  B31, 2416.
Marzilli, L.G.: 1977, S.J. Lippard, (ed.), *Progress in Inorganic*
  Chemistry, John Wiley, New York.
Marzilli, L.G., Epps, L.A., Sorrell, T., and Kistenmacher, T.J.:
  1975, *J. Amer. Chem. Soc.*, 97, 3351.
Marzilli, L.G., Kistenmacher, T.J., and Chang, C.H.: 1973, *J. Amer.*
  Chem. Soc., 95, 7507.
Marzilli, L.G., Kistenmacher, T.J., Darcy, P.E., Szalda, D.J., and
  Beer, M.: 1974, *J. Amer. Chem. Soc.*, 96, 4686.
Marzilli, L.G., Trogler, W.C., Hollis, D.P., Kistenmacher, T.J.,
  Chang, C.H., and Hanson, B.E.: 1975, *Inorg. Chem.*, 14, 2568.
Sorrell, T., Kistenmacher, T.J., Marzilli, L.G., and Epps, L.A.: 1976,
  to be published.
Sorrell, T., Marzilli, L.G., and Kistenmacher, T.J.: 1976, *J. Amer.*
  Chem. Soc., 98, 2181.
Sundaralingam, M. and Carrabine, J.A.: 1971, *J. Mol. Biol.*, 61, 287.
Szalda, D.J., Kistenmacher, T.J., and Marzilli, L.G.: 1975, *Inorg.*
  Chem., 14, 2623.
Szalda, D.J., Kistenmacher, T.J., and Marzilli, L.G.: 1976a, to be
  published.
Szalda, D.J., Kistenmacher, T.J., and Marzilli, L.G.: 1976b, to be
  published.
Szalda, D.J., Kistenmacher, T.J., and Marzilli, L.G.: 1976c, to be
  published.
Szalda, D.J. Marzilli, L.G. and Kistenmacher, T.J.: 1975a, *Biochem.*
  Biophys. Res. Commun., 63, 601.
Szalda, D.J., Marzilli, L.G., and Kistenmacher, T.J.: 1975b, *Inorg.*
  Chem., 14, 2076.

DISCUSSION

A. Pullman: (1) To what extent does the presence of the various ligands already bound to the cation influence the site of binding to the base? Would for instance the possibility of H bonding of this ligand to the NH_2 group of adenosine favor the positioning of the cation towards N_7 rather than towards N_1 ?

(2) You seemed to imply that the basicity of the nitrogens of the bases is involved in the binding to the cation. In that case, what would be the reason for adenine binding the cation through N_7 which is not the most basic site?

Kistenmacher: (1) Yes, certainly. That is the main thrust of our work, site specificity due to ligand-ligand interactions.

(2) We do not think that basicity is an overwhelming criterion in many instances.

Laniv: (1) What are the conditions for the preparation at the Co(III) complexes?

(2) We have prepared a series of Co(III) nucleotide complexes. It was found by Dr. A. Danchin and it was supported by NMR measurements that at least in part of them, O_2^- is bound as the sixth ligand at the Co(III). I wonder what is the sixth ligand in the case of Co(III)-purines and pyrimidines complexes prepared in aqueous solution, without adding ligands such as NO_2^- or Cl^- ions.

Kistenmacher: (1) $[Co(acac)_2(NO_2)_2]^-$ + nucleic acid base $\xrightarrow{H_2O}$ $[Co(acac)_2(base)NO_2]^o$ + NO_2^-

(2) Probably H_2O or OH^- depending on pH.

Werber: What would you expect would be the metal ion binding sites in adenine nucleotides, in particular in the case of Co(III)?

Kistenmacher: Very probably N(7) or N(1).

Saenger: (1) How does the interaction of the Cu(II) or Co(III) complexes with double helical, base paired DNA or RNA look like?

(2) What do you mean by 'mild' probe? Does the interaction between your probes and the nucleobases depend om the ligands you are using or on the type of cation?

Kistenmacher: (1) Our own experience is still pretty limited in this area. (2) We hope that many of the probes we are employing will not seriously alter the structure of the polynucleotide we are probing.

B. Pullman: In relation to Professor Kistenmacher's reference to the electrostatic molecular potential maps of the purine and pyrimidine bases, constructed by Mme Pullman and coworkers, as possible guides towards the discussion of cation-binding sites, we would like to report that we have studied recently in our laboratory explicitly the interaction between cations and these bases. Truly, the computations which involve in each case the interaction between a single base and a single cation cannot be considered more than model investigations for this particular situation which may rarely if ever be encountered. Real cation-ligand interactions in this field (and *a fortiori* with nucleosides, nucleotides or nucleic acids) imply generally a much more complex scheme involving a larger number of competing binding sites, anionic forms of the bases, hydrated or chelated cations, cross linkings etc.). Nevertheless it appears that this model study is able

to provide some fundamental information about the intrinsic cation
binding tendency of the bases which may be related to some aspects of
the interactions occurring in the more complex real situations.

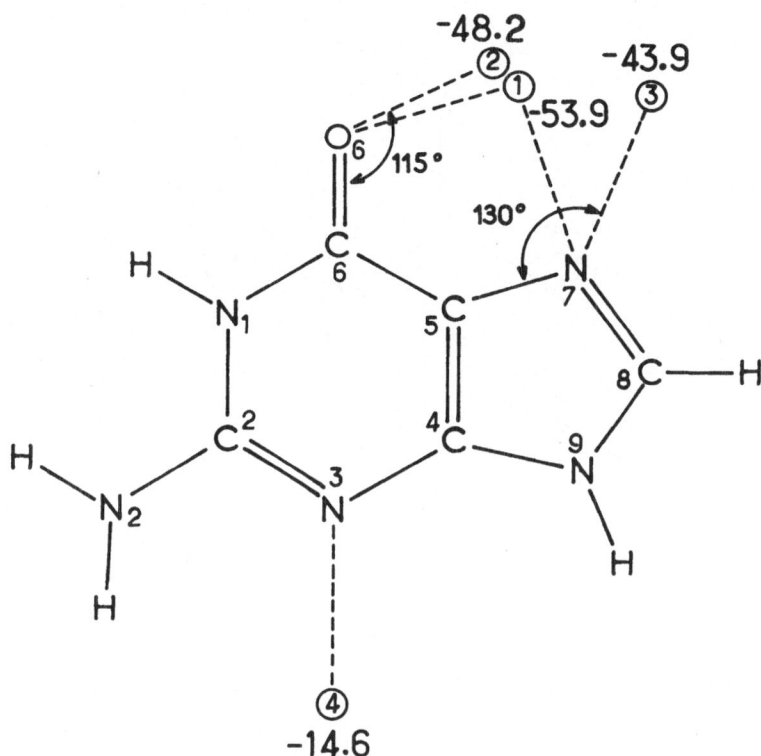

The computations have been performed by the SCF *ab initio* method and involved essentially the Na$^+$ cation. Figure 1 presents the summary of some of the essential results obtained. (For more details see Perahia, D., Pullman, A. and Pullman, B.,: *Theoret. Chim. Acta,* in press). It indicates the energies of interaction for the preferred binding sites. The principal results to underline, in particular in relation with the previous results on the electrostatic molecular potentials of the bases which indicated their preferential proton affinities. (See Pullman, A., in *Chemical and Biochemical Reactivity,* Proceedings of the 6th Jerusalem Symposium, E.D. Bergmann and B. Pullman (eds.), Academic Press, 1974, p. 1 and Bonaccorsi, A., Scrocco, E., Tomasi, J., and Pullman, A.,: 1975, *Theor. Chim. Acta* 36, 339) are:

    (1) The increased singificance, of the individual oxygen binding sites, in compounds containing both oxygen and nitrogen potential binding sites. Thus e.g. in cytosine, interactions energies at sites ② or ④ are greater than at site ③   The reverse was true for proton interaction energies.

    (2) The appearance in cytosine and guanine of 'bridged' sites ① , involving a nitrogen and a carbonyl oxygen, as the preferred binding sites. In the electrostatic potential energy maps for these compounds, these N and O atoms were generally associated with separate minima and the intermediate bridged position was less favorable.

    (3) In adenine, where only N binding sites are available and in

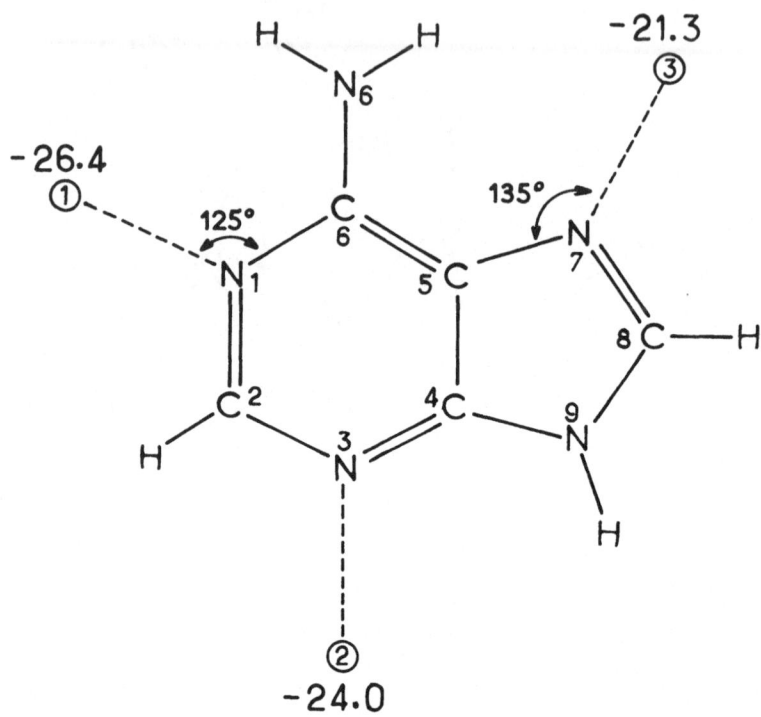

uracil where only O binding sites are present, the situation is
qualitatively similar to that revealed by the molecular electrostatic
potential map.

    Our theoretical results were obtained specifically as stated
above for the Na^+ cation and thus may not be used, at least without
caution for the interpretation of findings observed with other cations.
Nevertheless it seems evident that the findings described e.g. in (2)
above are related to Dr. Kistenmacher's observation about the combined
effects of N and O attractions in guanine and cytosine. Also, while
it is generally considered (see e.g. Izatt, R.M., Christensen, J.J.,
and Rytting, H.: 1971, *Chem. Rev.* 71, 439) that alkali metal ions bind
in solution exclusively to the phosphate moieties of ribonucleotides
and DNA (a situation understandable in view of the high interaction
energies involved, see the communication by Pullman, A. and Pullman, B.:
in these proceedings), a few recent crystallographic results indicate
Na^+ coordination to the carbonyl oxygens of uracil, cytosine and guanine.
Such is the case e.g. in the crystal structure of cytidine-5'-
diphospocholine (Viswamitra, M.A., Seshadri, T.P., Post, M.L. and
Kennard, O.: 1975, *Nature,* 258, 497), sodium β-cytidine-2',3'-cyclic
phosphate (Coulter, Ch.L.,: 1973, *J. Am. Chem. Soc.* 95, 570), the sodium
salt adenosyl-3',5'-uridine phosphate Rosenberg, J.M., Seeman, N.C.,
Po Kim, J.J., Suddath, F.L., Nicholas, H.B., and Rich, A.: 1973, *Nature,*
243, 150), thymidylyl-(5'-3')-thymidylate-5' (Camerman, N., Fawcett, J.K.,
and Camerman, A.: 1973, *Science,* 182, 1142) and disodium deoxyguanosine-
5-phosphate tetrahydrate Young, D.W., Tollin, P., and Wilson, H.R.: 1974,
Acta Cryst. B30, 2012).

RECENT STUDIES ON THE EFFECTS OF DIVALENT METAL IONS ON THE STRUCTURE AND FUNCTION OF NUCLEIC ACIDS

G.L. EICHHORN, J. RIFKIND, Y.A. SHIN, J. PITHA, J. BUTZOW,
P. CLARK, AND J. FROEHLICH
*Gerontology Research Center, National Institute on Aging,
National Institutes of Health, Baltimore City Hospitals,
Baltimore, Maryland 21224, U.S.A.*

Metal ions are required for many of the processes in which nucleic acids transfer genetic information (Eichhorn, 1973). Thus metal ions are essential for the proper functioning of the genetic apparatus. Nevertheless the presence of the wrong metal ions, or even the essential ones in the wrong concentration, can lead to errors. Let me illustrate such errors in DNA, RNA and protein synthesis.

The enzyme DNA polymerase I from *e. coli* requires divalent ions for its activity. When the ion is Mg^{2+} only the desired deoxynucleotides are incorporated into DNA, but when it is Mn^{2+}, the enzyme fails to discriminate between deoxynucleotides and ribonucleotides (Berg *et al.*, 1963). RNA polymerase from *e. coli* also requires activation by divalent ions. With this enzyme Mg^{2+} ions allow only the desired ribonucleotides to be incorporated into RNA, whereas Mn^{2+} ions again lead to a failure to discriminate between nucleotides that do or do not contain a 2' OH group (Steck *et al.*, 1968).

When polyuridylic acid is used as a message for *in vitro* protein synthesis, low concentrations of Mg^{2+} lead to the correct incorporation of phenylalanine into the protein chain. Higher concentrations result in a misreading of the genetic code so that leucine as well as phenylalanine are incorporated (Szer and Ochoa, 1964).

These phenomena illustrate not only the importance of metal ions in nucleic acid reactions, but also the sensitivity of these reactions to somewhat specific metal ion requirements.

Ultimately the explanation for both the necessary and the deleterious involvement of metal ions in genetic information transfer depends at least in part on an understanding of the way in which metal ions interact with the nucleic acids and their constituents.

In our laboratory we have worked for some years on the sites of interaction of metal ions with nucleic acids and constituents, and recently many other laboratories have been interested in this problem. Several papers are being presented here on the nature of the reacting sites. I shall summarize by pointing out that nucleic acids contain three classes of potential electron donor sites to which metal ions can bind: the heterocyclic bases, the phosphate groups, and the ribose hydroxyls. Experiments have shown that all potential sites can become actual interaction sites. Different metal ions have different binding preferences,

*B. Pullman and N. Goldblum (eds.), Metal-Ligand Interactions in Organic
Chemistry and Biochemistry, part 1 , 41 - 51. All Rights Reserved.
Copyright* 1977 *D. Reidel Publishing Company, Dordrecht-Holland.*

and the same metal ion can be made to adhere to one or another site by changes in reaction conditions.

Metals binding to the nucleic acids produce dramatic changes in their structure, and the nature of these changes depends upon the binding site. We shall consider the consequences of binding to the bases and to the phosphate groups.

1. EFFECTS OF BASE BINDING

One consequence of metal binding to the bases on a polynucleotide is the

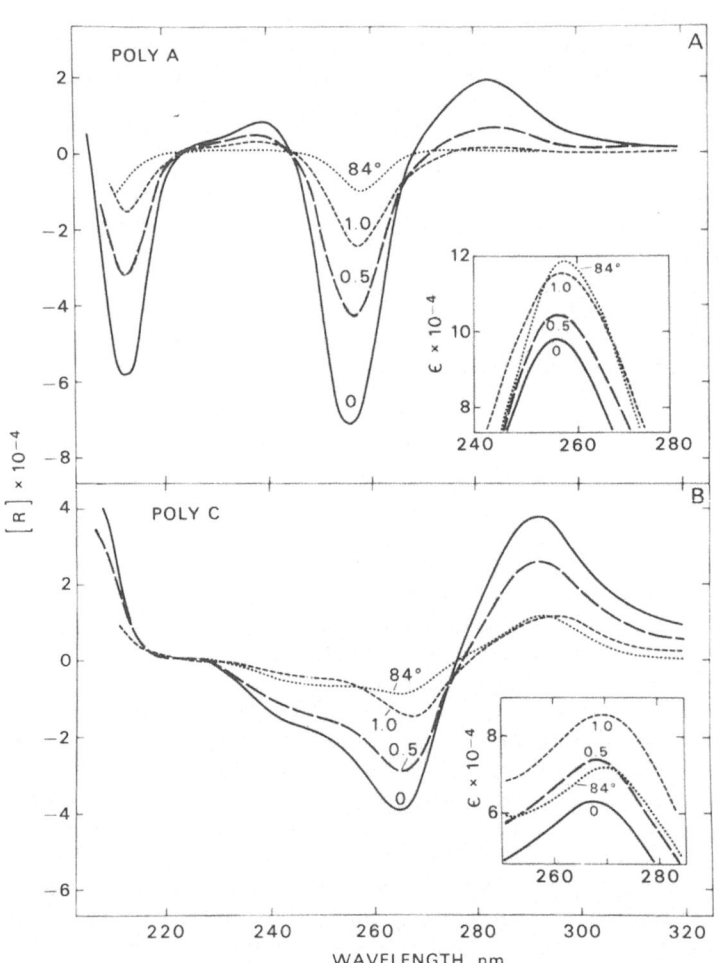

Fig. 1. Effect of Cu(II) on the ORD and UV spectra of single stranded poly(A) and poly(C). Nucleotide residue concentration 1.5×10^{-4} M, pH 7, 25°, copper/nucleotide residue given in this figure. Compared to spectra without Cu^{2+} at 84°.

destacking of the bases and the disordering of the ordered structure of the polymer. We have previously shown that the double helix of DNA is denatured by base-binding metal ions such as copper (II), and that metal ions form crosslinks between the DNA strands (Eichhorn and Clark, 1971).

We have recently shown that the stacked base structure of single stranded helical polynucleotides is also collapsed by copper ions (Rifkind *et al.*, 1976). Figure 1 reveals that copper (II) produces a dramatic decrease in the magnitude of the ORD spectra of poly(A) and poly(C)

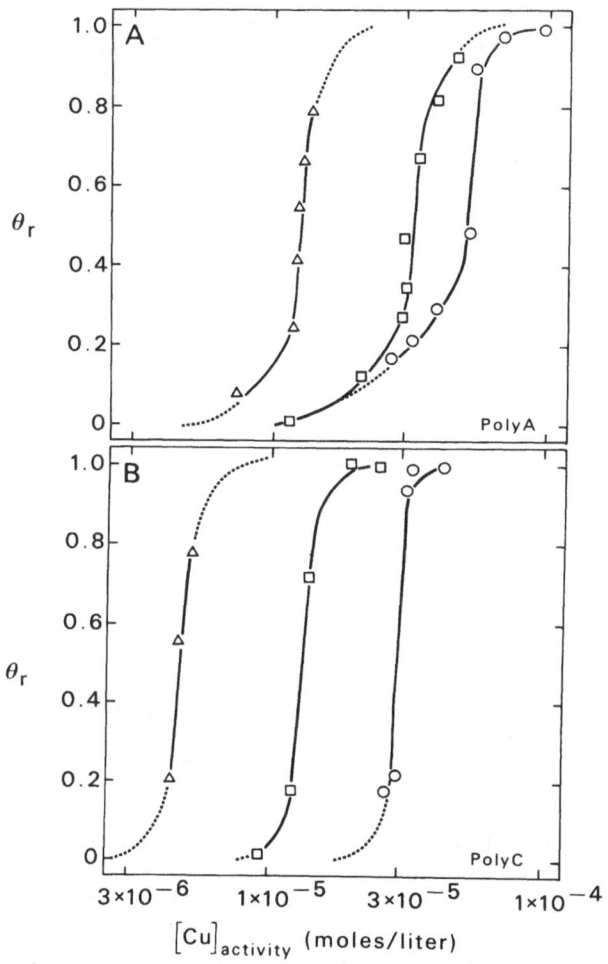

Fig. 2. Effect of polymer concentration on Cu(II) induced changes in rotation of single stranded poly(A) and poly(c) at pH 6 in 0.1 M $NaClO_4$ and 0.01 M cacodylate. θ_r is fractional change in rotation = $[\alpha] - [\alpha]_o / [\alpha]_d - [\alpha]_o$, where $[\alpha]$ = rotation at given Cu^{2+} activity, $[\alpha]_o$ = rotation in absence of Cu^{2+}, and $[\alpha]_d$ = rotation of fully disordered structure. Concentration of polymers, A-poly(A); B-poly(C): (Δ), 5×10^{-4}M; ($\square$), 5×10^{-5}M; (0), 5×10^{-6}M.

as well as an increase in the UV extinction coefficient. The latter
indicates destacking and the former a high degree of disordering.

The disordering of single stranded helical polynucleotides by
copper (II), unlike disordering by heat, is highly cooperative and dependent on the concentration of polymer, as demonstrated in Figure 2.
At high polymer concentration the disordering transition occurs at much
lower copper activities than at low polymer concentration. This polymer
concentration dependence indicates that Cu(II) binds not only to isolated polymer strands but also causes crosslinking between strands. Evidence has also been obtained for intramolecular crosslinks, or the formation of loops within single polynucleotide chains. Figure 3 demonstra-
tes the chain length dependence of the disordering transition; poly(A)
is disordered at much lower $[Cu^{2+}]$ than hexa(A). From the cooperative

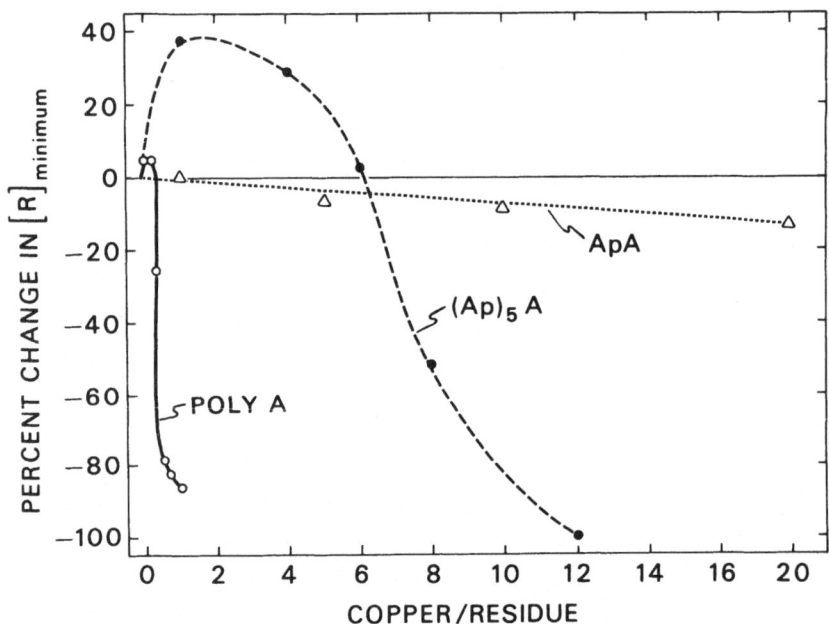

Fig. 3.   Effect of chain length on Cu(II) induced changes in
rotation for single stranded poly(A) and oligo(A) in water at
25°. Presented as % change in mean residue rotation at trough
(see Figure 1).

binding of poly(A) it is possible to calculate that the number of bases
disordered in one continuous sequence is 4-5. Thus if only intermolecular
crosslinking were involved, the disordering induced by copper in a hexamer
should be similar to that in a polymer, but it is found to be much less.
The difference between the polymer and hexamer results therefore indicates
the existence of a type of disordering that is possible in polymer but
not in hexamer, and that is presumably intramolecular crosslinking. Thus
Cu(II) forms both intermolecular and intramolecular crosslinks.

In spite of the fact that base-binding metal ions diminish the stack-

ing properties of the bases in polynucleotides, such metal ions can induce the stacking of mononucleotides (Rifkind and Eichhorn, 1972).

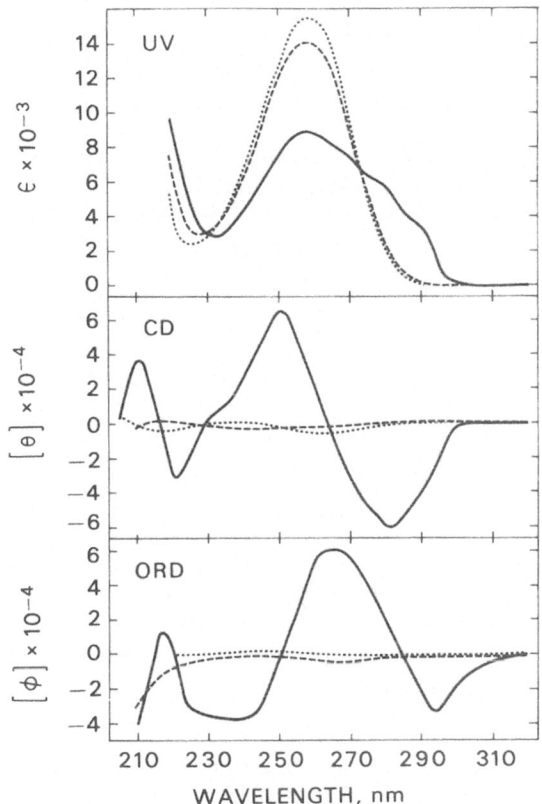

Fig. 4.   ..., UV, CD, and ORD spectra of a 1 x 10^{-3} M solution
of 5'AMP, pH 7.
----, in the presence of 2×10^{-3} M poly-L-lysine.
———, in the presence of 2×10^{-3} M poly-L-lysine and 1×10^{-3} M
Zn(II).

Figure 4 shows the effect of zinc on the UV, ORD, and CD spectra of 5'AMP
in the presence of polylysine or some other polymer which keeps the zinc
nucleotide complex in solution. The resulting intense conservative CD
spectrum accompanied by a large absorbance decrease presumably indicates
extensive base stacking. The question then arises why metal ions disor-
ganize an ordered polymer structure but cause monomeric nucleotides to
stack and become oriented in solution. The answer is probably that in
the polymer the sugar-phosphate backbone determines and restricts the
geometry of the base stacking and that metal binding to the bases causes
a perturbation in this geometry. When metal ions bind to the monomers,

they can produce an entirely new backbone for the bases which are not
already constrained by the sugar-phosphate.

Metal binding to nucleic acids can alter the specificity of enzymes
that act upon them. This phenomenon has been demonstrated by the action
of bovine pancreatic DNAse I on DNA bound to either Cu(II) or Hg(II), as
shown in Table I (Clark and Eichhorn, 1974). Cu(II) binds preferentially
to guanine sites, and as a result guanine sites are protected extensively

TABLE I:
End group analysis of DNase produced soluble fragments
in the presence of various metal ions.[a]

Metal ion	End groups (% of total nucleoside)			
	A	T	C	G
Cu^{2+}	18	55	18	9
Mg^{2+}	19	40	18	23
$Hg^{2+} (+Mg^{2+})$	52	16	10	21

[a]Boxes show the major decreases in base content of end
groups in the reaction with copper and mercury. Standard
deviations of the mean are as follows: Cu^{2+}, A, 1.0; T, 1.4;
C, 1.5; G, 1.8; Mg^{2+}, A, 0.5; T, 1.9; C, 1.8; G, 1.1; Hg^{2+},
A, 2.5; T, 0.3; C, 3.4; G, 3.1.

from the action of the enzyme. Hg(II) on the other hand binds preferentially to thymine sites, and as a result thymine sites are protected.
The reason for this specificity is presumably that the enzyme recognizes
a certain configuration around a nucleotide, and the change in configuration produced by metal binding obfuscates the recognition. Metal binding specificity is therefore translated into enzyme specificity.

Another way in which metal binding to nucleic acid bases can alter
an important enzymatic reaction is illustrated by RNA synthesis through
the action of *e. coli* RNA polymerase on a DNA (poly dAT) template that
has been modified by binding to cis-Pt(NH$_3$)$_2$Cl$_2$ (Froehlich and Eichhorn,
1974). The higher the amount of platinum bound to the DNA template, the
shorter the size of the RNA pieces becomes (Table II). We believe that
the enzyme cannot get past the barrier posed by the platinum and that
therefore the platinum serves as a sort of stop sign for RNA synthesis.

2. EFFECTS OF PHOSPHATE BINDING

Metal ions such as Mg^{2+} that bind to nucleic acid phosphate groups can
also have dramatic impact upon the structure and function of the genetic
molecules. In contrast to the disordering effect of base-binding metal

TABLE II

The effect of cis $[Pt(NH_3)_2 Cl_2]$ on RNA chain length

Measured by incorporation of $\gamma-{}^{32}P$ ATP[a] and ${}^{3}H$ UMP[b]

by E.coli RNA polymerase[c]

$\dfrac{\text{Pt (moles)}}{\text{Nucleotide (moles)}}$	${}^{3}H$ UMP/$\gamma-{}^{32}P$ ATP
0	1924
1:1000	374
1:500	242
1:200	126
1:100	104
1:40	53
1:10	26

[a] Measures end groups.
[b] Measures total incorporation.
[c] Using poly dAT template.

ions, those that bind phosphate stabilize ordered structures, whether single or double helical. The stabilization is due to the neutralization of the array of mutually repulsive negative charges on the phosphate groups on the surface of the polynucleotides.

The ability of divalent metal ions to stabilize the interaction between polynucleotide strands, as in a double helix, can bring about the mispairing of nucleotide bases (Eichhorn *et al.*, 1973). At low concentrations of divalent metal ions such strand interaction is relatively unstable, and therefore only those base pairs with the most stable hydrogen bonds, i.e. complementary base pairs, are formed. At high metal ion concentrations the strand interactions are stabilized with the result that hydrogen bonding between less stable base pairs is permitted. High metal ion concentration is therefore expected to lead to 'mispairing'

Such 'mispairing' at high Mg^{2+} ion concentration has been demonstrated by studying the reaction of various synthetic polynucleotides, such as poly($I_{0.5}$, $U_{0.5}$) and poly(A). At low Mg^{2+} concentration only the complementary A-U pairs are formed, but at high $[Mg^{2+}]$, A-I as well as A-U base pairs are produced, as shown schematically in Figure 5. Solely complementary base pairing corresponds to 2:1 poly(I,U):poly(A) stoichiometry, while 'mispairing' leads to 1:1 stoichiometry, and Figure 6 shows that increasing $[Mg^{2+}]$ produces a transition from the ratio expected for complementarity to that predicted for mispairing. This mispairing phenomenon has been confirmed by NMR studies (Eichhorn *et al.*, 1973) in which the resonances of hydrogen-bonded bases are broadened and the resonances of the bases remain sharp.

This transition from solely complementary base pairing to 'mispairing' occurs between 10 and 20 mM Mg^{2+} concentration. The concentration range for this transition is similar to the concentration range during which errors are introduced in protein synthesis *in vitro*. In the

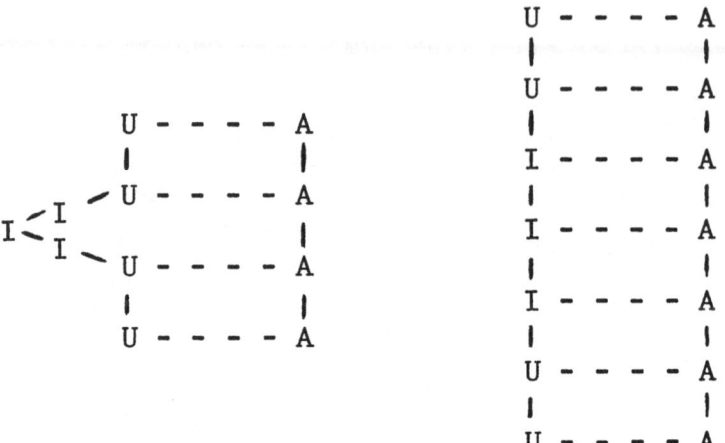

Fig. 5. Scheme for reaction of poly(I, U) and poly(A). Left,
complementary reaction; right, mispairing reaction.

lower $[Mg^{2+}]$ range only the 'correct' amino acids are incorporated, while
in the higher range a simultaneous incorporation of correct and incorrect
amino acids takes place (Szer and Ochoa, 1964). This correlation leads
to the possibility that misincorporation of amino acids into protein at
high metal ion concentration is caused by the mispairing of bases in the
codon-anticodon interaction. Theoretically metal ion induced mispairing
of bases could lead to errors in DNA and RNA synthesis as well as in protein synthesis.

Another consequence of metal binding to phosphate groups is the
cleavage of the phosphodiester bonds that results in the degradation of
RNA molecules. The mechanism of this reaction requires the presence of
a 2'OH group, so that the degradation occurs only with RNA and not with

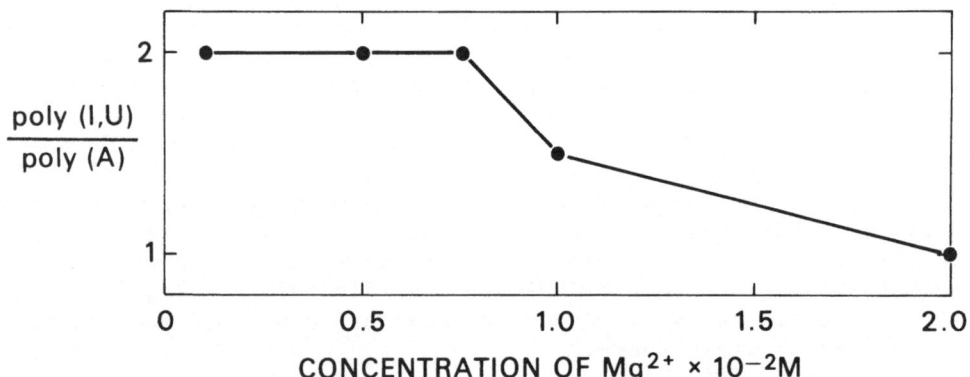

Fig. 6. Stoichiometry of reaction of 9×10^{-5}M poly(I,U) and
poly(A) as a function of Mg(II) concentration.

DNA (Butzow and Eichhorn, 1971). In fact one can speculate that one of
the reasons for the evolutionary selection of DNA rather than RNA as the
most common bearer of the primary genetic information is the greater
stability of DNA toward degradation by metal ions (Butzow and Eichhorn,
1975).

3. EFFECTS ON NUCLEOPROTEINS

Thus far we have illustrated the ways in which metal ions can alter structure and function of nucleic acid molecules without regard to other molecules to which they may be attached. It remains for us to consider the
influence of metal ions on nucleic acids that are associated with proteins
or polypeptides, as they are in the nucleus of eukaryotic cells. We therefore examine the effect of metal ions on the DNA-polylysine complex and
again become aware of the importance of the nature of the metal binding
site – phosphate or base.

Polylysine is bound to DNA in a manner that bears some resemblance
to the way in which histones, which contain a high proportion of lysine,
are bound to DNA in the cell nucleus. The polylysine has the effect of
greatly increasing the magnitude of the circular dichroism of DNA (Figure
7) so that the region around 270 nm is characterized by high negative
ellipticity (Jordan et al., 1972). This high ellipticity is due to
'handed' association, or anisotropic packing, of

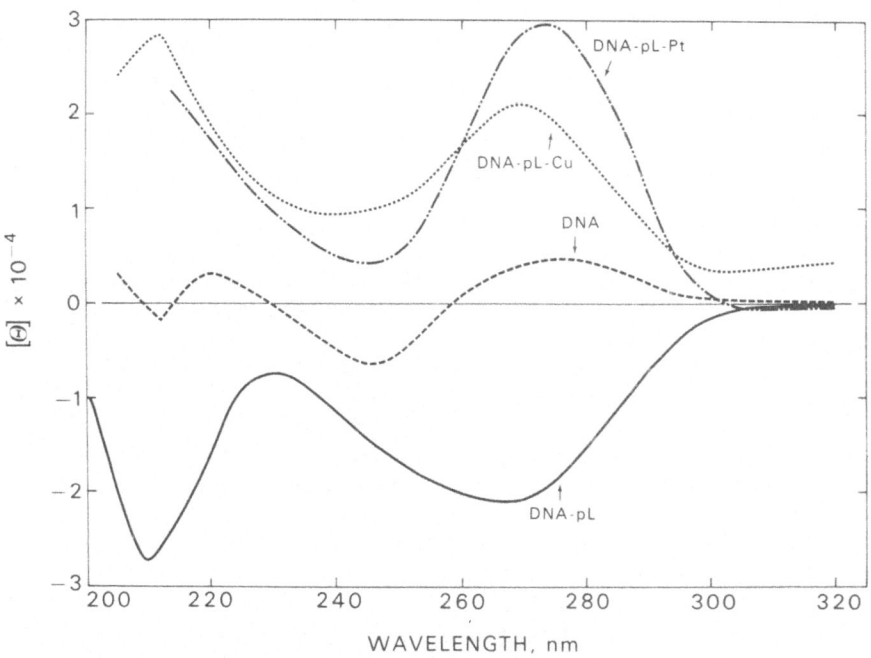

Fig. 7.   Effect of $CuCl_2$ and cis-$Pt(NH_3)_2Cl_2$ (1×10^{-4}M) on CD
spectra of complex of DNA and poly(Lys) in 1:0.5 ratio (DNA
concentration 1×10^{-4}M).

the DNA-polylysine complexes. Histone Hl in association with DNA gives a similar CD spectrum.

Metal ions binding to phosphate, such as Mg^{2+}, Mn^{2+}, Co^{2+}, or Zn^{2+}, enhance the negative ellipticity. This enhancement indicates that these metal ions increase the degree of intermolecular association of the DNA-polylysine complex (Shin and Eichhorn, 1972).

Metal ions that bind to the bases elicit an entirely different result (Shin and Eichhorn, 1976). This is illustrated by the CD spectrum obtained when the DNA-polylysine complex is heated with Cu(II). The magnitude of the ellipticity is like that of DNA-polylysine in the absence of divalent metal, but the sign of the ellipticity is now positive instead of negative (Figure 7). The reaction of cis[Pt(NH$_3$)$_2$Cl$_2$] with DNA-polylysine also produces such high positive ellipticity.

We interpret these CD spectra with positive bands in the 270 nm region, which have also been obtained with certain DNA-histone complexes in the absence of divalent ions, as indicative of a conformational change in the DNA structure which in turn causes the direction of the twist of the intermolecular packing to reverse. Removal of Cu^{2+} ions with EDTA converts this spectrum into one characteristic of DNA-polylysine in the absence of base binding metals. Obviously divalent metal ions strongly influence the manner in which a DNA-polypeptide complex is packed in a supramolecular association.

Divalent metal ions binding to either phosphate or nucleotide base groups thus exert remarkable effects on the structure and function of nucleic acids and nucleic acid-polypeptide complexes. It is conceivable that some of these reactions, which have been studied *in vitro*, may represent processes that are deleterious to the propagation of the genetic information in the cell. In this connection it may be noted that the concentrations of various metal ions in various tissues varies with the age of the organism.

REFERENCES

Berg, P., Fancher, H., and Chamberlain, M.: 1963, in H. Vogel (ed.), *Symposium on Informational Macromolecules*, Academic Press, New York, N.Y., 1963, p. 467.
Butzow, J.J., and Eichhorn, G.L.: 1975, *Nature* 254, 358.
Butzow, J.J., and Eichhorn, G.L.: 1971, *Biochemistry* 10, 2019.
Clark, P., and Eichhorn, G.L.: 1974, *Biochemistry* 13, 5098.
Eichhorn, G.L.: 1973, 'Complexes of Polynucleotides and Nucleic Acids', in G.L. Eichhorn (ed.), *Inorganic Biochemistry*, Elsevier, Amsterdam, p. 1210.
Eichhorn, G.L., Berger, N.A., Butzow, J.J., Clark, P., Heim, J., Pitha, J., Richardson, C., Rifkind, J.M., Shin, Y., and Tarien, E.: 1973, 'Some Effects of Metal Ions on the Structure and Function of Nucleic Acids', in Sanat K. Dhar (ed.), *Metal Ions in Biological Systems*, Plenum Publishing Corp., New York, N.Y., p. 43-66.
Eichhorn, G.L., and Clark, P.: 1965, *Proc. Natn. Acad. Sci. U.S.* 53, 586.
Fasman, G.D., Schaffhausen, B., Goldsmith, L., and Adler, A.: 1970, *Biochemistry* 9, 2814.

Froehlich, J.P., Richardson, C., and Eichhorn, G.L.: 1974, 'Platinum Complexes as Probes of the Mechanism of RNA Polymerase', in *Abstracts of Papers*, Biochemistry/Biophysics 1974 Meeting, Minneapolis, Minn., June, 1974.

Jordan, C.F., Lerman, L.S., and Venable, J.H.: 1972, *Nature New Biology* 236, 67.

Rifkind, J.M., Shin, Y.A., and Eichhorn, G.L.: 1976, *Biopolymers*, in press.

Rifkind, J.M., and Eichhorn, G.L.: 1972, *J. Amer. Chem. Soc.* 94, 6526.

Shin, Y.A., and Eichhorn, G.L.: 1976, *Biopolymers*, in press.

Shin, Y.A., and Eichhorn, G.L.: 1972, *Biophysical Society Abstracts* 12, 248a.

Steck, T.L., Caicuts, M.J., and Wilson, R.G.: 1968, *J. Biol. Chem.* 243, 2769.

Szer, W., and Ochoa, S.: 1964, *J. Mol Biol.* 8, 823.

DISCUSSION

Zundel: With poly(A) Mg^{2+} ions turn the phosphate groups toward the base residues and induce in this way a monohelical structure (Kôlkenbeck, K., and Zundel, G.: 1975, *Biophys. Struct. Mechanism* 1, 203). This could be the reason for the more cooperative melting.

The interaction of Cu^{2+} with poly(nucleotide) cannot be used as a model for the Mg^{2+} effects, since the nature of interaction of Cu^{2+} ions with these compounds is very different from the Mg^{2+} poly(nucleotide)-interaction.

Eichhorn: In much of our work on metal ion reactions with nucleic acids we have pointed out the opposite characteristics of Cu^{2+} and Mg^{2+}. We first did so by showing that Mg^{2+} raises the T_m of DNA, while Cu^{2+} lowers it (*Nature* 194, 474, 1962). Therefore, the last thing that we would want to do is to use Cu^{2+} interactions as models for Mg^{2+} effects. The effects of Cu^{2+} noted in this lecture were the effects of binding to bases, which Cu^{2+} does and Mg^{2+} does not.

For this reason also I do not believe that the effect of Mg^{2+} that you have shown is likely to occur with Cu^{2+}. The cooperative phenomena that we have demonstrated with Cu^{2+} do not take place with Mg^{2+}. Incidentally, I assume that your studies involved the conversion of double helix to single helix, while ours involved the disordering of single helix.

Werber: Is Pt(II) the only metal ion that inhibits DNA transcription? Could other 'inert' metal ions cause the same effect?

Eichhorn: We have demonstrated the decrease in size of RNA products only with Pt complexes. The effect is very much greater with cis[Pt$(NH_3)_2Cl_2$] than with the corresponding *trans* complex. One can conclude, therefore, that metal binding alone is not sufficient for maximal effect, but that it requires a certain geometry. I would guess that other metals binding to DNA bases can produce the necessary geometry, but we have no experiments bearing on this matter.

CRYSTALLOGRAPHIC STUDIES ON COPPER COMPLEXES OF NUCLEIC ACID COMPONENTS

EINAR SLETTEN
*Dept. of Chemistry, University of Bergen, 5000 Bergen,
Norway*

1. INTRODUCTION

The biological functioning of the nucleic acids is supposed to involve
the participation of metal ions. Certain metal ions stabilize the DNA
double helix while others have a strong destabilizing effect. When DNA
is denatured by heating in the presence of *e.g.* Cu^{2+} ions, the double
helix is reversibly renatured by cooling and subsequent addition of
electrolyte (Eichhorn and Shin, 1968). In the absence of metal ions the
cooling process does not produce a reversible rewinding of the two
strands. This phenomena is explained by copper ions forming crosslinks
between the two strands in the unwound form so as to keep them in register, making a reversible rewinding possible (Geiduschek, 1962). Many
metal ions are able to bind both phosphate and base sites, and their
effect on DNA depends on their relative affinity for the two types of
binding sites.
    Structural data on metal complexes of nucleic acid components have
been published by several groups the last few years (*e.g.* de Meester *et
al.*, 1974a, b, c; Sletten and Fløgstad, 1976; Aoki, 1975). These data
combined with information from spectroscopic investigations (*e.g.* Kotowycs
and Suzuki, 1973) have enabled us to draw some definite conclusions about
the coordination chemistry of these biologically important ligands. In
the following I will summarize some of the work on copper complexes which
we have carried out in our laboratory.

2. COPPER COMPLEXES OF NUCLEOSIDE ANALOGUES

The crystallographic work started on complexes of simple purine bases,
which were found to bind to copper at N(3) and N(9) forming dinuclear
complexes (Sletten, 1969; Sletten, 1970). Since N(9) binding is not possible in nucleosides and nucleotides these structures have limited biological interest. Attempts to crystallize copper complexes of nucleosides
were not successful. Instead a series of copper complexes of nucleoside
analogues where N(9) is blocked by a methyl group, were synthesized and
obtained in crystalline form.
    The coordination geometry of the first complex of this series,
(hypoxanthine)$_2$(H$_2$O)$_2$CuCl$_2$, is shown in Figure 1 (Sletten, 1974). The
copper ion is situated at a center of symmetry and has octahedral (4+2)

*B. Pullman and N. Goldblum (eds.), Metal-Ligand Interactions in Organic
Chemistry and Biochemistry, part 1 , 53 - 64. All Rights Reserved.
Copyright   1977 D. Reidel Publishing Company, Dordrecht-Holland.*

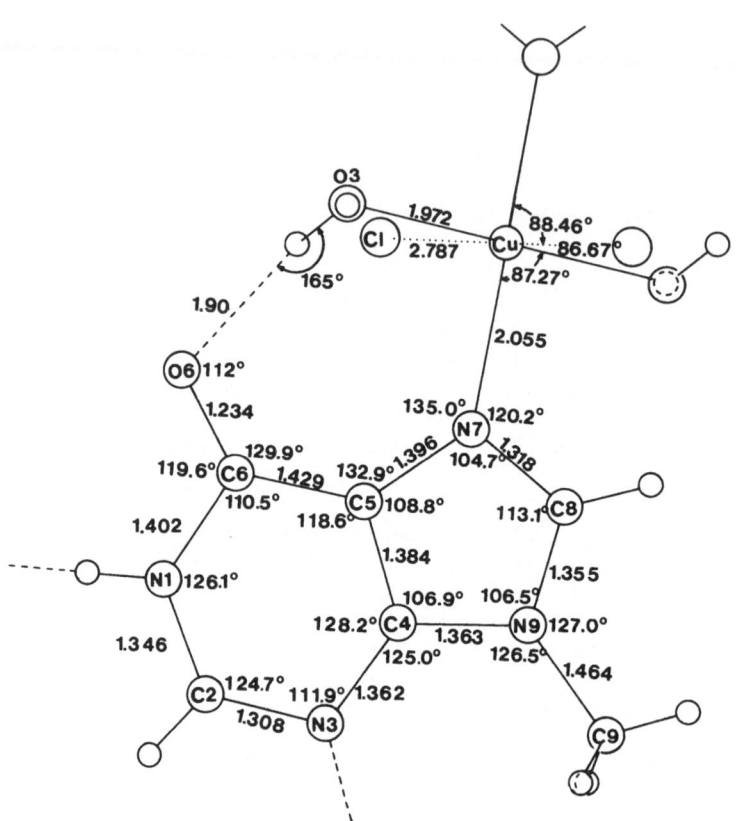

Fig. 1.   The molecular dimensions of [(9-methylhypoxathine)$_2$ (H$_2$O)$_2$CuCl$_2$]·3H$_2$O.

coordination. The hypoxanthine ligands are attached to copper in equatorial positions at imidazole nitrogen N(7). The chloride ions are situated in axial positions at a fairly long Cu-Cl distance of 2.785 Å. A
water molecule in equatorial position is involved in a relatively strong
hydrogen bond to the carbonyl group. This strong intramolecular hydrogen
bond between the substituent on carbon C(6) and a ligand in the coordination sphere is a characteristic feature in these purine complexes. In
the sulphate complex of the adenosine analogue this intramolecular hydrogen bond is established between the amino group and the sulphate ion
coordinated to copper (Figure 2) (Sletten and Thorstensen, 1974). In the
structure of (9-methyladenine)$_2$(H$_2$O)$_4$CuCl$_2$ (Figure 3) the chloride ion
is expelled from the coordination sphere. The tetra-aqua complex has a
very rare (2+4) coordination with inverted Jahn-Teller distortion, the
axial Cu-N(7) bonds being the shorter (2.008 Å) and the equatorial Cu-O
bonds being the longer (2.162 Å) (Sletten and Ruud, 1975). The complex
unit lies on a two-fold axis and a mirror plane, and the unusual coordination geometry may be a result of crystallographic symmetry requirements
The intramolecular hydrogen bond arrangement is established between a

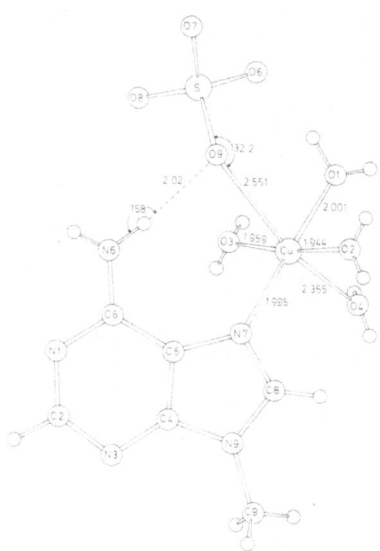

Fig. 2. The coordination geometry of [(9-methyladenine)$_2$
(H$_2$O)$_4$CuSO$_4$]·H$_2$O.

bifurcated amino hydrogen atom and two water molecules in the coordination sphere. The chloride ion is interacting with the base through
a C — H···Cl contact which may be considered a hydrogen bond.
      In Figure 4 the copper complex of the guanosine analogue, (9-methyl-
guanine)$_2$(H$_2$O)$_4$CuSO$_4$, is shown (Sletten and Fløgstad, 1976). The guanine
ligands have *syn*-configuration with a water molecule in the coordination
sphere serving as a hydrogen bond bridge between the two carbonyl groups.
The sulphate ion is interacting with guanine through a strong di-hydrogen
bond involving N(1)-H and the amino group. Guanosine is unique among the
purine and pyrimidine nucleosides, being the only one having two hydrogen bond donors. Furthermore the 'bite' distance between the two donor
H atoms at N(1) and the amino group (2.3 Å) is compatible with the distance between the acceptor atoms in oxyanions as *e.g.* SO$_4^{2-}$, ClO$_4^-$, NO$_3^-$,
PO$_4^{3-}$. The strong di-hydrogen bond established between guanine and sulphate may prevent the anion from interacting directly with the metal.
The H-bond donor properties of the purines together with steric requirements may determine whether the anion will enter the coordination sphere
or interact with the purine ring. Usually, the anionic effect has not
been dealt with explicitly in investigations of metal-nucleic acid interaction. The present result indicates that oxyanions may have a profound
effect on guanine rich polynucleotides.
      To find out whether oxyanions in general tend to form di-hydrogen
bonds with guanine, the corresponding nitrate complex was investigated
(Sletten and Erevik, 1976). The formation of a di-hydrogen bond was indeed confirmed (Figure 5), however, in addition, the nitrate ion is simultaneously coordinated to copper. The smaller size of NO$_3^-$ compared
to SO$_4^{2-}$ apparently makes it easier to accommodate both a metal bond and
the di-hydrogen bond in the NO$_3^-$ compound.

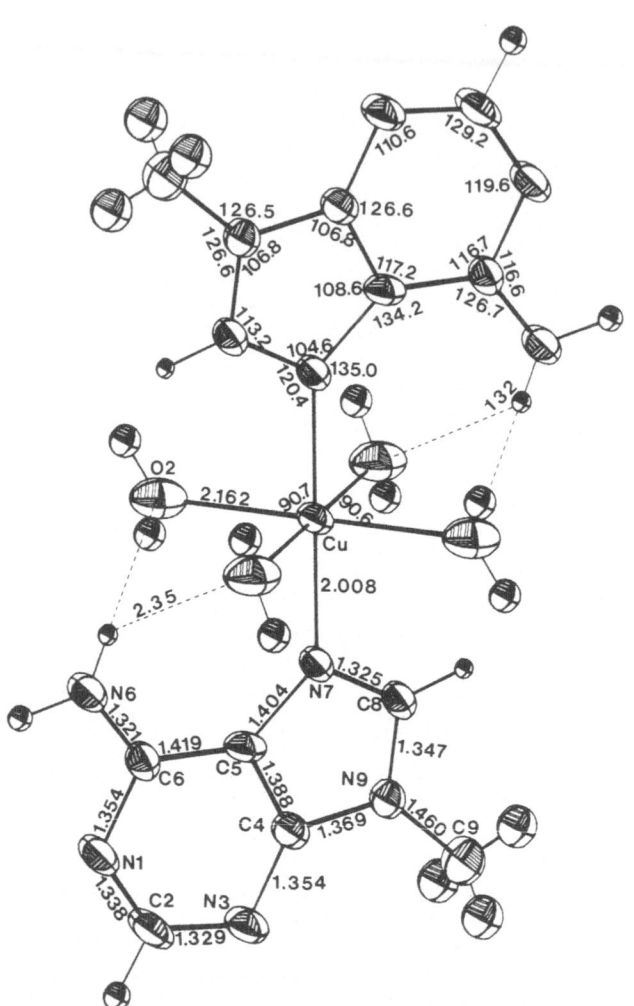

Fig. 3. Interatomic distances and angles in the complex unit of $[(9\text{-methyladenine})_2(H_2O)_4Cu]Cl_2 \cdot 2H_2O$.

For purines like adenine and hypoxanthine with only one H-donor, direct interaction with oxyanions should probably be less favourable. However, in $(9\text{-methylhypoxanthine})_2CuSO_4$ (Figure 6) the sulphate ion is positioned between the two bases so as to acquire a maximum anion-base interaction (Sletten and Kaale, 1976). The C(2)-H···O contact of 3.15 Å together with the N(1)-H···O hydrogen bond constitute an arrangement very similar to the di-hydrogen bond between purine and sulphate, described above.

Thus far only complexes of analogues of naturally occurring nucleosides have been described. They have all been shown to be mono-dentate, the binding site being imidazole nitrogen N(7). The sulphur analogue of hypoxanthine, 6-thiopurine, is known as a potent anticarcinogenic drug,

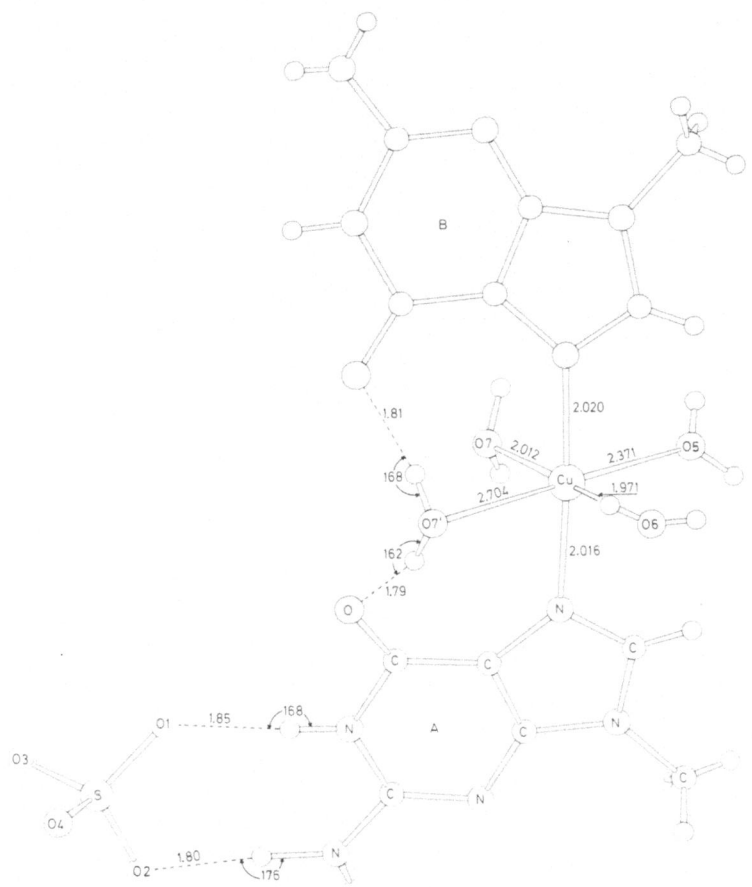

Fig. 4. The copper-guanine and sulphate-guanine bonding in
[(9-methylguanine)$_2$(H$_2$O)$_3$Cu]SO$_4$·3H$_2$O.

and its activity may be related to its metal binding properties (Furst,
1960). As shown in Figure 7 the geometry of (9-methyl-6-thiopurine)CuCl$_2$
is indeed quite different from the other copper-purine complexes (Sletten
and Apeland, 1975). Both sulphur and nitrogen N(7) are bonded to copper
forming a chelate ring. The 'bite' distance between S and N(7) is reduced
by 0.3 Å relative to the free ligand. The molecular dimensions of the
6-thiopurine ligand make it more suitable for chelation than the oxygen
analogue. Also sulphur has a greater ability than oxygen to bind to copper.

3. METAL-NUCLEOTIDE COMPLEXES

In the last couple of years structure determinations of nucleotide complexes of Zn(II), Mn(II), Ni(II) and Co(II) have been reported. The first

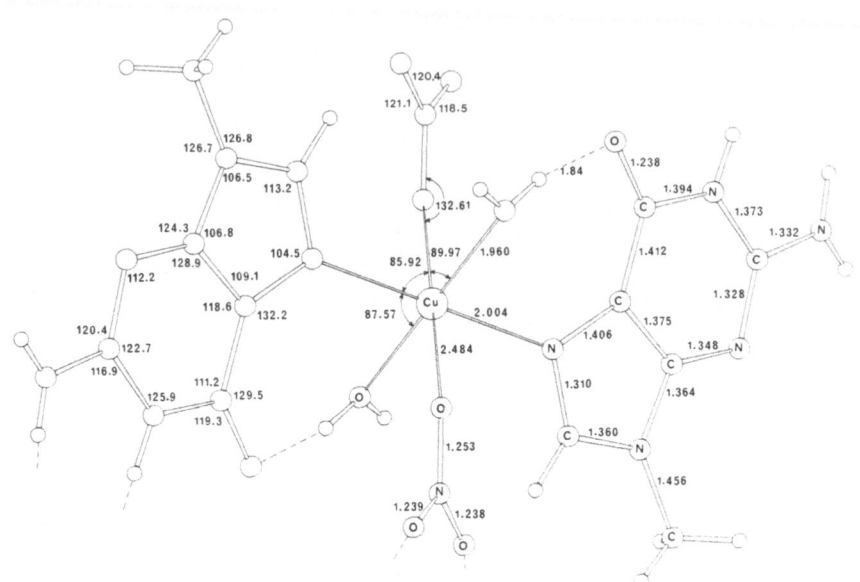

Fig. 5.   Interatomic distances and angles in (9-methylguanine)$_2$ (H$_2$O)$_2$Cu(NO$_3$)$_2$.

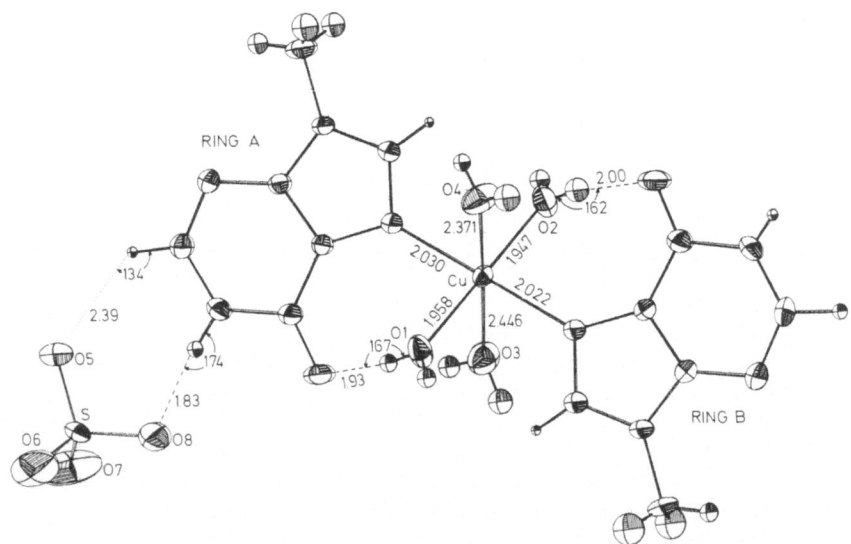

Fig. 6.   The copper-hypoxanthine and sulphate-hypoxanthine bonding in [(9-methylhypoxanthine)$_2$(H$_2$O)Cu]SO$_4$.

one, that of Zn(5'-IMP), is shown to be polymeric with the base attached at the N(7) position to a zink atom, which in turn is bonded to phosphate

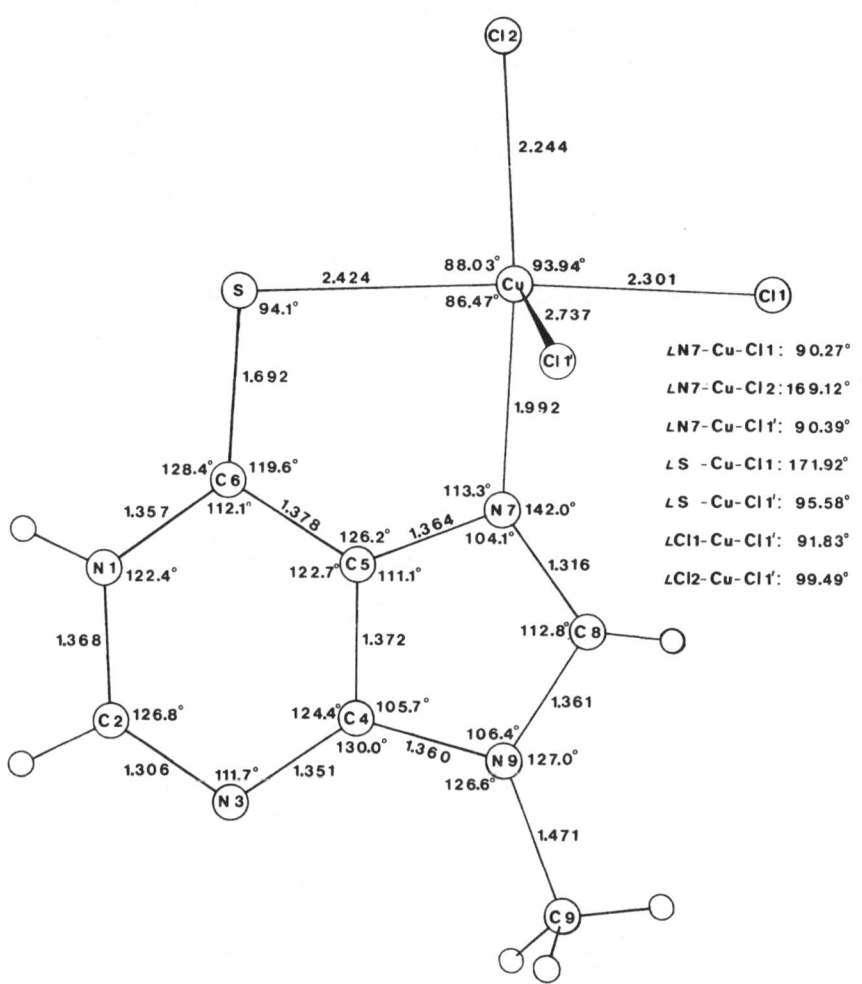

Fig. 7. The molecular dimensions of [(9-methyl-6-thiopurine) CuCl$_2$]·H$_2$O.

groups of three neighbouring 5'-IMP moieties (de Meester *et al.*, 1974a). The metals in the other metal-nucleotides are all found to be bonded to N(7) and highly hydrated. Hydrogen bonds are established between water molecules in the coordination sphere and the phosphate group (Aoki, 1975; de Meester *et al.*, 1974b, c).

Copper complexes of nucleotides have been studied by NMR techniques (Kotowycz and Suzuki, 1973; and references therein). In p.m.r. spectra of 5'-nucleotides, the H(8) signal is broadened preferably to the H(2) signal, while in ^{13}C spectra C(4), C(5) and C(8) resonances are equally affected. However, these results are not quite conclusive as to the binding site of the paramagnetic ion since both N(7) coordination and/or phosphate binding with the nucleotide in *anti* conformation might produce similar effects.

The stabilizing effect of metal ions on DNA is generally assumed to be connected with metal-phosphate binding, while destabilizing is explained by disruption of the complementary hydrogen bonding scheme. This simple picture is not corroborated by the type of coordination geometry observed in the structures of the metal-nucleotide complexes mentioned above. Furthermore, the mechanism by which certain metal ions bring about reversible rewinding of DNA is not readily explained by the type of bonding observed.

We have recently carried out a structure determination of Cu(5'-GMP) which has been found to crystallize with three copper-nucleotide complexes in the asymmetric unit (Figure 8) (Sletten and Lie, 1976). Each

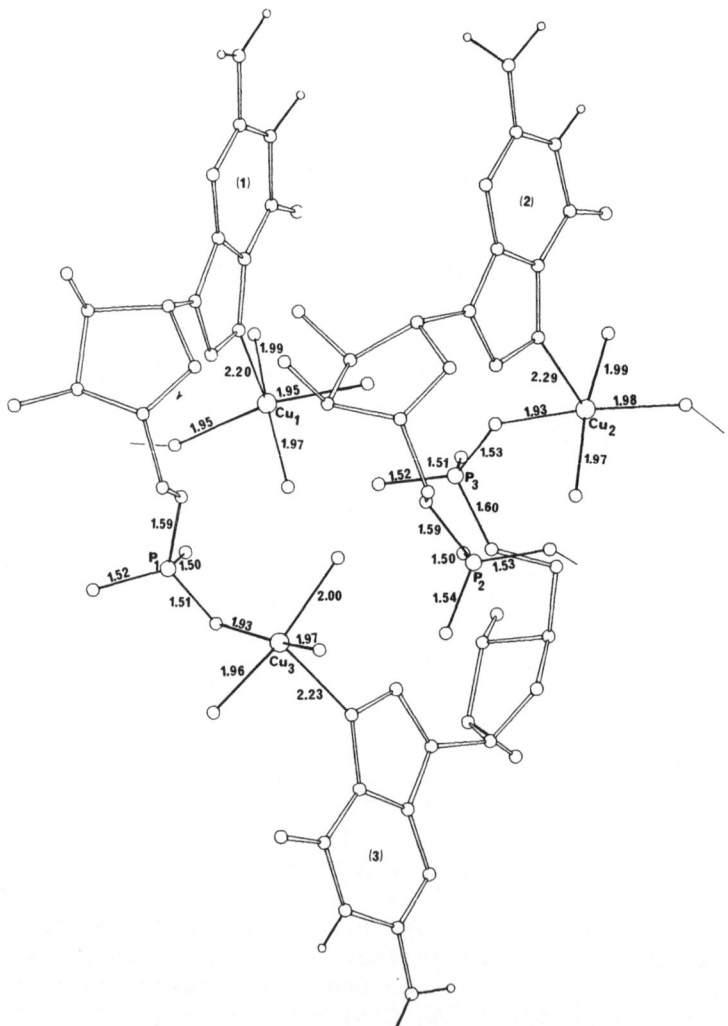

Fig. 8.   The copper-nucleotide repeat unit in the polymeric helical chain of [(5'-guanosinemonophosphate)$_3$(H$_2$O)$_8$Cu$_3$]·5H$_2$O.

guanine base is attached at the N(7) position to a copper atom, which in turn is bonded to a phosphate group of a neighbouring GMP. The copper ions have slightly irregular square pyramidal (4+1) coordination with the base nitrogen in axial position and the equatorial positions occupied by phosphate and water oxygens. The four equatorial ligands are approximately coplanar, with copper displaced out of plane towards the axial ligands. In the complexes of nucleoside analogues copper was found to be octahedrally surrounded and with an average Cu–N(7) distance of 2.016 Å. The shift to square pyramidal five coordination has lengthened the Cu–N(7) distance by approximately 0.2 Å. The characteristic intramolecular hydrogen bond between the coordination sphere and the substi-

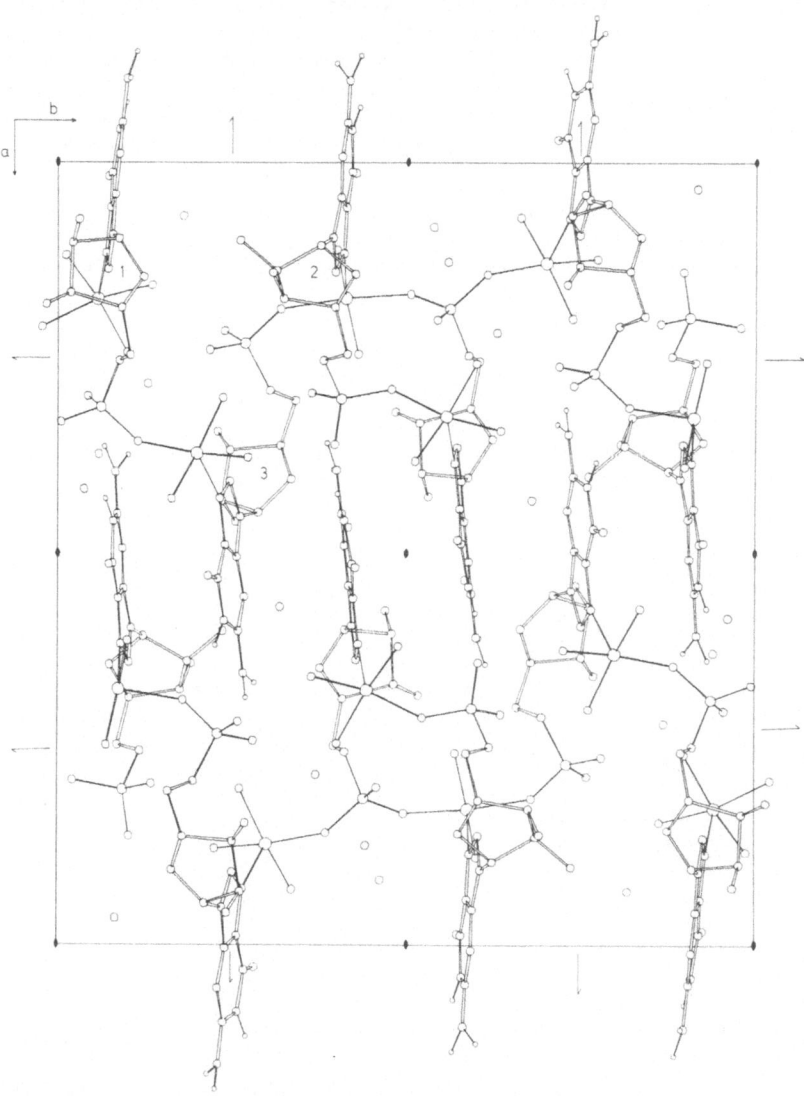

Fig. 9. The crystal packing of Cu(5'-GMP) as viewed along the *c*-axis.

tuent on C(6) is retained in the copper-nucleotide even though the ligand
configuration is drastically changed.

In phosphate transfer reactions, where the presence of divalent metal is known to be crucial, the functioning of the metal might be due to
polarizing effects excerted on the P-O bond to which it is attached. This
polarization would increase the positive charge on the phosphorous atom,
thereby facilitating a nucleophilic attack and subsequent breaking of
the P-O bond. Looking at the P-O distances in Cu(5'-GMP) there is no
systematic lengthening of the coordinated bond. The average P-O bond
distance within each group is 1.53, 1.54 and 1.54 Å, respectively. This
is in excellent agreement with the value of 1.54 Å proposed by Cruickshank
for the constant average P-O bond distance in phosphate groups
(Cruickshank, 1961). Thus there seems to be no structural evidence for
the bond polarization theory. The involvement of metal may instead be
explained by its ability to activate water, forming incipient hydroxyl
ions and simultaneously holding them in suitable position for attack on
phosphorous (Spiro, 1973). A reaction mechanism involving bidentate phosphate activation has also been proposed (Farell *et al.*, 1969). In
Cu(5'-GMP) the vacant sixth position in the coordination sphere of Cu(1)
and Cu(2) is approached by a second phosphate oxygen, and one may envis-
age a six-coordinated intermediate where bidentate phosphate activation
occurs.

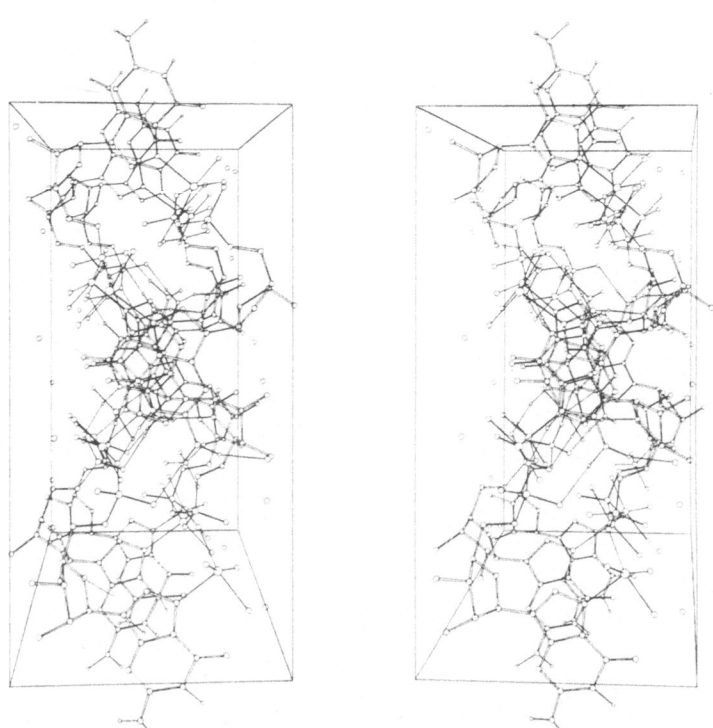

Fig. 10. Stereoscopic drawing showing the helical arrangement
of the polymeric metal-nucleotide chain in Cu(5'-GMP).

The packing arrangement in the crystal is shown in Figure 9. The nucleotides are linked together by alternating base-metal-phosphate bonds
forming an infinite helical chain about a crystallographic two-fold screw
axis. The individual nucleotides in the trimer are positioned so as to
make the screw pitch of the helix 1/3 of the *b*-axis. The base planes of
two adjacent helical chains are interleaved so as to form a stacking pattern comparable to that in DNA. The bases are almost parallel with an
average interplanar distance of 3.4 Å. The two strands are held together
via strong di-hydrogen bonds between guanine and phosphate. As already
pointed out, only guanine has the ability to form such di-hydrogen bonds
to oxyanions. In Figure 10 a stereo picture of the cell viewed along the
screw axis gives an impression of how the nucleotide chain is wound in a
right handed helical arrangement. The distance across the double stack
is approximately 20 Å which is almost identical to the corresponding
dimension for double-helical DNA.

The bonding pattern in Cu(5'-GMP) may be relevant for discussing
the mechanism of reversible unwinding and rewinding of DNA. The unwinding
process proceeds through a disruption of complementary hydrogen bonds
brought about by heating. The copper ions may in this initial stage be
coordinated to N(7) which is accessible. By subsequent unwinding of the
two single strands a phosphate group in one strand may be positioned so
as to bind to a copper ion already coordinated to N(7) in the other
strand. In this way cross-links between base and phosphate may keep the
two strands in register, making possible a reversible rewinding.

REFERENCES

Aoki, K.: 1975, *Bull. Chem. Soc. Japan* 48 1260-1271.
Cruickshank, D.W.J.: 1961, *J. Chem. Soc.* 5486-5504.
de Meester, P., Goodgame, D.M.L., Jones, T.J., and Skapski, A.C.: 1974a,
    Biochim. Biophys. Acta 353,392-394.
de Meester, P., Goodgame, D.M.L., Skapski, A.C., and Smith, B.T.: 1974b,
    Biochim. Biophys. Acta 340 113-115.
de Meester, P., Goodgame, D.M.L., Jones, T.J., and Skapski, A.C.: 1974c,
    Biochem. J. 139,791-792.
Eichhorn, G.L., and Shin, Y.A.: 1968, *J. Amer. Chem. Soc.* 90,7323-7328.
Farell, F.J., Kjellstrøm, W., and Spiro, T.G.: 1969, *Science* 164,320-321.
Furst, A.: 1960, in *Metal Binding in Medicine* (ed. by M.J. Seven and
    L.A. Johnson), Lippincott, Philadelphia, pp. 336-344.
Geiduschek, E.P.: 1962, *J. Mol. Biol.* 4 467-487.
Kotowycz, G., and Suzuki, O.: 1973,*Biochemistry* 12, 5325-5328.
Sletten, E.: 1969, *Acta Cryst.* B25,1480-1491.
Sletten, E.: 1970, *Acta Cryst.* B26,1609-1614.
Sletten, E.: 1974, *Acta Cryst.* B30,1961-1966.
Sletten, E., and Apeland, A.: 1975, *Acta Cryst.* B31,2019-2022.
Sletten, E., and Erevik, G.: 1976, to be published.
Sletten, E., and Fløgstad, N.: 1976, *Acta Cryst.* B32,461-466.
Sletten, E., and Kaale, R.: 1976, Acta Cryst., in press.
Sletten, E., and Lie, B.: 1976, Acta Cryst., in press.
Sletten, E., and Ruud, M.: 1975, *Acta Cryst.* B31,982-985.
Sletten, E., and Thorstensen, B.: 1974, *Acta Cryst.* B30,2438-2443.

Spiro, T.G.: 1973, in *Inorganic Biochemistry*, vol. 1 (ed. by G.L.
  Eichhorn), Elsevier, Amsterdam, pp. 568-573.

DISCUSSION

Saenger: Guanosine-5'-phosphate forms gels in the presence of sodium
ions. Did you observe similar gels with Cu^{2+}-ions?

Sletten: As a matter of fact, we did carry out quite a number of experiments to synthesize Cu(5'-GMP), which all failed, giving colloidal pre-
cipitation. In the only one successful, crystals appeared over night,
at pH $\simeq$ 5.5 using nitrate as electrolyte.

Truter: Did you find the donor-hydrogen...acceptor angles to be more than
$140°$ for *both* the acceptors in the bonds you describe as bifurcated?

Sletten: The N-H $\overset{\diagup O}{\underset{\diagdown O}{}}$ arrangement has a crystallographic mirror plane,

thus the two N-H...O angles are equal, being $132°$. The H···O distances
are 2.35 Å. In my opinion this clearly indicates the presence of a bifurcated hydrogen bond.

Eichhorn: We have previously postulated stacking of 5'-AMP and 3'-AMP by
Cu(II), based on NMR studies that indicated line broadening of the H8
resonance before the H2 resonance for these isomers, but not for 2'-AMP
or cyclic AMPs. It was then found that only the 5' and 3' isomers had
their N-7 group and phosphate groups in a position favorable for stacking.
I wonder therefore whether the structure that you have obtained for
5'-GMP can also be constructed for other isomers.

Sletten: We have not actually checked this by model building, however,
it looks like the helical arrangement observed for Cu(5'-GMP) with alternating base-metal-phosphate binding is dependent on a certain minimum
distance between N(7) and phosphate so as to get the bases into a proper
stacking interaction which probably is very important for stabilizing
the structure.

QUANTUM-MECHANICAL STUDIES ON CATION BINDING TO CONSTITUENTS OF NUCLEIC ACIDS AND PHOSPHOLIPIDS

BERNARD PULLMAN and ALBERTE PULLMAN
*Institut de Biologie Physico-Chimique, Laboratoire de Bio-
chimie Théorique associé au C. N. R. S., 13, rue P. et M.
Curie — 75005 Paris — France*

1. INTRODUCTION

We already had the opportunity of presenting some aspects of our studies on cation-binding to biomolecules in a preceding Jerusalem Symposium [1]. The work reported concerned, however, essentially binding to peptides and esters in connection in particular with the structure and properties of antibiotic cyclic depsipeptides. (For other papers in relation with this subject see references [2 — 6]. In the present communication we wish to present results relevant to other types of fundamental biomolecules: constituents of nucleic acids and phospholipids.

The reason for which we group together these two rather different types of compound is due to the fact that they both contain a common component, an anionic phosphodiester group, which moreover is one of the essential if not the essential center for interaction with cations in both of them. The study of the intrinsic cation binding properties of an appropriate model of this group is thus of utility for both of them. Beyond this analogy, the two types of biomolecule present, of course, their own problems and particularities.

In this paper, we would like to present results referring to two subjects:

(1) The cation binding properties of the phosphate group and the influence of the binding on the conformational properties of this group itself.

(2) The influence of cation-binding to the phosphate group on the conformational properties of the polar head of phospholipids.

A third subject, the cation binding abilities of the purine and pyrimidine bases of the nucleic acids are presented separately.

Underlying all these subjects are complex technical problems concerned with the selection of the appropriate methodology and within the methodology of the appropriate computational procedures. A general discussion of the principal aspects of this problem has been presented in references [8] and [9]. These technical considerations cannot be described in this short communication and have to be looked for in the above papers. We would like simply to indicate that our choice will constantly be the SCF *ab initio* procedure and that the computations described here

*B. Pullman and N. Goldblum (eds.), Metal-Ligand Interactions in Organic
Chemistry and Biochemistry, part 1 , 65 - 76. All Rights Reserved.
Copyright © 1977 by D. Reidel Publishing Company, Dordrecht-Holland.*

TABLE I

Gaussian exponents and contraction coefficients utilized on the
P. O, H atoms of DMP⁻

Atom	Type	Exponent	Coefficient
	s	22 566.5	0.001 5310
	s	3 380.80	0.011 793
	s	766.417	0.058 861
	s	214.964	0.208 183
	s	68.970 3	0.447 369
	s	23.919 5	0.390 968
	s	5.269 29	0.441 137
P	s	1.962 97	0.667 165
	s	0.350 435	0.557 322
	s	0.131 021	0.609 385
	p	109.959	0.028 840
	p	25.129 2	0.175 866
	p	7.511 27	0.461 765
	p	2.395 83	0.490 526
	p	0.531 089	0.391 025
	p	0.150 160	0.729 140
	d	0.380 000	1.0
	s	1 113.12	0.013 221
	s	172.26	0.087 629
	s	42.800 8	0.296 295
	s	13.371 0	0.492 042
	s	4.839 7	0.258 935
	s	1.073 8	0.497 086
	s	0.316 9	0.566 094
	p	6.922	0.148 880
	p	1.426 1	0.516 709
	p	0.321 2	0.558 700
	s	391.445	0.022 22
	s	69.735 8	0.132 968
	s	16.224 7	0.384 690
	s	5.334 6	0.458 385
	s	2.009 95	0.154 547
C	s	0.502 323	0.534 240
	s	0.155 139	0.524 992
	p	4.316 13	0.108 451
	p	0.873 682	0.461 166
	p	0.202 860	0.630 436
	s	6.480 53	0.070 480
H	s	0.981 039	0.407 890
	s	0.217 979	0.647 669

have been carried out using a combination of Clementi's type contracted (7 s, 3 p) basis set on O and C, a (3 s) basis on H, contracted to minimal with ζ = 1.2, and a Siegbahn-Roos (10 s, 6 p) basis set on phosphorus contracted to minimal and augmented by 6 d orbitals. The gaussian exponents and contraction coefficients are given in Table I. The basis used for the cations is our 3 G$^+$ basis optimized for each of them with deletion of the p orbitals of the empty valence shell (see reference [7] for details).

2. THE BINDING OF CATIONS AND ITS INFLUENCE ON THE CONFORMATION OF THE PHOSPHODIESTER LINKAGE

The problem was studied using the dimethylphosphate anian (DMP$^-$, Figure 1) as a model compound. Because of the fixation of the metal cations in the two most plausible sites, as indicated by preliminary computations

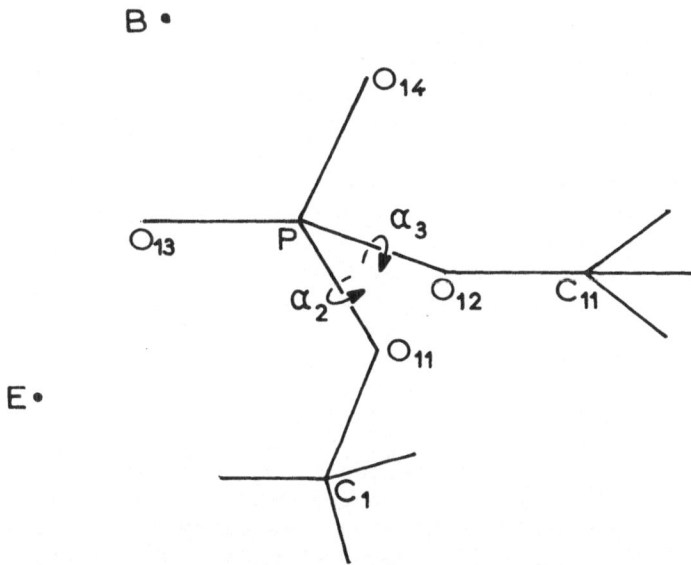

Fig. 1. DMP$^-$ and the location of the cation binding sites B and E. The conformations gg, gt (tg) and tt are considered with respect to the torsion angles α_2 (C_1-O_{11}-P-O_{12}) and α_3(O_{11}-P-O_{12}-C_{11}). This notations correspond to standard notations in the study of phospholipids [10, 11]. In the study of the nucleic acids, these two torsion angles are frequently referred to as ω, ω' [12, 13].

[9]. These sites are B on the bissectrix of the O_{13}-P-O_{14} angle (designated as B_{13} in reference [9] and E along the direction making an angle of 120° with the O_{13}P axis on the external site of the O_{13}-P-O_{14} angle (designated as E_{13} in reference [9]. For the same reasons of economy, the exploration of the conformational consequences of cation binding

was limited to the evaluation of the effect of such attachments upon
the relative energies of the three fundamental types of conformation
resulting from possible rotations about the P-O$_{ester}$ bonds : gauche-
gauche (*gg*), gauche-trans (*gt*) or trans-gauche (*tg*) and trans-trans
(*tt*). In free DMP⁻ the preferred conformation is *gg*, followed by *gt* (*tg*),
followed in turn by *tt* [10, 11]. With the basis set utilized in this
study, the *gt* (*tg*) form lies 2.3 kcal mole⁻¹ and the *tt* 5.7 kcal mole⁻¹
above the global one.

The principal results relevant to the energies of binding and the
influence of the binding upon the stability of the conformers are summed up in Table II. The main conclusions which can be drawn from this
study seem to be the two following ones:

(1) The strongest binding corresponds to the B site. Na⁺ and Mg⁺⁺
bind more strongly than K⁺ and Ca⁺⁺, respectively; the divalent cations
bind much more strongly than the monovalent ones. The ordre of decreasing
binding is Mg⁺⁺ > Ca⁺⁺ > Na⁺ > K⁺ as predictid by theoretical considerations [7].

(2) The fixation of the cations at the B site does not influence
the order of conformational preferences with respect to the torsion
about the P-O$_{ester}$ bonds existing in the free DMP⁻ : the *gg* conformation
remains the most stable one, followed by the *gt* (*tg*) one, followed in
turn by the *tt* one. In fact, the cation binding in its preferred B site
increases even the relative stability of the *gg* form with respect to
the two others. On the contrary, the binding of the cations to the external E site is able to produce modifications in the previous order of
conformational stabilities, the perturbations being stronger with the
divalent Ca⁺⁺ than with the monovalent Na⁺; with the monovalent Na⁺
the *tg* forms becomes the most stable one, followed, in decreasing order
of stability by *gg*, *tt* and *gt*; with the divalent Ca⁺⁺ both the *tg* and the
tt conformers become more stable than the *gg*, only the *gt* being less stable than it.

The essential conclusion which can be drawn from these results is
that according to the site of binding, the cation may or may not perturb
the intrinsic conformational preferences of DMP⁻. There are no direct
experimental results which may be compared with the theoretical predictions. X-ray crystallographic results on related systems show a conser-
vation of the *gg* conformation in some of them (barium diethylphosphate
[14], magnesium diethylphosphate [15] and glycerylphosphorylcholine
CdCl$_2$ trihydrate [16] but not in others (silver diethylphosphate [17]
The crystal data correspond, however, generally to complex interactions,
involving more than two interacting entities and cannot be easily compared with the theoretical ones, which must thus be considered for the
while being as primarily model studies on essential possibilities.

3. THE INFLUENCE OF CATION BINDING TO THE PHOSPHATE GROUP ON THE CONFOR-
   MATION OF THE POLAR HEAD OF PHOSPHOLIPIDS

We have demonstrated previously [18, 19] that the preferred conformation
of the polar head of phospholipids (whether phosphatidylethanolamines or
phosphatidylcholines) in its free state, corresponds to a highly folded

arrangement representing a strong electrostatic interaction between its cationic terminal group and its anionic phosphate group. NMR studies confirm the presence of such structures in non polar solvents [20]. With the usual notations for the torsion angles in phospholipids [10, 11] this arrangement is associated in the basis set utilized in the present work with a *gg* orientation about the α_2 and α_3 (the P-O bonds) and values of $\alpha_4 = 270°$ and $\alpha_5 = 30°$ (Figure 3). We have also shown (19) that the interaction of the polar head with water should produce an extension of the structure, α_5 remaining gauche but α_4 tending towards $180°$.

We now present the results of a preliminary exploration of the effect of cation binding to the phosphate group upon the conformation of the polar head, represented by the model compound EP (Figure 2). The effect was studied for Na$^+$ and Ca^{++} attached at the *B* and *E* binding sites.

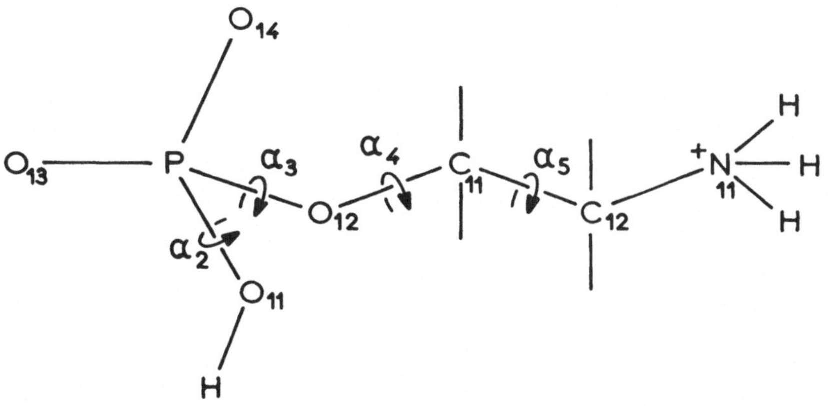

Fig. 2.   Ethanolamine phosphate (EP), model compound for the polar head of phospholipids. The torsion angles α_4 and α_5 are defined $\alpha_4 (P-O_{12}-C_{11}-C_{12})$ and $\alpha_5 (O_{12}-C_{11}-C_{12}-N^+)$ [10[5], 11].

The results of this exploration for Na$^+$ are summed up in Figs. 4-7 which are of two kinds. Figures 4 and 5 represent the interaction energies of the cations, fixed at sites *B* and *E*, with EP, as a function of α_4 and α_5, taken with respect to the energy of the most stable conformation of free EP. Figures 6 and 7 represent the conformational energy maps of the adducts, as a function of α_4, α_5, with respect to the most stable conformation of the adduct taken as energy zero. Because of the high cost of the computations only selective points of the α_4, α_5 map have been evaluated. They seem, however, to lead to straightforward conclusion.

Thus it is seen in Figure 4 that for the adduct at *B* a number of combinations of $\alpha_4 - \alpha_5$ correspond to greater interaction energies than the one associated with the minimum of the free EP. As a result one observes, in Figure 6, that the most stable conformation of this adduct is substantially different from the most stable conformation of free EP. In particular, the preferred conformation of the adduct while still corresponding to an essentially *gauche* arrangement with respect to α_5 is

TABLE II

Cation binding to DMP⁻

Site of cation	The cation to DMP⁻	Energy of cation binding : ΔE with respect to DMP⁻ and cation at infinite separation				Difference in energy between conformers		
		gg	gt	tg	tt	$\Delta E(gt\text{-}gg)$	$\Delta E(tg\text{-}gg)$	$\Delta E(tt\text{-}gg)$
Free DMP⁻						2.3	2.3	5.7
B	Na⁺	-150.3	-148.3	-148.3	-146.0	4.3	4.3	10.0
	Mg⁺⁺	-340.1	-335.2	-335.2	-330.0	7.2	7.2	15.9
	K⁺	-128.3	-126.3	-126.3	-124.1	4.3	4.3	9.9
	Ca⁺⁺	-285.0	-280.3	-280.3	-275.2	7.0	7.0	15.4
E	Na⁺	-118.6	-115.8	-125.6	-122.7	5.1	-4.7	1.6
	Ca⁺⁺	-227.0	-220.3	-240.0	-233.7	8.5	-10.8	-1.3

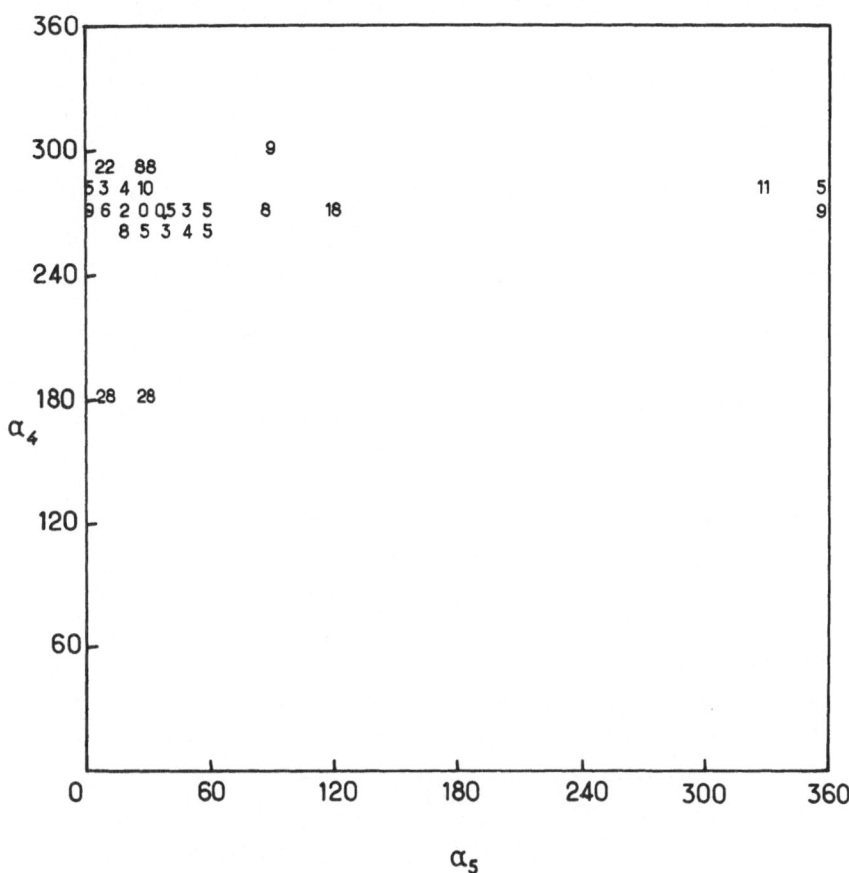

Fig. 3.   Conformational energy map of PE, computed by SCF *ab initio* method with the basis set indicated in the text. Energies in kcal/mole with respect to the most stable conformation taken as energy zero.

is associated with α_4 = 180° (instead of 270° in free EP). The fixation of a cation at B has thus the effect of extending the structure of the polar head with respect to α_4. This effect is thus comparable to the influence of polyhydration [19].

   On the other hand cation binding at the E site has quite different consequence. Figure 5 indicates that the binding energies conserve their highest values in the vicinity of (α_4, α_5) = (270°, 30° − 90°) i.e. in the vicinity of the most stable conformation of free EP and are appreciable lower in the other explored regions. The examination of the molecular models shows that the low interaction energies found for the two indicated values of α_5 (0°, 330°) and for α_4 = 180° should be conserved for other conformations with or in the vicinity of this last value of α_4. As a result, the most stable conformation of the E adduct is gener-

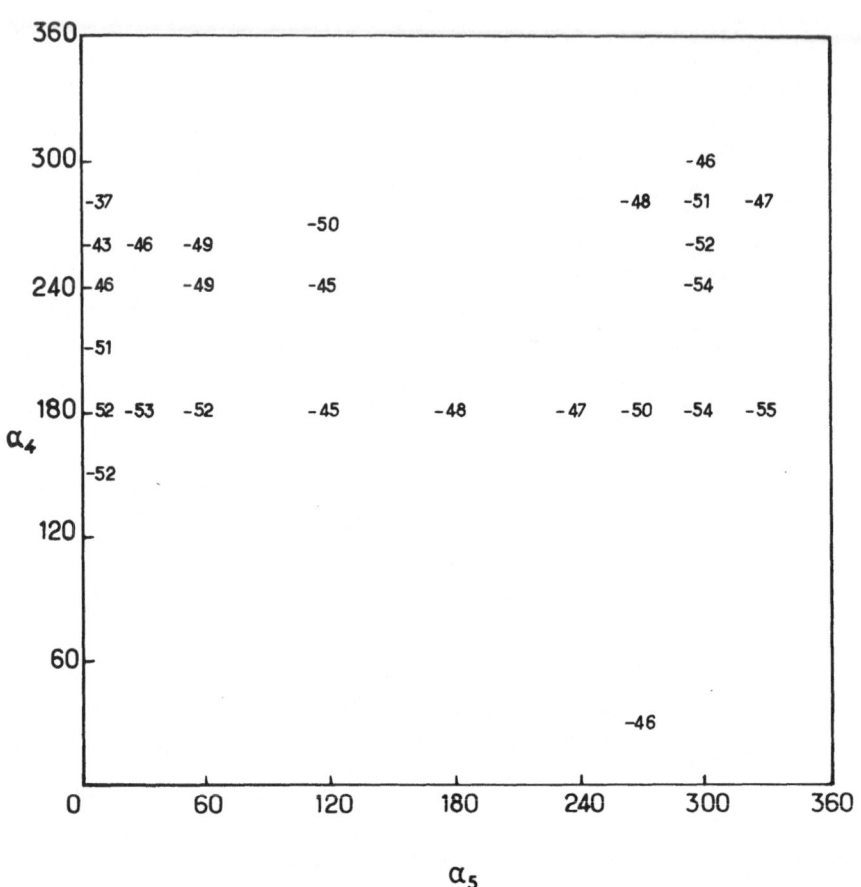

Fig. 4. Interaction energies (kcal/mole) for the formation
of the EP...Na⁺ adduct (cation at site B), with respect to
the sum of the energies of the free constituents.

ally the same as that for free EP, although its location is less strin-
gent (Figure 7). The minimum acquires a substantial degree of fluidity
with respect to α_5, no significant variation of the conformational energy occurring for $30° < \alpha_5 < 90°$.

Computations performed for the effect of the Ca⁺⁺ ion, the details
of which will be presented elsewhere, show essentially similar results.
The fixation of this cation at site B favors again the elongated conformation associated with $(\alpha_4, \alpha_5) = (180°, 300°)$, which moreover is now
stabilized by 39 kcal/mole with respect to the conformation prevalent
for free EP $(\alpha_4, \alpha_5) = (270°, 30°)$. The fixation of the cation at site
E labilizes the stable conformation with respect to α_5. In fact the conformation of the cation-bound ligand at $(\alpha_4, \alpha_5) = (270°, 90°)$ is now
about 6 kcal mole -1 more stable than that at $(\alpha_4, \alpha_5) = (270°, 30°)$.

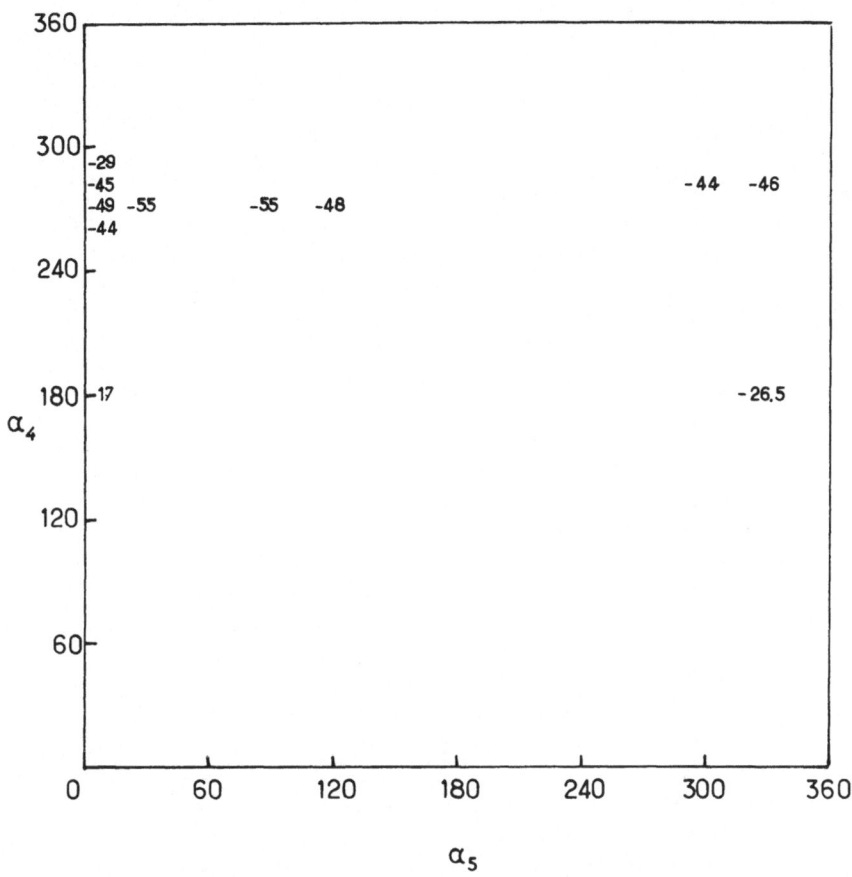

Fig. 5. Same as Figure 4 but the cation at site *E*.

4. CONCLUSION

The present work indicates the high affinity of the phosphate monoanion for the binding of cations and illustrates the possibility that such an interaction may produce conformational changements in the ligand for some sites of binding but not necessarily for all of them. The necessity of using refined *ab initio* treatments of the problem is evident from the observation that the results presented here on the absence of any effect on the preferred conformation of DMP‾ by cation binding to the *B* site are in contradiction with results obtained for the same situation using the CNDO/2 method [21] which predicts that the *tt* conformation becomes the most favorable one in DMP‾...Na$^+$ and DMP‾...Mg^{++} complexes of this sort. The indications of the CNDO method are certainly artifacts of the procedure. An analysis of the causes of this artifact is given in our paper on cation binding to purine and pyrimidine bases where similar discrepancies occur [22].

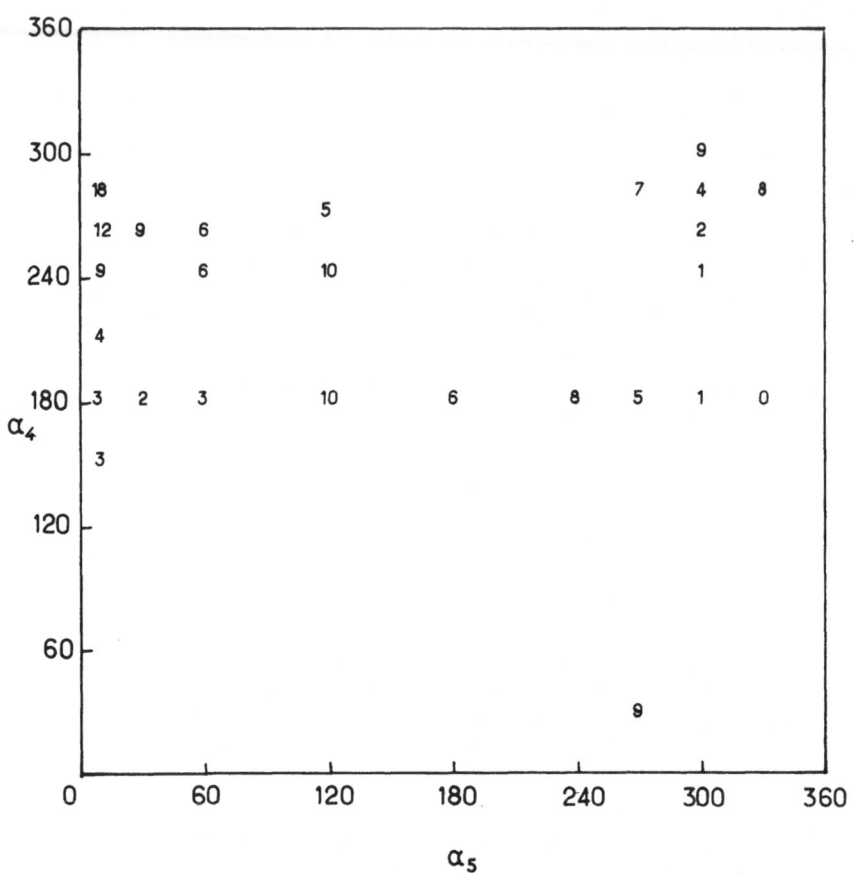

Fig. 6.   Conformational energy map of the EP...Na[+] adduct
(cation at site B). Energies (kcal/mole) with respect to the
global minimum taken as energie zero.

The marked effect that cation binding exerces on the conformational pref
erences of the polar head of phospholipids illustrates further, after
the demonstration of the effect of water in this respect [19], the conformational flexibility of this chain of phospholipids under the influence of environmental perturbations. It must, of course, be realized
that the theoretical studies performed so far are essentially model investigations which treat interactions between one ligand molecule and
one (or more) external water molecules or cations. The results are thus
still far away from dealing with real situations in which multiple interactions may occur between a number of ligand molecules themselves
and in which water molecules or cations may bridge different ligands
in different ways. We are, however, moving closer and closer to the examination of these complex situations and the progressive development
of the studies, from free molecular species, through reduced binary interactions to more complex situations has at least the advantage of en-

abling a better understanding of the interplay of intrinsic and environmental effects.

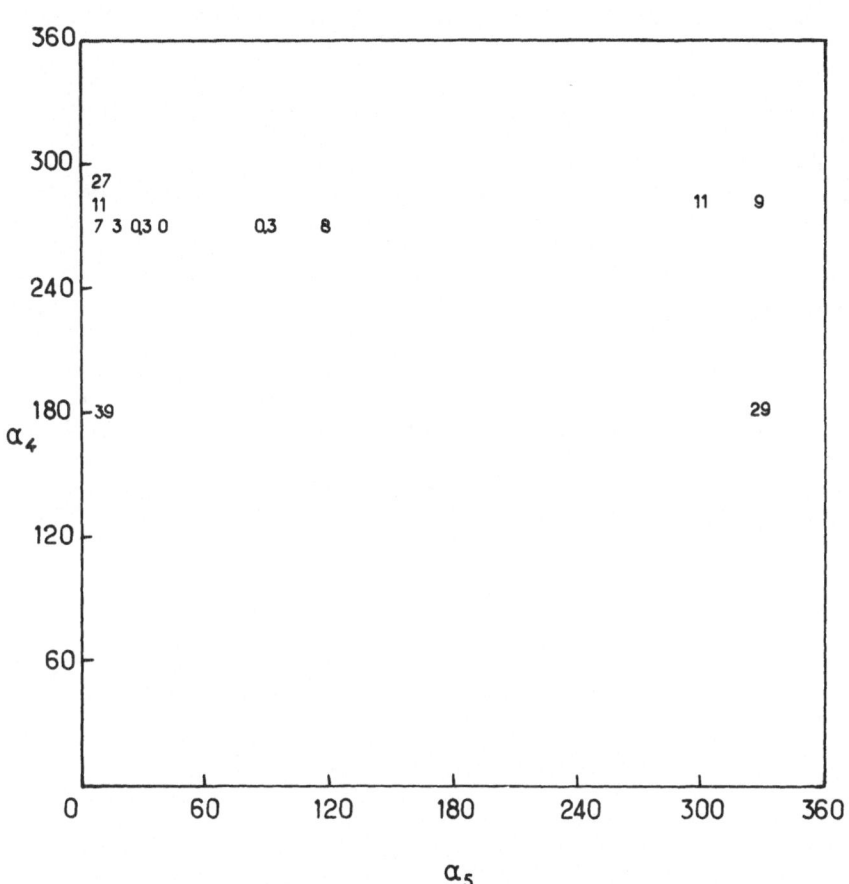

Fig. 7. Same as Figure 6 but the cation at site *E*.

REFERENCES

1. Pullman, A.: in E.D. Bergman and B. Pullman (Eds.), *Molecular and Quantum Pharmacology*, Proceedings of the 7th Jerusalem Symposium, 1974, p. 401.
2. Pullman, A.: *Int. J. Quantum Chem. Quantum Biol. Symp.* 1, 33 (1974).
3. Perricaudet, M. and Pullman, A.: *FEBS Letters* 34, 222 (1973).
4. Armbruster, A.M. and Pullman, A.: *FEBS Letters* 49, 18 (1974).
5. Pullman, A., Giessner-Prettre, C., and Kruglyak Yu, R. *Chem. Phys. Lett.* 35, 156 (1975).
6. Schuster, P., Marius, W., Pullman, A., and Berthod, H.: *Theoret. Chim. Acta*, 40, 323 (1975).

7. Pullman, A., Berthod, H., and Gresh, N.: *Inter. J. Quant. Chem. Quant. Chem. Symp.* 10, in press.
8. Pullman, B., Pullman, A., Gresh, N., and Berthod, H.: *Theoret. Chim.. Acta*, in press.
9. Pullman, B., Gresh, N., and Berthod, H.: *Theoret. Chim. Acta*, 40, 71 (1975).
10. Sundaralingam, M.: *Ann. N.Y. Acad. Sci.* 195, 324 (1972).
11. Pullman, B. and Saran, A.: *Intern. J. Quant. Chem. Quant. Biol. Symp.* 2, 71 (1975).
12. Sundaralingam, M.: *Biopolymers* 7, 821 (1969).
13. Pullman, B. and Saran, A.: *Progress in Nucleic Acid Research and Molecular Biology*, in press.
14. Kyogoku, Y. and Iitaka, Y.: *Acta Cryst.* 21, 49 (1966).
15. Ezra, F.S. and Collin, R.L.: *Acta Cryst.* B29, 1398 (1973).
16. Sundaralingam, M. and Jensen, L.H.: *Science*, 150, 1035 (1965).
17. Hazel, J.P. and Collin, R.L.: *Acta Cryst.* B28, 2951 (1972).
18. Pullman, B. and Berthod, H.: *FEBS Letters* 44, 266, 1974.
19. Pullman, B. and Berthod, H.: *FEBS Letters* 43, 199 (1975).
20. Henderson, T.O., Glonek, T., and Myers, T.C.: *Biochemistry* 13, 523 (1974).
21. Nanda, R.K. and Govil, G.: *Theoret. Chim. Acta*, 38, 71 (1975).
22. Perahia, D., Pullman, A., and Pullman, B.: *Theoret. Chim. Acta*, 42, 23 (1976).

THE CONFORMATIONAL SENSITIVITY OF DNA TO IONIC INTERACTIONS IN AQUEOUS SOLUTIONS

Sue Hanlon, Barry Wolf, Stephen Berman, and Aurelia Chan
Dept. of Biological Chemistry, College of Medicine, Univ. of Illinois at the Medical Center, Chicago, Illinois U.S.A. 60612.

ABSTRACT. The circular dichroism (CD) properties of calf thymus DNA have been examined at 27 °C and pH 7 over the wavelength range of 215 to 300 nm in aqueous solutions of varying concentrations of NaCl, KCl, LiCl, CsCl, NH_4Cl, $MgCl_2$, $CaCl_2$, $MnCl_2$, $ZnCl_2$, and $CoCl_2$. Regression analyses of the data were used to determine the minimum number of spectral components in each electrolyte set, their interchangeability between sets, and their relationship to the standard A, B and C crystallographic structures of DNA. The transformations of these spectra as the concentration of electrolyte increases have been interpreted in terms of a conversion of a form resembling the B conformation to the C structure for most of the sets examined. The actual 'C' spectral end point in $ZnCl_2$ and $CoCl_2$, however, is similar but not identical to the C spectrum determined for the other salts. This is probably due to changes in the spectral properties of the purine bases interacting with Zn^{++} or Co^{++}, rather than a dramatic difference in conformation. In LiCl, NH_4Cl and CsCl, an A form appears at the higher concentrations of these salts and, finally, at the very highest concentration of LiCl, a fourth as yet unidentified component is required to account for the data. These transconformational reactions can be rationalized in terms of cationic association and the hydration of the DNA in these solutions.

GLOSSARY OF SYMBOLS

Symbol	Meaning
r.h.	relative humidity
DNA	deoxyribonucleic acid
A	secondary A structure of DNA
B	secondary B structure of DNA
C	secondary C structure of DNA
K	generalized secondary structure of DNA
$[\theta]$	molecular ellipticity (in $deg_1 cm^2/decimole$)
E	extinction coefficient (in $M^{-1} cm^{-1}$)
λ	wavelength (in nm)
J	wavelength index
i	solution index
f	fraction of nucleotide residues in a given secondary structure

B. Pullman and N. Goldblum (eds.) Metal-Ligand Interactions in Organic Chemistry and Biochemistry, part 1, 77-106.

1. INTRODUCTION

A number of studies have confirmed the fact that the conformational properties of linear double stranded DNA are profoundly sensitive to the
ion and water content of the medium surrounding the nucleic acid. The
secondary structure of the Li salt of DNA in fibers, for instance, will
transform from the Watson Crick B form (Langridge *et al.*, 1960) to the
C structure when the relative humidity (r.h.) drops below 66% (Marvin
et al., 1961). The Na salt, in contrast, will transform from the B to
the A form upon decreasing the r.h. (Fuller *et al.*, 1965) and the point
at which the transition occurs is dependent on the ion content of the
hydrated fiber. Similar results have been found for films of DNA examined by IR spectroscopy (Brahms *et al.*, 1973; Pilet and Brahms, 1973).
    In solution, the result of investigations on the effects of various
types of electrolytes on the optical rotatory dispersion (Cheng, 1965;
Tunis and Hearst, 1968) and the circular dichroism (CD) properties of
DNA in solution (Nelson and Johnson, 1970; Hanlon *et al.*, 1972; Studdert
et al., 1972; Ivanov *et al.*, 1973; Luck and Zimmer, 1972; Zimmer and Luck
1973; Zimmer *et al.*, 1974; Hanlon *et al.*, 1975) have been interpreted
in terms of conformational changes similar to those undergone by DNA in
fiber form. These transitions have been attributed to two factors. The
initial effect at low ion concentration is caused by binding the cations
of the medium (Zimmer *et al.*, 1974; Hanlon *et al.*, 1975). At higher electrolyte concentrations, the dehydration of the DNA which is a consequence
of the lower activity of water of aqueous solvents, causes further change
    (Hanlon *et al.*, 1972; Zimmer and Luck, 1973; Wolf and Hanlon, 1975).
With the advent of the identification of the CD spectra appropriate for
the B, the C and the A forms of DNA by Tunis-Schneider and Maestre (1970)
the changes in most of the electrolytes examined have been attributed
to the production of the C form of DNA at the expense of a structure
resembling the B (Hanlon *et al.*, 1972; Studdert *et al.*, 1972; Zimmer and
Luck, 1973; Zimmer *et al.*, 1974; Hanlon *et al.*, 1975). In some electrolytes, an A form appears, as well (Hanlon *et al.*, 1975). There is evidence from additional techniques that an accompanying change in tertiary
and quaternary structure of DNA in these solutions may also be occurring
(Zimmer *et al.*, 1974; Maniatis *et al.*, 1974; Wolf and Hanlon, 1975; Wolf,
Berman, and Hanlon, submitted to *Biochemistry*, 1976).
    Despite the apparent unanimity in the above description of the cur-
rent state of research on the conformational sensitivity of DNA to the
presence of electrolytes, there have been no attempt, to date, to formu-
late a quantitative picture of the role of ion binding and hydration in
modulating the conformational properties of DNA in aqueous solutions.
What we wish to do in the present work is to present such a picture for
a specific DNA (calf thymus) and describe the approach and the analytical procedures whereby we have arrived at our conclusions.

2. EXPERIMENTAL

The calf thymus DNA sample employed in these experiments was a commercia
preparation obtained from Sigma (lot 802184) whose properties have been

previously described (Johnson *et al.*, 1972). All solutions of DNA were prepared in solvents in which the pH ranged between 6.8 and 7.2. The manner in which the circular dichroism and absorption spectra as well as the sedimentation coefficient of the DNA component of these solutions were obtained has been reported elsewhere (Hanlon *et al.*, 1975; Wolf and Hanlon, 1975) and will be only briefly covered here. Circular dichroism spectra were obtained at 27°C in 1 cm cells with a Cary 60 spectropolarimeter equipped with a 6001 CD unit. The concentration of DNA in these experiments, 2×10^{-4} M (nucleotide residues), was determined by measuring the absorption spectra, obtained in a Cary 14 CMR spectrophotometer at 25 $^{\circ}$C, of appropriate dilutions of a stock solution of DNA in 0.01 m salt, pH 7. An extinction coefficient of 6600 M^{-1} at 259 nm, previously determined (Johnson *et al.*, 1972), was employed in these calculations.

Sedimentation velocity experiments were conducted at 24 to 26 $^{\circ}$C with solutions of DNA at 1×10^{-4} M in top loading cells with a Spinco Model E analytical ultracentrifuge equipped with absorption optics. The median sedimentation coefficient obtained in these experiments was corrected to the conventional anhydrous standard state of a medium whose density and viscosity is that of water at 20 $^{\circ}$C by applying the appropriate viscosity, density, and preferential solvation factors. The values of this latter correction in the various electrolyte solutions were obtained from the data of Hearst and co-workers (Hearst and Vinograd, 1961a, 1961b; Hearst, 1965; Tunis and Hearst, 1968a) as well as Cohen and Eisenberg (1968).

3. SPECTRAL OBSERVATIONS

Figures 1 and 2 present characteristic patterns of the transformation of the CD spectral properties of DNA at 27 $^{\circ}$C as a function of increasing concentrations of mono and divalent cations. Figure 1 displays the spectra of DNA in increasing concentration of NaCl (A), CsCl (B), NH_4Cl (C) and LiCl (D). A similar set of spectra were also obtained in KCl. Figure 2 shows the spectra of the same sample of DNA in a lower concentration range of $CaCl_2$ (A), $MnCl_2$ (B) and $ZnCl_2$ (C). Two other sets (not shown) in $MgCl_2$ and $CoCl_2$ at the same concentration of salt, were also obtained. The set in $MgCl_2$ was found to be similar in shape to that in $CaCl_2$ while the set in $CoCl_2$ was similar to the set obtained in $ZnCl_2$.

Despite the differences in the shapes of the spectra, one consistently finds that as the concentration of the electrolytes increases, the rotational strength of the positive band above 255 nm decreases. The negative band below 255 nm is affected only minimally. In some electrolytes, a substantial depression of the positive band occurs at fairly low concentrations of salt. This point is demonstrated by the spectra of DNA in 0.01 molal solutions of various electrolytes, shown in Figure 3, and the behavior of the molecular ellipticity at 275 nm, $[\theta]_{275}$ (deg cm^2/ decimole), as a function of electrolyte concentration, shown in Figure 4. These effects are not attributable to denaturation since changes in the absorbance spectra, even for the divalent ion Zn^{++}, Ca^{++} and Mn^{++}, are minimal. The upper part of Figure 4 shows the behavior of the extinction coefficients at 259 nm (E_{259}) (1/mole cm) over the same concentration

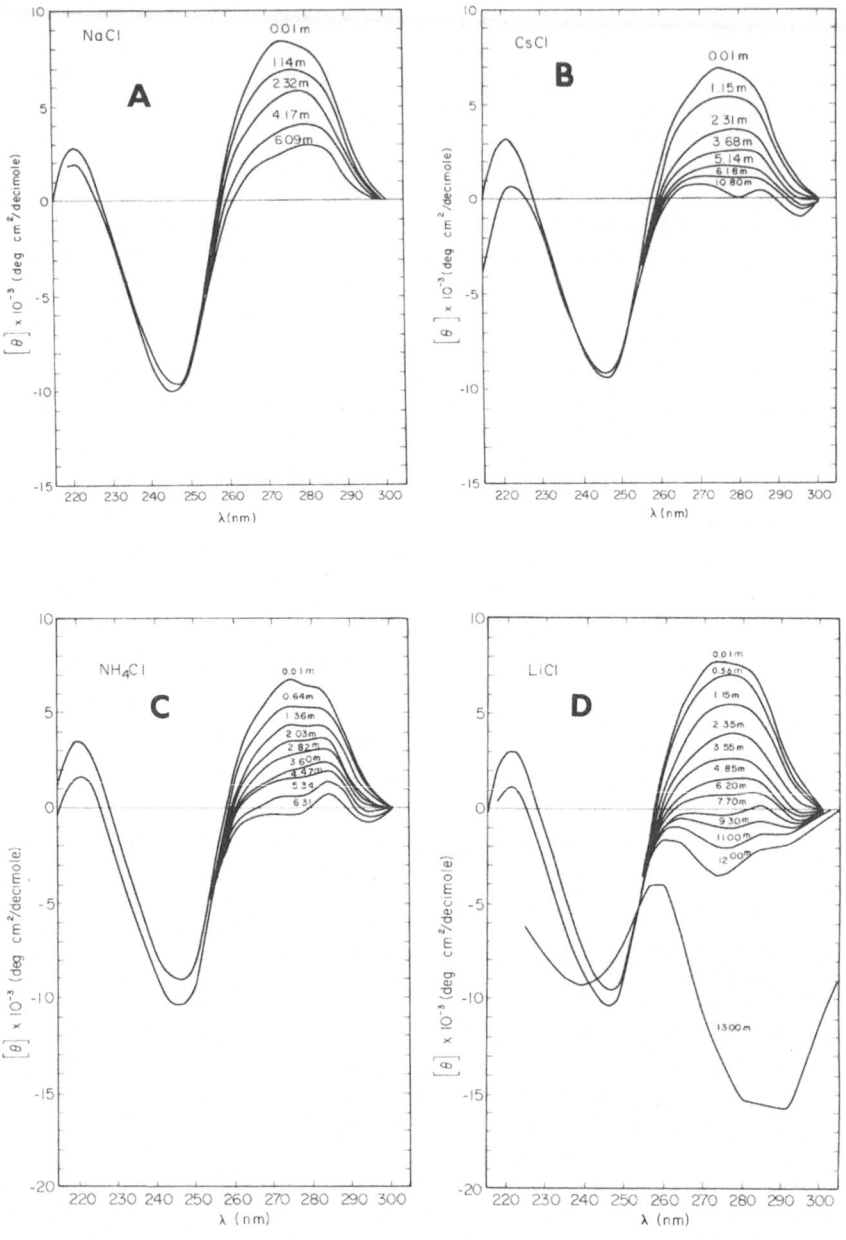

Fig. 1. The circular dichroism spectra of calf thymus DNA at pH 7 (27°C) in aqueous solutions of salts with univalent cations. Spectra are shown in NaCl (A), CsCl (B), NH$_4$Cl (C) and LiCl (D) at the molal concentrations indicated.

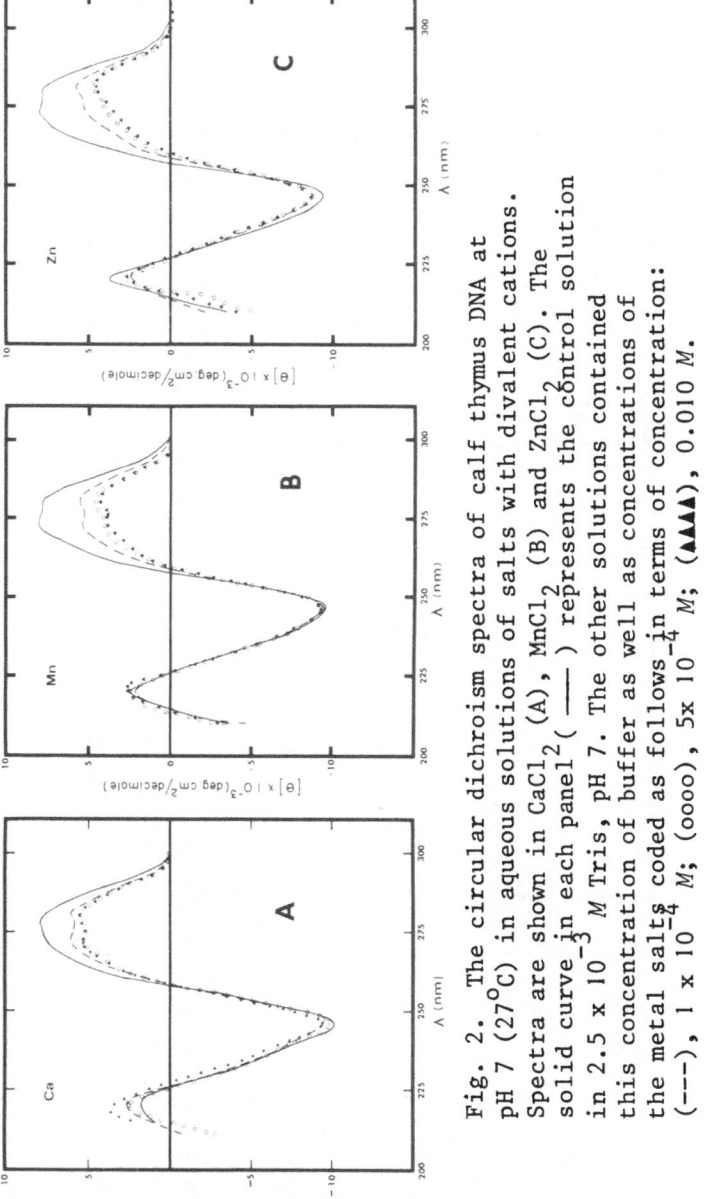

Fig. 2. The circular dichroism spectra of calf thymus DNA at pH 7 (27°C) in aqueous solutions of salts with divalent cations. Spectra are shown in $CaCl_2$ (A), $MnCl_2$ (B) and $ZnCl_2$ (C). The solid curve in each panel ($\underline{\hspace{1cm}}$) represents the control solution in 2.5×10^{-3} M Tris, pH 7. The other solutions contained this concentration of buffer as well as concentrations of the metal salts coded as follows in terms of concentration: (---), 1×10^{-4} M; (oooo), 5×10^{-4} M; (▲▲▲▲), 0.010 M.

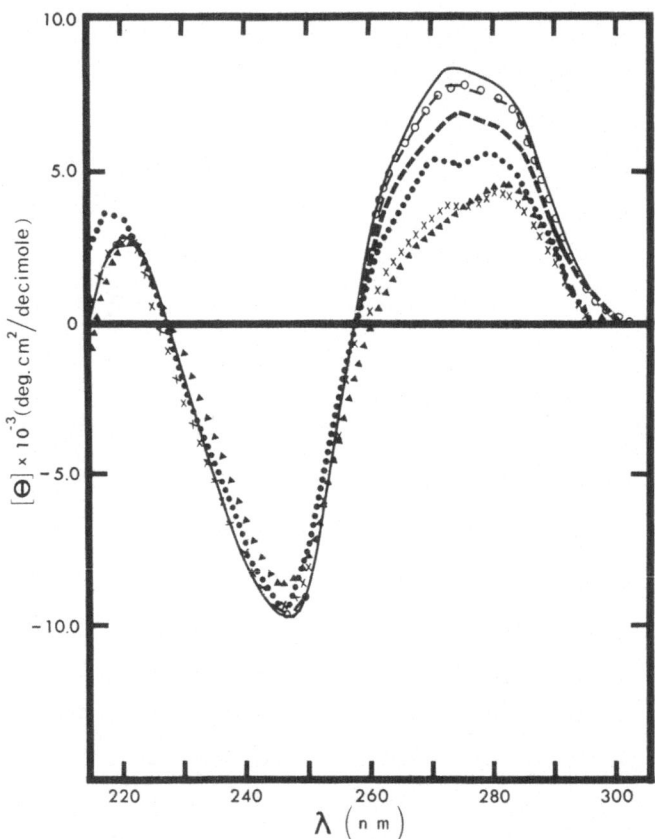

Fig. 3. The circular dichroism spectra of calf thymus DNA at pH 7 (27°C) in 0.01 M solutions of various salts. (————), NaCl; (o-o-o), KCl and LiCl; (———), CsCl and NH₄Cl; (●●●) CaCl₂; (××××), MnCl₂; (▲▲▲▲), ZnCl₂. All solutions contained additionally 2.5 x 10⁻³ M Tris. The mole ratio of metal ion to nucleotide was 40:1 in all cases.

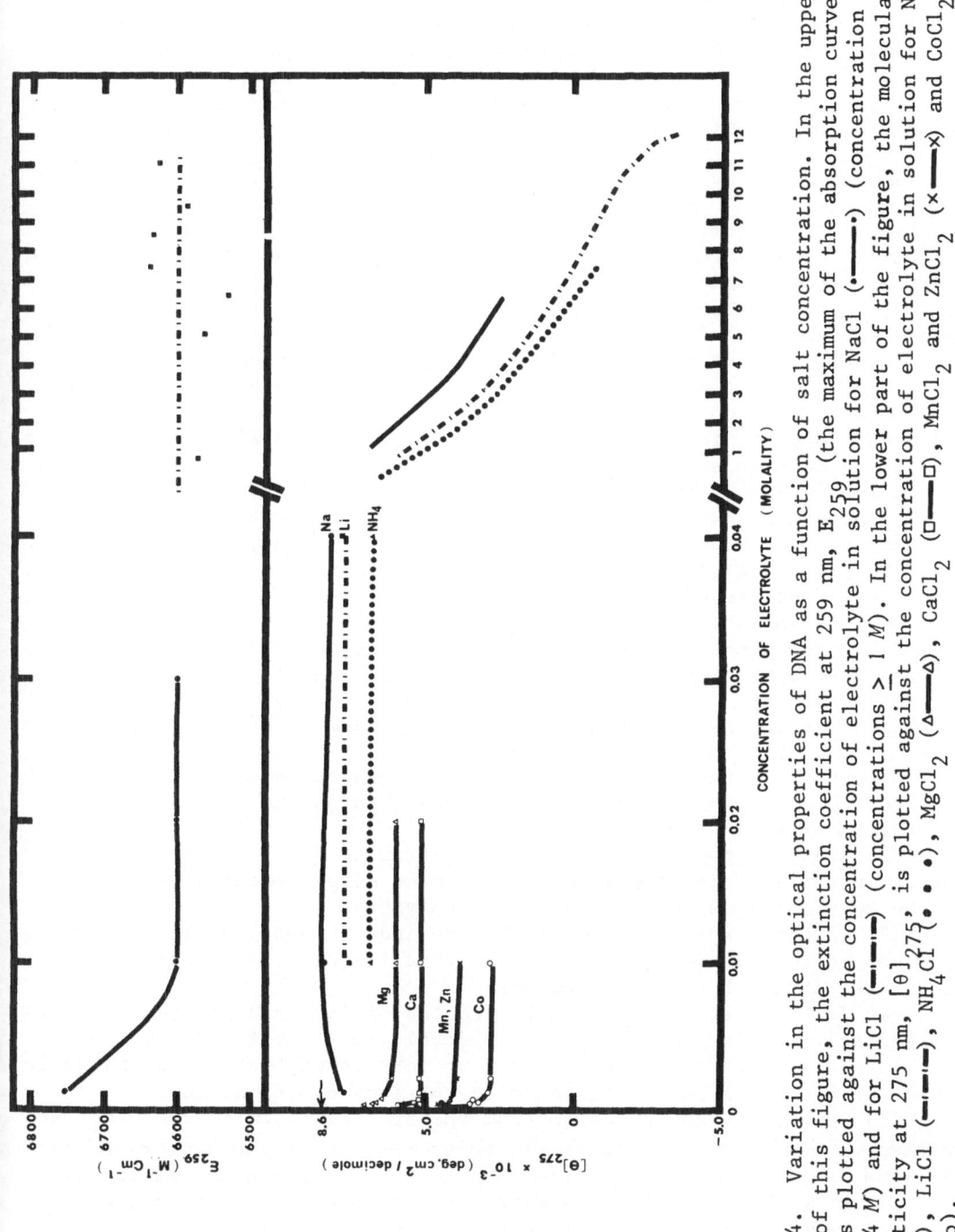

Fig. 4. Variation in the optical properties of DNA as a function of salt concentration. In the upper part of this figure, the extinction coefficient at 259 nm, E_{259} (the maximum of the absorption curve) of DNA is plotted against the concentration of electrolyte in solution for NaCl (•——•) (concentration $\leq 0.04\ M$) and for LiCl (■——■) (concentrations $\geq 1\ M$). In the lower part of the figure, the molecular ellipticity at 275 nm, $[\theta]_{275}$, is plotted against the concentration of electrolyte in solution for NaCl (——), LiCl (■——■), NH_4Cl (• • •), $MgCl_2$ (△——△), $CaCl_2$ (□——□), $MnCl_2$ and $ZnCl_2$ (×——×) and $CoCl_2$ (○——○).

range of electrolyte. (The only solution in which there appeared to be
detectable unstacking was that in 0.001 m NaCl. The data point, $[\theta]_{275}$,
in the lower part of the figure shows that this effect is accompanied
by a decrease in $[\theta]_{275}$).

4. ANALYSIS OF SPECTRAL DATA

4.1. *General Strategy*

As the data in Figure 4 demonstrate, the spectral changes appear to occur in two stages. The changes at low concentration of salt suggest that
a direct interaction of the cation of the salt with the nucleotide residues of DNA is responsible for this initial lowering of the positive
band spectrum above 255 nm. This process continues, however, well past
the saturation point of the ion binding to DNA and, hence, additional
aspects of the system must be invoked to explain the subsequent changes.
    Before questions of the origin and nature of these effects can be
answered, however, it is necessary to investigate certain properties and
quantitate certain aspects of these transitions. For instance, a simple
visual inspection of the spectra shows certain significant differences
in the pattern of changes observed when low and high salt concentrations
are compared and when sets from two different electrolytes are compared.
Thus, it is not at all apparent that the transformations within a set
and between sets involve the same spectral forms.
    In pursuing this general problem of the nature and origin of the
spectral changes observed in these experiments, however, it is important
not to lose sight of the forest because of differences in the individual
trees. In order to avoid the latter temptation we have found it useful
to follow a strategy in which, sequentially, we attempt to answer the
five major questions outlined below:
    (1) What are the minimum number of independent spectral components
required to account for the set of spectra obtained by varying the concentration of a given electrolyte?
    (2) Are the spectral components required to describe one set of
spectra equally satisfactory, within allowed experimental error, to describe sets obtained in different electrolytes?
    (3) If these spectral components are indeed identical for all sets,
can they be identified with specific crystallographic structures of DNA,
(i.e., the A, the B and the C forms)?
    (4) If they are not identical, are the differences between two sets
of electrolytes attributable to (1) radically different conformations
or (2) to differences in the spectral properties of the purine and pyrimidine chromophores in the presence of the ions of the medium?
    (5). Can the transformation from one spectral form to the other be
related directly and quantitatively to properties of the macromolecule,
DNA, and its interaction with the components of its aqueous environment?
    These questions must be answered in the sequence given, since the
answer to each in order provides the basis of the approach to the subsequent one. Questions (1) and (2) are relatively easy, and can be answered in an unequivocal fashion without recourse to additional experi-

mental data. The approach for so doing is described under Topic 4.2 below   (Component Analysis). Questions (3), (4) and (5), however, require information input from other types of experiments and a strong reliance on data available in the literature. In fact, the accuracy of our conclusions depends crucially on the accuracy and correctness of the cited literature results.

4.2 *Component Analysis*

There are a number of statistical procedures available for determining the minimum number of independent components, or factors, and their interchangeability, which appropriately describe a given set of data of the type shown in Figures 1 and 2. The approach which we have favored is a multiple linear regression analysis of the set of spectra obtained by varying the concentration of given electrolyte. It can be demonstrated that if a given set of spectra for i solutions (in terms of the optical signal, $[\theta]^{OBS,i}_{\lambda j}$, vs. wavelength, λj) can be represented by a linear combination of the fractional contributions of k spectral components in each solution,

$$[\theta]^{OBS,i}_{\lambda j} = \sum_{k=1}^{q} f_k \, [\theta]^{k}_{\lambda j} \tag{1}$$

in which $\Sigma f_k = 1$, then a given spectrum from this set may also be expressed as a linear combination of fractional amounts of other observed spectra from the same set.
That is,

$$[\theta]^{OBS,i=p}_{\lambda j} = \sum_{i} f'_i \, [\theta]^{OBS,i\neq p}_{\lambda j} \tag{2}$$

The minimum number of observed spectra substituted on the right hand side of Equation (2) required to make the coefficients, f'_i, sum to 1 will be equal to q, the number of independent spectral components present in these spectra.
    In order to minimize the experimental errors, a residual

$$R = \sum_{j} \left([\theta]^{OBS,i=p}_{\lambda j} - \sum_{i} f'_i [\theta]^{OBS,i\neq p}_{\lambda j} \right)^2 \tag{3}$$

is differentiated with respect to each f'_i. These partials are then set equal to 0, thus minimizing the sum of the square. The resulting set of simultaneous equations may then be solved for unique values of f'_i and a sum, $\Sigma f'_i$, obtained.
    We have used this procedures in a systematic fashion, testing a given concentration range of a specific set of spectra first with one, two, three and, finally, four substitutions for the right hand side of

TABLE I
Results of Component analysis

Salt	Minimum number of components	Concentration range of salt (molal)	Component interchangeability
Monovalent cations			
NaCl	2	0.01 to 6.09 m	Yes
KCl	2	0.01 to 4.73 m	Yes
LiCl	2	0.01 to 2.32	Yes, except
	3	0.01 to 9.6	for 4 th
	4	0.01 to 11.04 m	component
CsCl	2	0.01 to 1.15	Yes
	3	0.01 to 10.8	Yes
NH_4Cl	2	0.01 to 3.67	Yes
	3	0.01 to 7.38	Yes
Divalent cations			
$MgCl_2$	2	1×10^{-4} to 1 m	Yes
$CaCl_2$	2	1×10^{-4} to 1 m	Yes
$MnCl_2$	2	1×10^{-4} to 0.01 m	Yes
$ZnCl_2$	2	1×10^{-4} to 0.01 m	No
$CoCl_2$	2	1×10^{-4} to 0.01 m	No

Equation (2). When two different sets of spectra in two different electrolytes yield the same number of spectral components, the $i \neq p$ spectra
serving as 'bases' have been interchanged, in order to ascertain whether
the spectral components appropriate for one set are equally applicable
to the other.

The results of these analyses for the sets of spectra in the ten
different salts are summarized in Table I. Column 2 in this table gives
the number of experimental spectra in a given series which must be substituted in order to bring the $\Sigma f_i' = 1.00 \pm 0.05$ for the other spectra
in the series within the concentration range given in column 3. The last
column indicates whether the experimentally observed spectra used as
'bases' are equally suitable for the analysis of sets of spectra obtained
in other electrolytes.

As these results show, essentially only two spectral components are
required to account for the spectral transition in the lower concentration ranges of all salts and in all concentrations up to saturation in
NaCl and KCl. Except for two cases, $ZnCl_2$ and $CoCl_2$, these spectral components are interchangeable. In three salts, LiCl, CsCl and NH_4Cl, a
third component comes in at the upper end of the concentration scale.
And, finally, in LiCl at the very highest concentration, a fourth component appears. This last is unique to the LiCl set and is not formed in
the spectra obtained in other electrolytes.

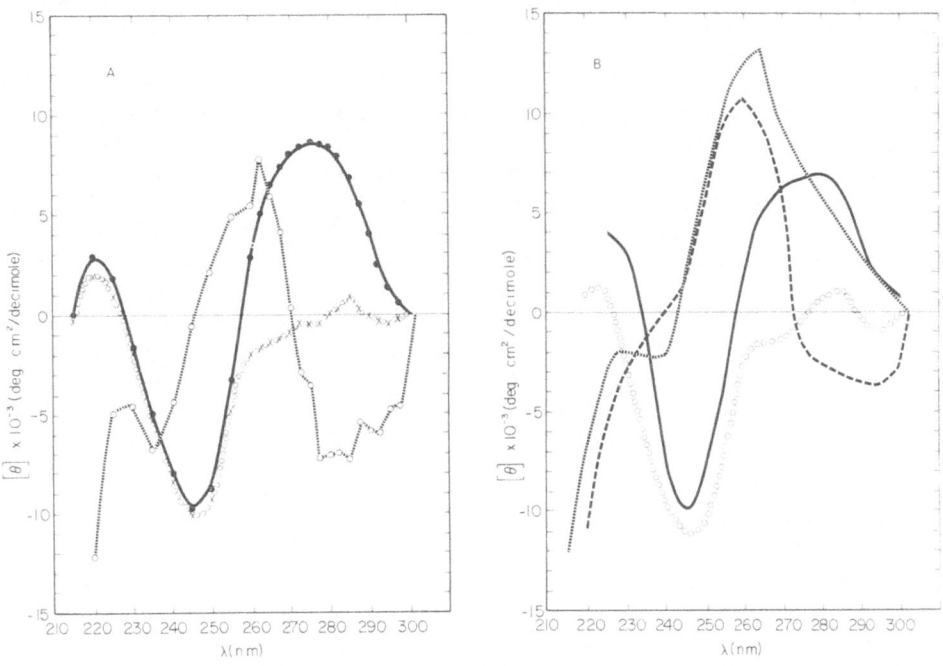

Fig. 5. Circular dichroism spectra of DNA in the B, C and A secondary structures. (A) The circular dichroism spectra of calf thymus DNA obtained by the methods described in this work: (●━━●), Spectrum of DNA extrapolated to 0 concentration of NaCl and used as the aqueous B spectrum; (xooxoo), the average C spectrum obtained by averaging data in NaCl, KCl, and NH₄Cl (in the two component range); (o■o■o), the 'A' spectrum of DNA in the LiCl solutions. (B) Circular dichroism spectra of DNA taken from Tunis-Schneider and Maestre (1970). (——), the Na salt of a film of calf thymus DNA at 92% r.h., taken from Figure 2 of the cited paper and identified as the B form spectrum; (oooo), the Li salt of a film of calf thymus DNA at 75% r.h. taken from Figure 5 of the cited paper and identified as the spectrum of the C form of calf thymus DNA; (■■■■), the Na salt of calf thymus DNA at 75% r.h. taken from Figure 3 of the cited paper, and identified as the spectrum of the A form of calf thymus DNA; (----), the Na salt of a film of E. coli DNA at 75% r.h. taken from Figure 4 of the cited paper and identified as the spectrum of the A form of E. coli DNA.

4.3. *Identification of the Spectral Components*

(A) Sets With Two Components

(a) Monovalent Cations. At this stage in the analysis, one may begin to make tentative assignments of the spectral components to specific confor-

mations and delineate spectral 'end points'. Obviously, it is useful to
begin with the sets of spectra which require only two components to describe their properties over a wide range of electrolyte concentrations.
This would include the spectra in NaCl, KCl and most of the NH$_4$Cl sets.

Since the transformations in the spectral properties are brought
about by an increase in the electrolyte concentration, it is reasonable
to assume that one of the two spectral components can be arrived at by
extrapolating the spectrum which seems least affected by the presence
of the salt, i.e., that set in NaCl, to 0 concentration of electrolyte,
under condition in which the DNA is still fully hypochromic.

The results of this extrapolation is shown in Figure 5B as the solid curve. In the adjacent panel, Figure 5A, we have reproduced the spectra of films of DNA obtained by Tunis-Schneider and Maestre (1970), identified by these authors as being the spectra appropriate for the B
(—— curve), the C (ooo curve) and the A form (--- curve) of DNA. Our
spectral limit in NaCl resembles the shape, but not the magnitude, of
the B spectrum obtained by Tunis-Schneider and Maestre. Based on the
results of wide angle X-ray scattering experiments, Bram (1971) has suggested that the conformation of DNA at these lower concentrations of
salt is characterized by a winding angle smaller than that characteristic of the Watson Crick B form. Thus, it is not surprising that the CD
spectrum which we obtain is not identical with that exhibited by the
Watson Crick B structure.

There are a number of approaches which can be employed in order to
ascertain the shape of the other spectral component in a two state transition. A simple visual inspection of the data for NaCl or KCl indicated
that a spectrum similar to the one exhibited by DNA in the C structure
(see Figure 5B) might be appropriate. This was confirmed by a factor analysis of the spectral data in NaCl at λ_j values greater than 255 nm.
One eigen value for the matrix accounted for ca. 98% of the variation
in the spectral data. Since the component analysis indicated that two
components were present, this meant that one of the spectral components
had values of $[\theta]_{\lambda_j}$ near 0. The C spectrum of Tunis-Schneider and Maestre
fit that requirement.

Using the C spectrum of these authors as a starting approximation,
the 'C' spectrum appropriate for each i solution in a set was calculated
by a linear regression analysis. Equation (1), in specific terms, was

$$[\theta]_{\lambda_j}^{OBS} = f_B[\theta]_{\lambda_j}^{B} + f_C[\theta]_{\lambda_j}^{C} \tag{4}$$

It was rearranged to cast $[\theta]_{\lambda_j}^{C}$ as the dependent variable which was to
be smoothed by minimizing the λ_j residual, R, in the expression

$$R = \sum_{\lambda_j} \left\{ \frac{1}{f_C}[\theta]_{\lambda_j}^{OBS,i} - \left(\frac{f_B}{f_C}\right)[\theta]_{\lambda_j}^{B} - [\theta]_{\lambda_j}^{C} \right\}^2 \tag{5a}$$

or

$$R = \sum_{\lambda_j} \left\{ \alpha[\theta]_{\lambda_j}^{OBS,i} - \beta[\theta]_{\lambda_j}^{B} - [\theta]_{\lambda_j}^{C} \right\}^2 \tag{5b}$$

using spectral data above 255 nm. The two simultaneous equations gener-
ated by setting $(\partial R/\partial \alpha)_\beta = 0$ and $(\partial R/\partial \beta)_\alpha = 0$ could then be solved for
α and β (or f_C and f_B and the best fitting values of $[\theta]_{\lambda_j}^{C,i}$ for the given
solution, i. For the rest of the spectrum below 255 nm, values of
$[\theta]_{\lambda_j}^{C,i}$ were calculated as

$$[\theta]_{\lambda_j - 255 \text{ nm}}^{C,i \leq} = \left\{ \frac{[\theta]_{\lambda_j}^{OBS,i} - f_{B_i} [\theta]_{\lambda_j}^B}{f_{C_i}} \right\} \tag{6}$$

The complete C spectrum from all solutions in which $(f_{C_i} + f_{B_i}) = 1$ were
then mean averaged to yield an average for the given electrolyte series. These, in turn, were averaged for three salts, NaCl,
KCl and NH$_4$Cl, to yield the spectrum displayed in Figure 5a.

An alternate method of analysis was also employed for extracting
the C reference spectrum. It involves a double summation over all the
available data, in a given electrolyte, yielding a set of nonlinear simultaneous equations which can be solved if approximations of values
of some of the 'unknowns' are available.

The general expression to be minimized is

$$G = \sum_{i=1}^{n} \sum_{j=1}^{m} \left\{ [\theta]_{\lambda_j}^{OBS,i} - \sum_{k=1}^{q} f_{k_i} [\theta]_{\lambda_j}^K \right\}^2 \tag{7}$$

A set of $q(m+n)$ simultaneous equations is generated by setting the partial derivatives of G with respect to the coefficients, f_k, and the variables, $[\theta]_{\lambda_j}^k$ to 0. (The number of such equations can be reduced and the
subsequent procedure for extracting solutions can be simplified appreciably if restrictions such as $\sum f_{k_i} = 1$ and/or some of the K reference spectra are not assigned as variable components, but are substituted as error free quantities). Values of $[\theta]_{\lambda_j}^K$ above 255 nm which are known only
approximately are substituted as starting approximations in the expression for the partial derivatives and the equations are solved simulta-
neously for the best fitting values of $f_{k,i}$ for each i solution and $[\theta]_{\lambda_j}^K$
for each spectral component at wavelength, λ_j.

For the two component spectral sets, we employed the restriction
that $f_B + f_C = 1$. The corresponding expression for the residual is:

$$G = \sum_{i=1}^{n} \sum_{j=1}^{m} \left\{ [\theta]_{\lambda_j}^{OBS,i} - f_{B_i} [\theta]_{\lambda_j}^B - (1-f_{B_i})[\theta]_{\lambda_j}^C \right\}^2 \tag{8}$$

This method resulted in average C and B reference spectra which differed
insignificantly from that obtained by the simpler linear regression analysis. For these simple two component systems in which one spectral end
point can be determined experimentally, there is no marked gain in employing this more elaborate regression analysis. When the spectral components exceed two, however, it is advantageous to employ either this
method or a less elaborate version of it (Berman, S. and Hanlon, S., manuscript in preparation).

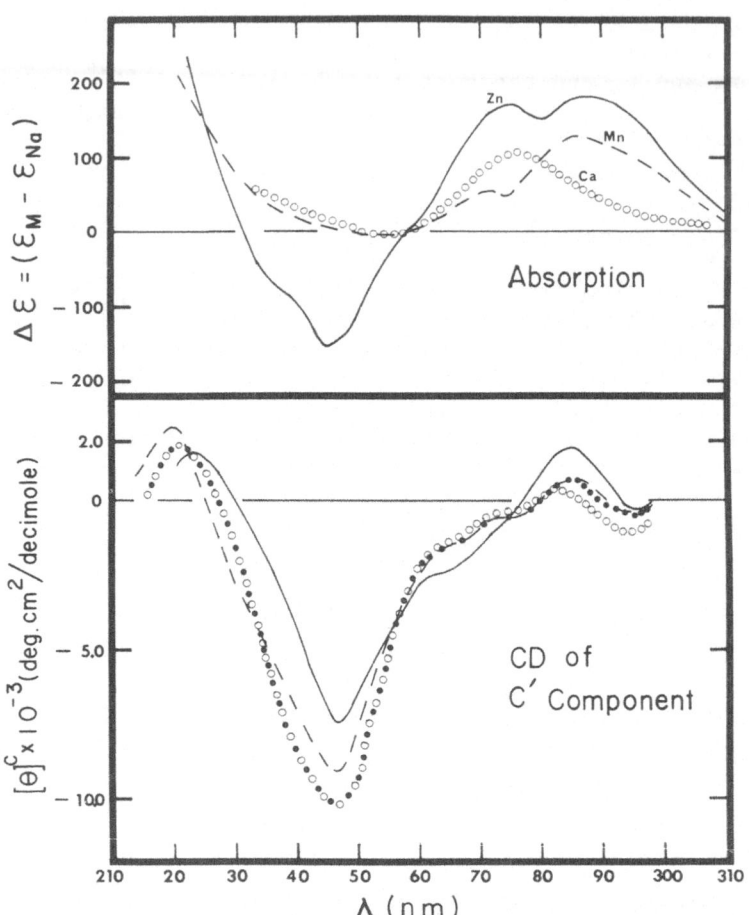

Fig. 6. Comparison of the circular dichroism spectra of the
'C' spectral component and the difference spectra of DNA in
the presence of divalent cations. The upper part of the spectrum displays the difference absorption spectra at pH 7 (25 $^{\circ}$ C)
in terms of the difference in extinction coefficient between
a solution of DNA at 2.3 x 10^{-4} M in the presence of 5 x 10^{-5} M
concentrations of divalent cations (Zn^{++}, Mn^{++} and Ca^{++}) and
in 0.01 M NaCl. The lower portion of the figure displays the
average 'C' reference spectra obtained by analyses of the spectral data sets in the presence of these same divalent cations.
The codes for the identity of the divalent cation are the same
in both parts of the figure: (——) Zn^{++}, (----) Mn^{++} and
(oooo) Ca^{++}. The C spectrum appropriate for the univalent
salts (••••) is also displayed for comparison.

b. Divalent Cations. When the C reference spectrum obtained by the analysis of the spectra in NaCl, KCl and NH$_4$Cl was used with the 'B' spectrum
determined experimentally for the analysis of the spectral changes caused

by changes in the concentration of divalent cations, the fit of the data above 255 nm was good and the 'smoothed' C spectrum was reasonable close to that obtained in the univalent salt series for Ca^{++}, Mg^{++} and Mn^{++}. This was not the case for Co^{++} or Zn^{++}, however, whose 'C' spectra differed from that of the input C.

The remainder of the spectral end point in each electrolyte was calculated by Equation (6) and mean averaged over the spectral set. These complete 'C' spectra for three of the divalent ion sets (Ca^{++}, Mn^{++} and Zn^{++}) are shown in the lower part of Figure 6. The curve for the $MgCl_2$ series was essentially identical to that in $CaCl_2$ while that for the $CoCl_2$ solutions was very similar to the spectrum shown for the Zn set. The differences between the 'C' spectrum in $CaCl_2$ and that displayed in Figure 5a, obtained from the analysis of the univalent salt solutions, are not significant. The C spectrum from the $MnCl_2$ solutions differs significantly from the C spectrum in Figure 5a only in the 245 nm region, where the negative band falls. The C spectrum from the $ZnCl_2$ series, however, exhibited significant differences in both the positive and the negative bands.

The effect of these ions on the absorption spectrum of DNA is shown in the upper part of Figure 6. This figure presents the difference spectrum, $(E_{M/DNA} - E_{Na/DNA})$ in which the absorbance of a DNA solution in a $5 \times 10^{-4} M$ solution of the divalent cation salt (plus 0.0025 M Tris) is read against a DNA solution at the same nucleotide concentration in 0.01 M NaCl, 0.0025 M Tris. The ion concentration in all solutions employed in these experiments is sufficient to saturate the phosphate groups of the polynucleotide backbone, and hence the electrostatic repulsive interactions leading to base unstacking have been relieved. Thus, the spectral effects observed cannot be attributed to the denaturation of DNA in one or more of the solutions examined.

The difference spectrum of DNA in the presence of $5 \times 10^{-4} M$ $CaCl_2$ has the shape of the change in refractive index in an absorption band. This suggests that this difference spectrum is due to a small difference in light scattering between the Ca and the Na salts of DNA. The difference spectra of solutions of DNA in $MnCl_2$ and $ZnCl_2$ also have an apparent maximum in about the same wavelength region (ca. 270 nm) which could be explained in a similar fashion. In addition, however, the spectra of DNA in the presence of these two salts also have positive peaks at 290 nm and the Zn salt has a spectral depression in the 240 nm region. These effects are diagnostic of perturbations of the spectral properties of the chromophores and cannot be attributed to differences in light scattering.

These changes in the absorption spectra of DNA upon interacting with Zn^{++} and Mn^{++} have been observed previously by Zimmer *et al.* (1974) and Luck and Zimmer (1972), who also found that the magnitude of the effect was proportional to the GC content of the DNA sample employed in the experiment. These authors have concluded that these changes in the difference spectra, as well as in the CD spectra in the presence of Mn^{++}, Zn^{++}, Co^{++} and other metal ions are due to the formation of a metal ion chelate involving the N_7 position of guanine and the phosphate oxygens of the polynucleotide backbone. If this interpretation is correct, then the additional bands at 290 and 245 nm in the difference spectra of the Zn

salt and Mn salts of DNA represent red shifts in the spectral properties
of the GC base pairs. These perturbations have a relatively minor effect
on the CD properties of the Mn salt and one may conclude that the conformational change resulting from the interaction of this ion with Zn^{++} DNA are
very similar to that seen in the univalent electrolytes. The Zn^{++} salt
of DNA, on the other hand, seems to be distinctly more affected, both
with respect to its absorbance and its CD properties. Several spectral
bands have obviously been red shifted in the absorbance spectrum. Similar effects are seen in the CD spectrum, and this may be the reason for
the changed characteristics of the spectral end point found in solutions
of this salt and in $CoCl_2$. Although it is somewhat hazardous to draw any
conclusions about conformations in this instance, we tentatively suggest
that the secondary structure of this spectral end point in the presence
of Zn^{++} (and Co^{++}, as well) is similar to the C structure found in the
solutions containing the univalent salts.

(B) Sets With Three Components

Although two components were satisfactory for the univalent electrolyte
sets on NH_4Cl, $CsCl$ and $LiCl$ up to concentrations of 2 to 3 m, the spectra at concentrations above these required three components. The shapes
of the spectra at these higher electrolyte concentrations suggested that
this third component was an A form. This possibility could be tested by
simply expanding the two types of regression analyses described under
Section 4. 3(A), to include a third spectral component, $[\theta]^A_{\lambda_j}$, and its
coefficient f_A. Equations (5a) and (5b) for the multicomponent linear
regression analysis thus become

$$R = \sum_j \left\{ \frac{1}{f_A} [\theta]^{OBS,i}_{\lambda_j} - \frac{f_B}{f_A}[\theta]^B_{\lambda_j} - (\frac{f_C}{f_A})[\theta]^C_{\lambda_j} - [\theta]^A_{\lambda_j} \right\}^2 \qquad (9a)$$

or

$$R = \sum_j \left\{ \alpha' [\theta]^{OBS,i}_{\lambda_j} - \beta'[\theta]^B_{\lambda_j} - \gamma[\theta]^C_{\lambda_j} - [\theta]^A_{\lambda_j} \right\}^2 \qquad (9b)$$

and the residual, R, is minimized by setting the partial derivatives of
R with respect to α', β' and γ to 0. The 3 simultaneous equations can
then be solved for the coefficients and the smoothed values of $[\theta]^{A,i}_{\lambda_j}$.
This procedure demands accurate values of $[\theta]^B_{\lambda_j}$ and $[\theta]^C_{\lambda_j}$, and
approximate values of $[\theta]^A_{\lambda_j}$.
    With the addition of another component, the residual, G, for the
more complicated regression analysis becomes

$$(10) \qquad G = \sum_{i=1}^n \sum_{j=1}^m \left\{ [\theta]^{OBS,i}_{\lambda_j} - f_{B_i}[\theta]^B_{\lambda_j} - f_{C_i}[\theta]^C_{\lambda_j} - (1-f_{B_i}-f_{C_i})[\theta]^A_{\lambda_j} \right\}^2$$

Minimization of this residual will involve solving only $(m + 2n)$ non-
-linear equations, since the values of $[\theta]^B_{\lambda_j}$ and $[\theta]^C_{\lambda_j}$ are presumably

known accurately, and hence only the partials,

$(\partial G/\partial f_{B_i})$, $(\partial G/\partial f_{C_i})$, and $(\partial G/\partial [\theta]^A_{\lambda_j})$ are involved. This procedure also demands that approximate values of $[\theta]^A_{\lambda_j}$ are available.

The data for the A spectrum of calf thymus DNA from Tunis-Schneider and Maestre (1970) used as the starting approximation of the A form in the linear regression analysis (Equations (9a), (9b) of the spectral data, above 255 nm, yielded a sum of the coefficients $(f_A + f_B + f_C)$ which differed significantly from one. The A spectrum of *E. coli* DNA reported by these authors provided a somewhat better fit, but was obviously not perfect. Finally, a simplified version of the regression analysis employing Equation (10) above yielded a spectrum which gave a satisfactory fit with all the data. This A spectrum above 255 nm was further smoothed by Equations (9a, 9b) for each solution, i. The remainder of the spectrum below 255 nm was calculated by

$$[\theta]^{A,i}_{\lambda_j} = \left\{ \frac{[\theta]^{OBS,i}_{\lambda_j} - f_{B_i}[\theta]^B_{\lambda_j} - f_{C_i}[\theta]^C_{\lambda_j}}{f_{A_i}} \right\} \tag{11}$$

The values of $[\theta]^{A,i}_{\lambda_j}$, were then mean averaged over the appropriate concentration range λ_j in each electrolyte.

A spectrum obtained in this manner from the LiCl spectral set is shown in Figure 5a. Similar results were also obtained in NH_4Cl and CsCl. The shape of this spectral component resembled that assigned to the A form of *E. coli* DNA, shown in Figure 5b.

In principle, it is possible that this spectrum is not a conformational form, but rather reflects a perturbation of the C spectrum due to the specific binding of Li^+, Cs^+ and NH_4^+ to the bases at high concentration of these salts. We do not see a spectral component of this type, however, in the divalent ions series and, hence, we feel that this is a rather unlikely explanation. Rather, we think that this spectrum reflects the presence of an A conformation, in a minor fraction of the nucleotide population. Since it resembles the higher GC content *E. coli* DNA spectrum rather than that for calf thymus, it is possible that the heavy satellites of calf thymus DNA are assuming an A conformation because their high GC content prevents them from converting to an C form (Brahms et al., 1973).

Once the B. the C and the A reference spectra were obtained in the manner described, we calculated the fractional amounts of these secondary structures present in the DNA solutions at various concentrations of the several electrolytes employed in these experiments. (For the solutions of DNA in $ZnCl_2$ and $CoCl_2$, we used the 'C' spectrum appropriate for these electrolytes). For these calculations, we simply reversed the order of input into the simple multicomponent linear regression program, permitting us to smooth values of $[\theta]^{OBS,i}_{\lambda_j}$ (rather than $[\theta]^C_{\lambda_j}$ or $[\theta]^A_{\lambda_j}$). That is, the residual, R, was

$$R = \sum_{\lambda_j} \left\{ [\theta]^{OBS,i}_{\lambda_j} - f_B[\theta]^B_{\lambda_j} - f_C[\theta]^C_{\lambda_j} - f_A[\theta]^A_{\lambda_j} \right\}^2 \tag{12}$$

R was differentiated with respect to f_B, f_C and f_A and set equal to 0 to generate the appropriate simultaneous equations. Again, this procedure does not require that the coefficients sum to 1. This, in fact, is used as a criterion, together with the standard deviation,

$$\text{STD} = \sqrt{\frac{\sum\limits_{j=1}^{m}([\theta]_{\lambda_j}^{\text{OBS},i} - [\theta]_{\lambda_j}^{\text{CALC},i})^2}{m-1}} \qquad (13)$$

as a measure of the goodness of fit. The standard deviations for all spectra were usually $\pm 0.2 \times 10^3$ (deg cm^2/decimole). The sum of the coefficients, $(f_B + f_C + f_A)$, averaged 1.00 ± 0.04. Some representative data for the fractional distribution of the A, B and C secondary structures in the 10 salts are given in Tables II and III.

TABLE II

Fractional Amounts of the A, B and C conformations of DNA in solutions of salts with monovalent cations.

NaCl (m)	0.01	1.14	2.32	4.17	6.09	
f_B	0.99	0.82	0.68	0.49	0.35	
f_C	0.01	0.12	0.34	0.55	0.67	
Sum	1.00	0.94	1.02	1.04	1.02	

KCl (m)	0.01	0.56	1.38	3.50	4.73	
f_B	0.91	0.84	0.70	0.47	0.34	
f_C	0.09	0.19	0.28	0.51	0.69	
Sum	1.00	1.03	0.98	0.98	1.03	

LiCl (m)	0.01	0.77	2.62	3.56	5.52	8.53
f_B	0.93	0.76	0.44	0.35	0.21	0.04
f_C	0.07	0.19	0.51	0.61	0.67	0.77
f_A	0	$0 \le f_A \le 0.04$	± 0.04	0.05	0.09	0.15
Sum	1.00	0.95	0.96	1.01	0.97	0.96

NH$_4$Cl (m)	0.01	0.10	3.67	5.34	7.38	
f_B	0.80	0.77	0.30	0.14	0.02	
f_C	0.20	0.25	0.65	0.83	0.93	
f_A	0	0	$0 \le f_A \le 0.04$	0.05	0.09	
Sum	1.00	1.02	0.95	1.02	1.04	

CsCl (m)	0.01	2.31	5.14	10.8
f_B	0.80	0.46	0.26	0.11
f_C	0.20	0.51	0.72	0.73
f_A	0	$0 \le f_A \le 0.04$	0.10	0.16
Sum	1.00	0.97	1.08	1.00

TABLE III

Fractional amounts of the B and C conformations of DNA in solutions
of salts with divalent cations

Salt	\multicolumn{6}{c}{Molal concentration of salt}					
	1×10^{-4}	2×10^{-4}	3×10^{-4}	5×10^{-4}	0.01	0.10
$MgCl_2$						
f_B	0.85	0.83	0.80			0.73
f_C	0.08	0.07	0.19			0.24
Sum	0.93	0.90	0.99			0.97
$CaCl_2$						
f_B	0.72	0.66	0.63	0.64	0.65	0.58
f_C	0.25	0.48	0.43	0.26	0.25	0.41
Sum	0.97	1.14	1.06	0.90	0.90	0.99
$MnCl_2$						
f_B	0.65	0.57	0.55	0.53	0.49	
f_C	0.39	0.61	0.31	0.53	0.52	
Sum	1.04	1.18	0.86	1.06	1.01	
$ZnCl_2$						
f_B	0.63	0.56	0.50	0.50	0.47	
f_C	0.36	0.44	0.52	0.44	0.52	
Sum	0.99	1.00	1.02	0.94	0.99	
$CoCl_2$						
f_B	0.56	0.45	0.43	0.42	0.36	
f_C	0.45	0.54	0.50	0.57	0.62	
Sum	1.01	0.99	0.93	0.99	0.98	

(C) High Concentrations of LiCl with Four or More components

At the very highest LiCl concentrations (10.5 m and above) both component
analysis and factor analysis indicate that components in excess of three
are required to describe the transitions. In this concentration range,
Equation (1) is not adequate to describe the data. In addition, the solutions evidence more light scattering, and the effects seen in Figure
1d at 13 m LiCl are clearly attributable to scattering artifacts. Sedimentation measurements on DNA in these solutions reveal the existence
of two sedimenting species, a fast and a slow, at concentrations of LiCl
above 7 m. The sedimentation properties of these solutions are given in
Table IV. An increase in light scattering accompanied the increase in
the fraction of fast moving species as the concentration of LiCl increased
from 7 m to 13 m.
    These facts taken together lead us to the conclusion that large
micelle structures of DNA are forming at these higher LiCl concentrations.

TABLE IV
Sedimentation of DNA in various concentrations of
LiCl solution

Molality LiCl	s_{50} a)		f_f (b	f_s (b
0.00				
(0.1 m NaCl)	17.06			
0.77	17.54			
5.03	17.51			
6.41	17.08			
	$s_{50,f}$	$s_{50,s}$		
7.39	25.28	16.75	6.7	93.3
8.53	28.00	14.65	15.2	84.8
9.51	31.36	13.75	23.7	76.3
11.03	40.56	11.73	39.8	60.2
12.04	40.33	11.34	44.3	55.7
11.10	39.00	11.70	42.0	58.0
11.10 + 0.005 m Na EDTA	38.69	11.23	43.7	56.3

a) Corrected to anhydrous Li DNA
b) Corrected for radial dilution

The secondary structure of the DNA in these aggregates amounts to ca.
80% C and 20% A forms. The molecules are packed in some ordered arrangement, giving rise to a large negative band around 270 *nm*.

5. RELATIONSHIP OF THE CONFORMATIONAL CHANGES TO OTHER PROPERTIES OF DNA IN SOLUTIONS OF ELECTROLYTES

As the data in Table II and III demonstrate, there is a significant increase in the C character at the expense of the B form at low concentrations of some salts. NH_4Cl and CsCl are particularly effective among the
univalent cation salts, and all of the divalent cation salts are more
potent at the same concentration. Although this transition at these low
concentrations of electrolytes must obviously arise by virtue of a direct
interaction of the cation with the DNA structure, it is not at all obvious as to how to rationalize the order of effectiveness of the various
ions. Differences between the univalent and the divalent cation efficien-
cies are likely to be due to differences in the affinity of these two
sets of ions, for the phosphate binding sites on the DNA backbone. In
fact, the proportional change which occurs in the CD spectra in the presence of these cations in the extremely dilute (1×10^{-4} m) solutions of
the metal salts is explainable only in terms of interaction in which one

divalent positive ion neutralizes two monovalent negative charges (Zimmer *et al.*, 1974).

Within a given family of cations, however, the *magnitude* of the depression of the positive band, and hence, the extent of the transconformational reaction which this phenomenon reflects, in dilute solutions of a given salt, such as 0.01 *m*, bears no relationship to the expected affinity of the cation for the phosphate sites. The relative order of the affinity constants of the univalent cations for the DNA phosphates should be $Li^+>Na^+>K^+$, $NH_4^+>Cs^+$. The efficiency in effecting the B→(C+A) structural conversion, however, is Cs^+, $NH_4^+>K^+$, $Li^+>Na^+$. Thus, site binding of a cation to DNA phosphate groups, while important, clearly does not provide the entire explanation of the mechanism of structural conversion in these electrolyte solutions.

The fact that these transconformational reactions continue in these solutions well past the concentration of cation needed to saturate the phosphate sites of DNA lends support to the above conclusion. And it is this observation which provides a clue to an additional important aspect of the mechanism. As has been demonstrated by Hearst and co-workers (Hearst and Vinograd, 1961a, b; Hearst, 1965; Tunis and Hearst, 1968a) as well as by Cohen and Eisenberg (1968), the preferential solvation, or net hydration, of DNA is modulated by the water activity of the solu-

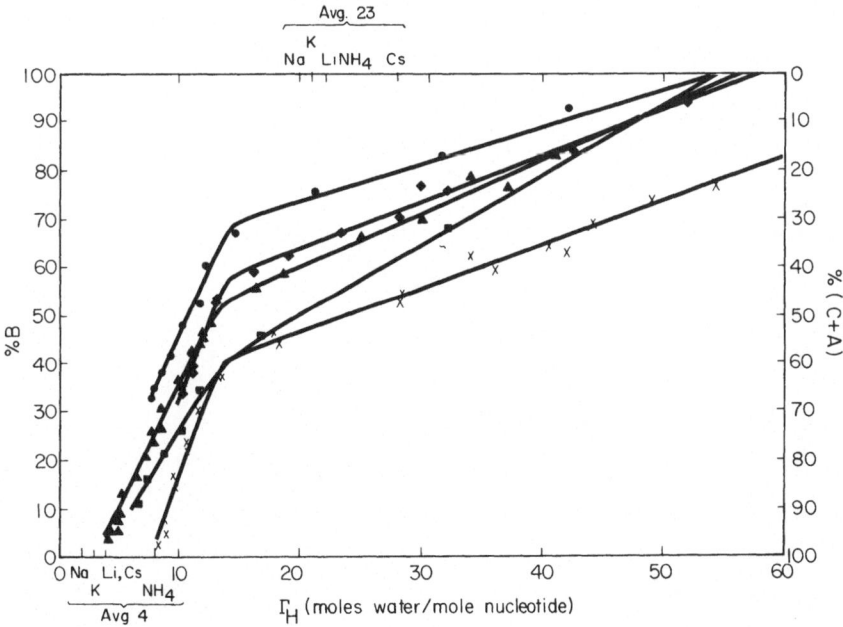

Fig. 7. Behavior of the fractional 'B' content of calf thymus DNA as a function of its net hydration in solutions of NaCl (●), KCl (◆), LiCl (▲), CsCl (■), and NH_4Cl (×). The fractional values of the secondary structures were calculated as described in the text. The net hydration, Γ_H, was taken from the data of Hearst and co-workers.

lion. As the concentration of electrolyte increases in solutions of salts of Na^+, Li^+, and Cs^+, the water activity decreases with a corresponding decrease in the net hydration. These workers have found that this effect is more or less independent of the nature of the cation (in the list given above).

Since we knew the molal concentration of univalent salts and, consequently, the water activity of each of the DNA solutions, we could calculate the net hydration of this macromolecule from the data available in the above cited literature. When we plotted the conformational information gleaned from our CD measurements against the net hydration, Γ_H, of DNA taken from the data of Hearst and co-workers, the results shown in Figure 7 were obtained. (Similar behavior was also observed when the net hydration data of Cohen and Eisenberg (1968) were plotted in the same fashion). Although there are specific cation effects on the structural properties, the general pattern in each electrolyte is similar. The data fall in two linear regions, one at lower net hydration values (Region I) and the other at higher hydration (Region II). The straight line relationships describing the behavior in these two regions intersect Γ_H of ca. 13 to 14 moles of water per nucleotide residue. This critical hydration value at which the B character begins to fall more precipitously is also that point at which the A form makes its appearance in solutions of NH_4Cl LiCl and CsCl. This is illustrated in Figure 8 which shows the variation

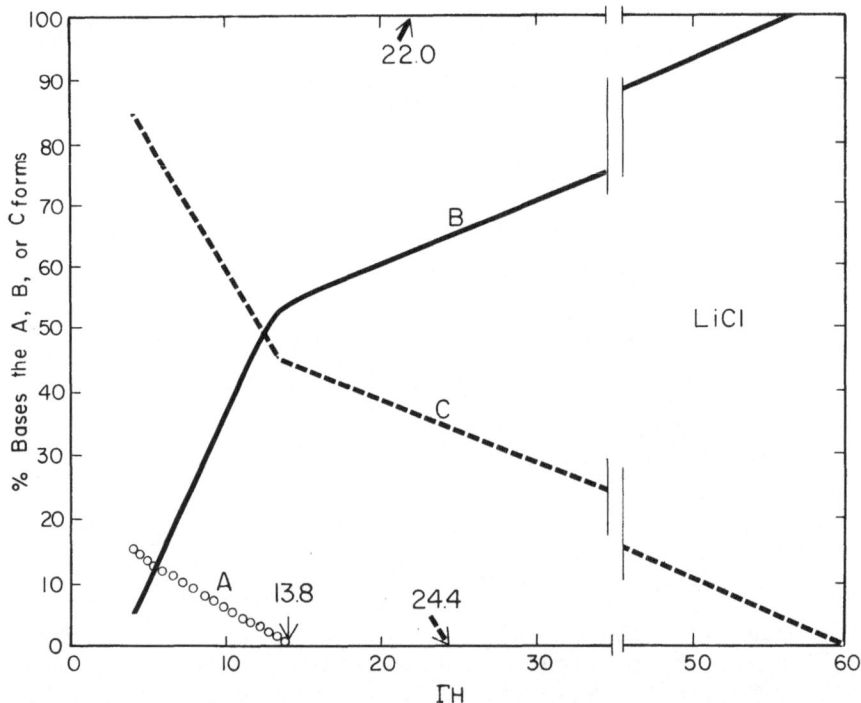

Fig. 8. Relationship between the fractional content of the 'B', the 'C' and the 'A' forms of calf thymus DNA in LiCl and the net hydration, Γ_H, of the macromolecule in these solutions. Γ_H values were taken from the data of Hearst and co-workers.

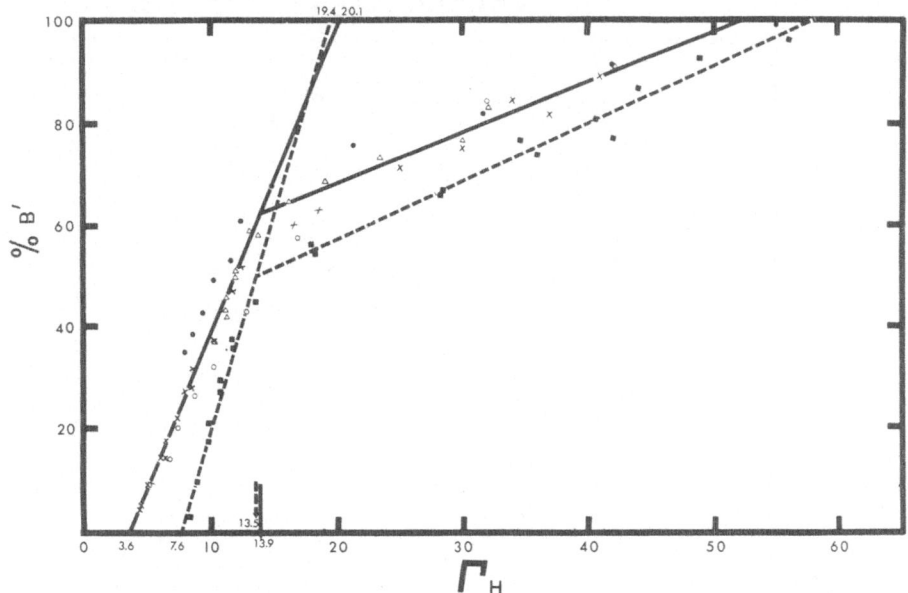

Fig. 9. Behavior of the normalized B content of calf thymus DNA as a function of its net hydration in solutions of NaCl (●), KCl (Δ), LiCl (×), CsCl (○) and NH$_4$Cl (■). The solid line represents the common linear regression of the alkali metal salts, and the dashed line that for NH$_4$Cl. The values of Γ_H are taken from the data of Hearst and co-workers.

in the fractional amounts of every structural component, individually, in solutions of LiCl as a function of the net hydration.

In Figure 9, we have normalized the fractional B content in the individual salt solutions by using the spectrum obtained in a given electrolyte at 0.01 m as the 100% B spectral end point appropriate for that salt. Although there is some scattering of points, the data for the alkali metal salts form a common curve. The NH$_4$Cl values, however, are still deviant.

Before interpreting these effects as well as those at low cation concentrations, we would like to review a picture of the hydration properties of DNA previously proposed (Wolf and Hanlon, 1975) which was based on the data and discussions of Hearst (1965), Tunis and Hearst (1968a, b), Cohen and Eisenberg (1968) and Falk *et al.* (1962, 1963a, b). As indicated schematically in Figure 10A, the hydration of DNA consists of a primary shell of ca. 20 water molecules per nucleotide adjacent to the surface of the DNA duplex, together with a more diffuse secondary shell of ca. 40 to 60 water molecules per nucleotide. The primary shell is the 'true' hydration layer, consisting of water molecules which are in a chemically different state compared to liquid water. Its OH frequencies are different and it is impermeable to electrolytes in the sense that ions cannot randomly move about in this layer at the same rate (or with the activa-

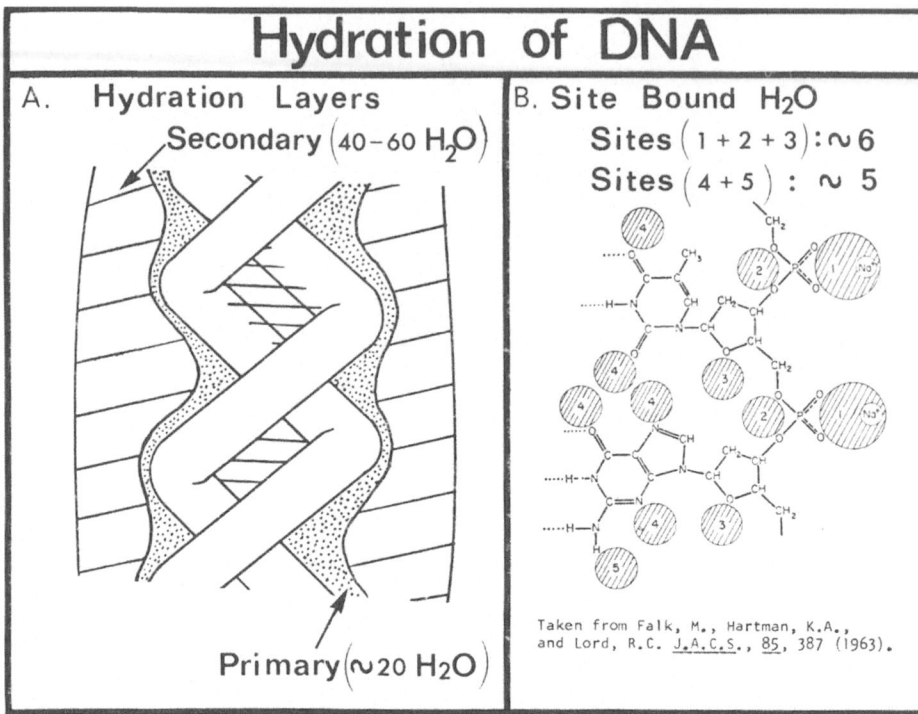

Fig. 10. The Hydration of DNA (A) Schematic version of the
secondary (cross hatched) and primary (stippled) hydration
layers of DNA. (B). The DNA hydration sites within the primary
hydration layer. This is a reproduction of a figure taken from
Falk, M., Hartman, K.A. and Lord, R.C., *Journal of the American
Chemical Society* 85, 387 (1963): The numbered cross hatched
areas indicate the sites of DNA hydration in the order of their
affinity for H_2O.

tion energy) that characterizes their motion in liquid water.
     Within this primary layer, water molecules are differentiated with
respect to whether they are directly bound to DNA sites, and their strengt
of binding to those sites. According to Falk *et al.* (1962, 1963a, b),
some 11 to 12 water molecules are bound to DNA at the sites indicated
in Figure 10B. We shall refer to these as Class A water molecules. Five
to six of these are associated with the base sites in the grooves of the
DNA helix and have a lower affinity than the remaining six. The four water molecules bound to the furanose oxygens and the phosphodiester linkages are somewhat more tightly bound than the ones associated with the
base sites. The remaining two water molecules associated with the ionic
phosphates have the highest affinity of this Class A type.
     The other class of water molecules (B) found in the primary hydra-
tion layer are distinguishable only insofar as the hydration layer of
DNA does not evidence the infrared spectrum of liquid water until ca.

20 water molecules have been adsorbed per nucleotide residue. Thus, 8 to 9 water molecules must be hydrogen bonded to the site bound water of Class A. One would assume these Class B molecules have the lowest affinity and would be the most readily removed water molecules of the primary shell.

With this model in mind, we further proposed that the maintenance of the 'B' conformation (that present in dilute NaCl solution at 0.01 m) was dependent on a combination of minimal interactions of the phosphate groups with the diffusible cations of the environment and the integrity of the primary hydration shell. The behavior of the data shown in Figures 7 through 9 was then rationalized in the following manner: As the hydration layers of DNA are reduced in Region II, it is primarily the secondary shell with a few of the weakly bound Class B water molecules of the primary shell which are removed. The loss of the B character, dependent on the removal of the primary shell, thus proceeds slowly in this region. When the net hydration has been reduced to 12 to 14 moles H_2O, only the tightly bound water molecules of the primary shell are left. As the latter are removed by the further lowering of the water activity, the % B character falls more rapidly, reflecting the true functional dependence of the conformational transitions on hydration. Since only the primary shell is being removed in this region, an extrapolation to 100% B at Γ_H<13 yields the magnitude of the primary layer minimally required to maintain the 'B' conformation. This value averages about 23 moles of water per nucleotide for the unnormalized data shown in Figure 7 and 19 to 20 moles for the normalized data in Figure 9. The difference between the intercepts of the two figures presumably reflects the efficiences of the various cations, compared to Na$^+$, in an initial displacement of water from the primary hydration layer in dilute solutions of the salt. These values agree reasonable well with estimate of the maximal amount of spectrally perturbed water bound to fibers of DNA (Falk *et al.*, 1962; 1963a, b).

Removal of all but ca. 4 of these water molecules from the primary shell of DNA in the presence of the alkali metal salts and 8 in the presence of NH_4Cl would, in principle, convert the DNA entirely to the (C + A) form. The difference between the hydration of these latter structures and that of the 'B' form probably reflects the loss of Class B water and Class A water molecules from the grooves of the helix.

At the very highest concentrations of LiCl, presumably the dehydration is so severe that DNA molecules must share the available water of hydration remaining on the phosphate sites if they are to remain in solution. It is conceivable that under these conditions, micelle structures form interlocking helices in which the water and Li$^+$ ions bound to the phosphate groups are shared intermolecularly.

As we thought about this interpretation over the past year, it seemed to us that we had missed a very crucial point: We had described this phenomenon of structural conversion in terms of a cation association and hydration dependent process, but had failed to explain why these events should be linked. In mulling over this problem, we have arrived at a revised point of view, as follows:

It is clear that the limiting 'B' conformation of calf thymus DNA present in 0.01 m NaCl at 25° to 27 ^{o}C (pH 7) represents a conformation

on the edge of stability. Electrostatic repulsive interactions are
screened to just that point which allows effective base stacking. Under
these conditions, the DNA structure should be one which maximizes the
interstrand distance between the phosphate groups without sacrificing
van der Waals contact between the purine and pyrimidine bases. Such a
structure can be achieved by decreasing the winding angle between the
base pairs in a B structure thereby increasing the size of the grooves
(Ivanov et al., 1973) In the absence of appreciable concentrations of
ions competing for the water in its hydration shells, the DNA molecule
will be extensively hydrated and the primary hydration shell should be
intact. We now feel that this latter hydration layer is probably important to the 'B' structure only insofar as it provides some degree of
screening and diffusion of electrostatic charge on the phosphate groups.
In the absence of this layer of dipole water molecules, DNA denatures,
as has been demonstrated by Falk et al. (1963a, b).

       As ions other than Na^+ bind to this 'B' structure, in dilute solu-
tions of electrolyte, the ones which are larger, or fit best into the
structure of the primary layer or, in fact, displace it to permit more
effective ion pairing, will be more effective in relieving the electrostatic repulsive interactions of the phosphate groups in a direction
along the helix axis. In the presence of these more efficient cations,
or at higher concentrations of Na^+, DNA relaxes into a form which permits
the phosphate groups of adjacent strands to approach more closely. The
groove size correspondingly decreases. In addition, these same ions,
either in their anhydrous or their hydrated state may be able to span
the distance between the N_7 sites on the purines in the large groove and
the oxygens of the phosphates in the polynucleotide backbone, thus
screening repulsive interactions in a horizontal direction as well. This
interaction, even though mediated by a water bridge would tend to pull
the bases off axis in the major groove which would correspondingly become
more shallow. The net effect of this movement coupled with the decrease
in the groove size would be to create a C like structure. If similar
ionic interactions between the base sites exposed in the minor groove
and the polynucleotide backbone ensued, an A structure would form. Ejec-
ting site bound water molecules from the grooves of the helix would facilitate these processes considerably. One would expect to see a correlation between the decrease in the net hydration and an increase in the
C and A character if the mechanism described above are operative.

       Thus the role of the primary hydration shell may be more of a pro-
hibitive rather than a stabilizing one for the B→(C+A) transconformational reaction. The presence of water in the grooves of the helix in the
B form prevents those cation interactions which promote the formation
of alternate secondary structures, such as the C and the A forms, whose
structural characteristics are such that their stability demands effective vertical and horizontal screening of repulsive electrostatic interactions between interstrand phosphate groups and between base dipoles
and negatively charged phosphate oxygens.

REFERENCES

Brahms, J., Pilet, J., Lan, T.T.., and Hill, L.R.: 1973, *Proc. Natl. Acad. Sci. U.S.A.* <u>70</u>, 3352.

Bram, S.: 1971, *J. Mol. Biol.* <u>58</u>, 277.

Cheng, P.Y.: 1965, *Biochim. Biophys. Acta* <u>102</u>, 314.

Cohen, G. and Eisenberg, H.: 1968, *Biopolymers* <u>6</u>, 1077.

Falk, M., Hartman, Jr., K.A., and Lord, R.C.: 1962, *J. Am. Chem. Soc.* <u>84</u>, 3843

Falk, M., Hartman, Jr., K.A., and Lord, R.C.: 1963a, *J. Am. Chem. Soc.* <u>85</u>, 387.

Falk, M., Hartman, Jr. K.A., and Lord, R.C.: 1963b, *J. Am. Chem. Soc.* <u>85</u>, 391.

Fuller, W., Wilkins, M.H.F., Wilson, H.R., and Hamilton, L.D.: 1965, *J. Mol. Biol.* <u>12</u>, 60.

Hanlon, S., Johnson, R., Wolf, B., and Chan, A.: 1972, *Proc. Natl. Acad. Sci. U.S.A.* <u>69</u>, 3263.

Hanlon, S., Brudno, S., Wu, T.T., and Wolf, B.: 1975, *Biochemistry* <u>14</u>, 1648.

Hearst, J.E. and Vinograd, J.: 1961a, *Proc. Natl. Acad. Sci. U.S.A.* <u>47</u>, 825.

Hearst, J.E. and Vinograd, J.: 1961b, *Proc. Natl. Acad. Sci. U.S.A.* <u>47</u> 1005.

Hearst, J.E.: 1965, *Biopolymers* <u>3</u>, 57.

Ivanov, V.I., Minchenkova, L.E., Schyolkima, A.K., and Poletayev, A.I.: 1973, *Biopolymers* <u>12</u>, 89.

Johnson, R.S., Chan, A., and Hanlon, S.: 1972, *Biochemistry* <u>11</u>, 4347.

Langridge, R., Marvin, D.A., Seeds, W.E., and Wilson, H.R.: 1960, *J. Mol. Biol.* <u>2</u>, 38.

Luck, G. and Zimmer, C.: 1972, *Eur. J. Biochem.* <u>29</u>, 528.

Maniatis, T., Venable, Jr., J.H., and Lerman, L.S.: 1974, *J. Mol. Biol.* <u>84</u>, 37

Marvin, D.A., Specer, M., Wilkins, M.H.F., and Hamilton, L.D.: 1961, *J. Mol. Biol.* <u>3</u>, 547.

Nelson, R.G. and Johnson, Jr., W.C.: 1970, *Biochim. Biophys. Res. Commun.* <u>41</u>, 211.

Pilet, J, and Brahms, J.: 1973, *Biopolymers* <u>12</u>, 387.

Studdert, D.S., Patroni, M., and Davis, R.C.: 1972, *Biopolymers* <u>11</u>, 761.

Tunis, M.J. and Hearst, J.E.: 1968a, *Biopolymers* <u>6</u>, 1218.

Tunis, M.J. and Hearst, J.E.: 1968b, *Biopolymers* <u>6</u>, 1325.

Tunis-Schneider, M.J.B. and Maestre, M.F.: 1970, *J. Mol. Biol.* <u>52</u>, 521.

Wolf, B. and Hanlon, S.: 1975, *Biochemistry* <u>14</u>, 1648.

Zimmer, C. and Luck, G.: 1973, *Biochim. Biophys. Acta* <u>312</u>, 215.

Zimmer, C., Luck, G., and Triebel, H.: 1974, *Biopolymers* <u>13</u>, 425.

DISCUSSION

Saenger: Fibre x-ray diffraction studies of DNA have shown that in the B-form the base pairs are approximately vertical to the helix axis but they are tilted by about $20°$ in A-DNA and by $-6°$ in C-DNA. Further, in

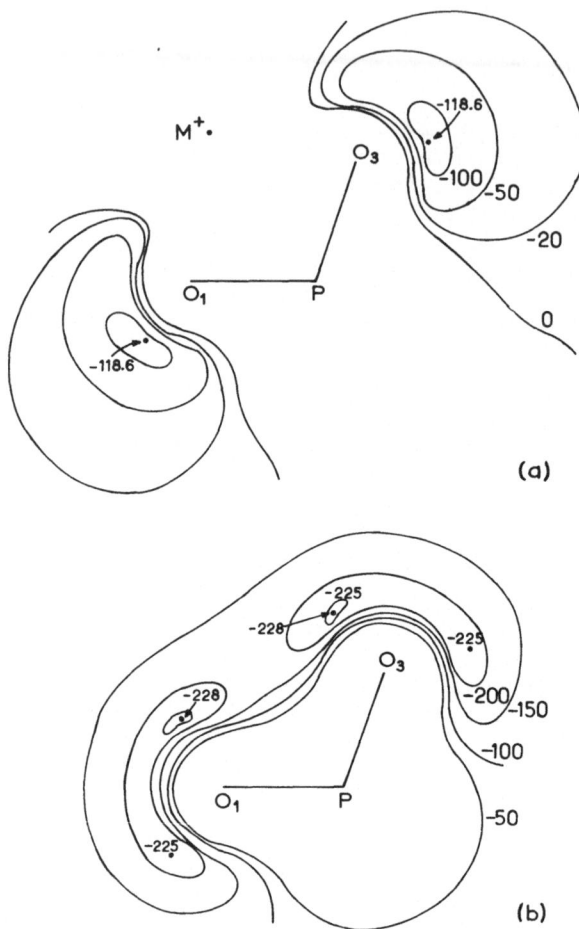

(a)

(b)

B-DNA the helix axis passes through the base pairs but it is located in
the large groove in A-DNA and in the narrow groove in C-DNA. In other
words, A- and C-DNA represent two extreme conformations (with also different sugar puckering). How do you explain the fact that both of these
conformations, or an intermediate version, can coexist in the same macromolecule?

Hanlon: We didn't mean to imply that an intermediate conformation (between A and C) can exist. However, it is possible that the two separate
conformations can coexist in the same macromolecular population. The
sample employed in these experiments is calf thymus DNA which has several heavy satellites of high GC content. We think that these fractions,
because of their high GC content are converted to the A conformation
rather than the C conformation as they are dehydrated in the concentrated
salt solutions.

A. Pullman: In relation to Dr Eichhorn's and Dr Hanlon's remarks on the
screening of the electrostatic potential of the phosphate group by mono-
and divalent cations and its significance for the stability of the DNA

TABLE I

The molecular potential (kcal/mole) at large distances (Ångström units) from the phosphate anion and its modification by a cation in position B.

O_3...cation distance	Phosphorus-cation distance	Molecular potential due to		
		DMP^-	$DMP^-...M^+$	$DMP^-...M^{++}$
5	5.70	−54.2	−5.5	+43.2
6	6.87	−46.3	−3.6	+39.1
7	7.85	−40.3	−2.4	+35.5

structure, I would like to report that we have carried out recently a quantitative evaluation of the screening effect (Pullman, A. and Berthod, H. *Chem. Phys. Letters,* in press). For this sake we have computed the molecular isopotential energy map of the phosphate group in the presence of a Na$^+$ cation placed at the most stable position, the bridge position, between the two anionic oxygens O_1 and O_3, (Figure 1a) and compared it to the map of the free phosphate group in the O_1PO_3 plane (Figure 1b). The modification of the potential by the ion indicates a strong decrease in the attractive character of the phos-

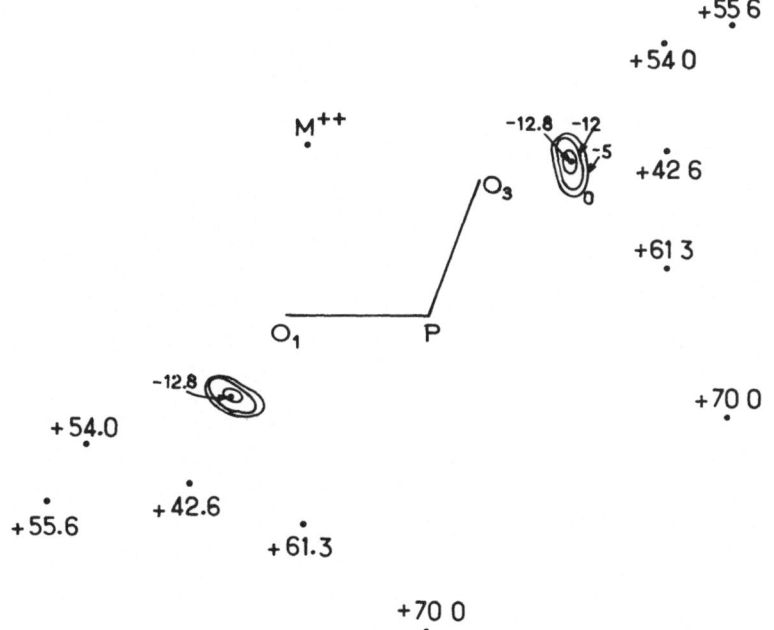

phate group: the value of the potential minimum is reduced to about one half of its former value and the dimension of the attractive zone is strongly decreased. A further illustration of the situation is provided in Table I where we given the values of the potential at three points situated at 5, 6 and 7 Å respectively from the O_3 oxygen, a range of distances which corresponds approximately to the distance between two adjacent phosphate groups in helical DNA. Whereas the values for the isolated phosphate group are appreciably negative, they become close to zero in the presence of the monocation.

We have done the same computation for a dipositive cation placed in the same bridge position. The results, presented in Figure 2, show that the attractive isoenergy lines have now practically disappeared; there remains only a very narrow zone of negative potential in the region of the external minimum of the isolated phosphate, while the whole O_1PO_3 plane is repulsive (see also Table I). The screening of the electrostatic potential of the negative phosphate is thus practically complete in this case.

INTERACTION OF GADOLINIUM (III), MANGANESE (II) AND COPPER (II) WITH CYCLIC NUCLEOTIDES

G.V. FAZAKERLEY, G.E. JACKSON, J.C. RUSSELL, and M.A. WOLFE
Dept. of Inorganic Chemistry, University of Cape Town, Rondebosch, 7700, South Africa

1. INTRODUCTION

Reactions of nucleic acids in biological systems are generally mediated by the presence of metal ions [1]. Because of the variety of sites available for metal binding (base and phosphate) this interaction can be complex as the metal ion does not necessarily occupy a unique site. As models for nucleic acids the cyclic nucleotides being phosphodiesters have been chosen. With monoesters there is the additional complication that first row transition metal ions may form a backbound chelate linking the base and the phosphate through a bridging water molecule. This has been suggested for Mn(II) and Cu(II) complexes with ATP [2, 3]. Although Mn(II) might be expected to bind predominantly to the phosphate and Cu(II) to base nitrogen donors the enhanced stability of the Cu(II) complex of 5' AMP relative to that of adenosine [4] and the observation of line broadening of the ^{31}P resonance of 5'AMP in the presence of Cu(II) [5] suggests that direct phosphate binding cannot be eliminated. Two furthers factors complicate the interpretation of n.m.r. line broadening studies. In employing high concentrations of ligand relative to paramagnetic ion multinuclear species may be present in solution and in, for example a 1:2 metal ligand complex the metal may be bound to different sites on the two coordinated ligands. Also as the n.m.r. data on species undergoing rapid comformational changes reflects the time averaged conformation it is necessary to determine what conformations exist in solution and their contribution to the overall effect to interpret line broadening studies. For the cyclic nucleotides rapid interconversion between syn and anti forms and rapid flipping of the ribose may be expected.

From analysis of coupling constant data the cyclic ribonucleotides show a preference for a ^{3}T (3'endo) conformation of the ribose whereas the deoxyribonucleotide (cTMP) is $_4$T (4'exo) [6]. The phosphate ring is locked in the chair conformation, Figure 1.

Potential energy calculations on 3',5' cyclic purine nucleotides [7, 8] show two broad energy minima corresponding to syn and anti conformations, Yathindra and Sundaralingam predict that the anti conformation is slightly favoured for cAMP and syn for cIMP.

B. Pullman and N. Goldblum (eds.), Metal-Ligand Interactions in Organic Chemistry and Biochemistry, part 1, 107-113. *All Rights Reserved.*
Copyright © 1977 D. Reidel Publishing Company, Dordrecht-Holland.

Lanthanide shift studies on cAMP at pH 5,5 have been interpreted in terms of a syn conformation [9]. Another study was interpreted in terms of contributions from both syn and anti conformations [10].

2. EXPERIMENTAL

Spectra were recorded on a Bruker WH-90 DS and a Varian XL-100 at 90 MHz and 100 MHz respectively. The ligands were 0.04 M in ^2H_2O. Measurements were made in the presence of 0.04 M praesodymium to shift the spectrum and allow observation of most of the proton resonances. Experiments in the absence of praesodymium showed that this had no observable effect upon the relative relaxation times of H2, H8 and H1' and thus does not influence the binding of Cu(II) and Mn(II). The Gd-EDTA, and Cu-ethyl-endiamine complexes were used for relaxation studies which limits the species formed to 1:1 metal substrate. For the Mn(II) titrations the nitrate was used.

3. RESULTS AND DISCUSSION

3.1. *Conformation of cAMP, cIMP and 6 chloro purine riboside 3'5 cyclic phosphate (c Cl PMP)*

The conformations were investigated by measurement of the relaxation times (T_1 and T_2) in the presence of Gd(III). The metal ion titrations gave good straight lines and the slopes $(fT_{2p})^{-1}$ and $(fT_{1p})^{-1}$, where f is the ratio of metal ion concentration to total ligand concentration, are shown in Table I. These are related to the metal ion- observed proton internuclear distances (r) by the Bloembergen and Solomon equations [11, 12] In the absence of scalar interactions the equations reduce to:

$$\frac{1}{fT_{2p}} = \frac{K}{r^6} \ , \ \frac{1}{fT_{1p}} = \frac{K'}{r^6} \tag{1}$$

where K and K' are constants during the experiment.

Tests with the nucleotides show that at the Gd(III) concentrations used in the experiments metal binding to adenosine and 6 chloropuriri-boside is negligible and to inosine very slight. The dominance of phosphate binding is further confirmed by the observation that the protons closest to the metal when bound on the phosphate (H3' and H5'u) are

TABLE I

Ligand	Metal	Hz	H8	H2	H1'	H2'	H3'	H5'u
3'5'cAMP	Gd(III)	$(fT_1p)^{-1}$	4100	4050	1580	2870	13300	12300
		$(fT_2p)^{-1}$	1220	1220	811	1063	9190	6280
	Mn(II)	$(fT_2p)^{-1}$	1390	1390	310	737	1778	2950
	Cu(II)	$(fT_1p)^{-1}$	1080	1870	81	55	56	24
3'5'cIMP	Gd(III)	$(fT_2p)^{-1}$	1450	232	390	743	—	2920
	Mn(II)	$(fT_2p)^{-1}$	1328	120	112	—	1638	1457
3'5'cClPMP	Gd(III)	$(fT_2p)^{-1}$	885	845	463	928	6286	4485
	Mn(II	$(fT_2p)^{-1}$	211	181	101	—	—	3195

most affected. The calculated internuclear distances agree well with those expected using data from x-ray studies on 3'5' cyclic phosphates [13, 14]. These are shown in Table II.

TABLE II
Observed and calculated internuclear distance referenced to H2'. The Gd(III) — 0 bond distance used was 2.3 Å.

	H1'	H2'	H3'	H5'u
3'5'cAMP	7.1	6.4	4.9	5.0
3'5'cIMP	7.2	6.4	5.0	5.2
3'5'cClPMP	7.1	6.4	4.7	4.9
Calculated	8.1	6.4	4.8	4.9

The relaxation data on cAMP show within experimental error an equal effect of the metal ion on H2 and H8. This would correspond to a position of the base midway between syn and anti conformations. From potential energy calculations such a conformation appears highly unlikely. Also in the crystalline state [13] two forms were observed, one corresponding to an anti conformation and the other to a syn conformation. It seems more likely that the data correspond to a time average of syn and anti conformations rather than a unique one. Using the atomic coordinates from the crystal structure with the ribose in the ^3T conformation the contributions from syn and anti forms required to fit the data were calculated. As we cannot expect the glycosyl torsion angles ϕ_{CN} [15] corresponding to syn and anti to be precisely defined searches were carried out over a number of permutations in the range $+120^{\circ} \pm 20^{\circ}$ (syn) and $-40^{\circ} \pm 20^{\circ}$ (anti). The corresponding populations are shown in Table III.

TABLE III
Observed contributions to the time averaged species.

	syn	anti	
3'5'cAMP	45%	55%	(± 5%)
3'5'cIMP	5%	95%	(± 2%)
3'5'cCLPMP	40%	60%	(±10%)

The effect of Gd(III) upon the H2 and H8 resonances of cIMP is markedly different. H8 is much more affected than H2 and qualitatively this must be explained in terms of a dominance of the anti conformation. Similar calculations to those made on cAMP were carried out and the results, Table III show only a small contribution from the syn conformation.

The data for c Cl PMP is similar to that observed for cAMP. Substantial contributions from both syn and anti conformations are required to explain this and the calculations show that the anti conformation is slightly favoured, (Table III).

3.2. *Manganese(II) binding*

The effect of Mn(II) upon the relaxation times of the protons of c Cl PMP is, within experimental error, the same as Gd(III). If binding to the base were to occur at all all the relaxation of either or both of H2 and H8 would be enhanced relative to the ribose protons. As this is not observed Mn(II) must bind wholly to the phosphate as does Gd(III). It is clear that the phosphate is a much better donor than monodentate base nitrogen donors.

The effect of Mn(II) upon the relaxation of the protons of cIMP differs from that observed with Gd(III). There is, relative to the ribose protons on approximately two fold enhancement on H8. This can only be explained in terms of some base binding. Studies on substituted hypoxanthines and inosines [16] have shown that the only site favoured for Mn(II) binding is with the metal ion coordinated to N7 and O6. Given the time averaged conformation of the molecule from the Gd(III) study and allowing binding to both phosphate and base a computer search reveals a substantial contribution from the base site viz. phosphate 70%, base 30%.

For cAMP the only base site that need be considered is a similar bidentate one through N7 and the amine group. If binding were to occur there then by analogy with cIMP the relaxation of H8 should rise relative to H2. This is not observed and binding to the phosphate must dominate.

3.3. *Copper(II) binding*

Although there have been many studies on Cu(II) binding reported it has recently been shown that T_2 may be dominated by the scalar term in the Bloembergen and Solomon equations [17]. However studies on model compuonds show that even with Cu(II) bound directly to aromatic systems T_1 is dominated by the dipolar term.

The data for cAMP chow clearly that base binding predominates and that the metal ion has considerable binding in the vicinity of H2. This could be accounted bor by binding to N1 and or N3. However as the effect upon H1' is very low and this proton would be very close to Cu(II) bound on N3 (binding can only occur when the molecule is in the anti conformation), N1 must be the most favoured site. That monodentate coordinaton is dominant rather than a chelate formed by the amino group and N7 is not unexpected as models shown that such a chelate would be quite strained. However the data cannot be explained by a unique site of coordination and a computer search was carried out allowing binding at N1, N3, N7 and the phosphate. The results are N1 40%, N3 <5%, N7 50%, phosphate 7%.

4. CONCLUSIONS

The three cyclic nucleotides studied exist in a dynamic syn/anti equi librium the position on that equilibrium being influenced by the 6-substitent. Mn(II) shows a strong preference for phosphate binding although with an oxygen donor as the 6-substituent base binding also occurs. The binding of Cu(II) to cAMP is complex involving all possible donor sites but with a strong preference for the base.

REFERENCES

1. Eichhorn, G.L., Berger, N.A., Butzow, J.J., Clark, C., Rifkind, J.M., Shin, Y.A., and Tarein, E.: 1971, Adv. Chem. 100, 135 and references therein.
2. Sternlicht, H., Shulman, R.G., and Andersen, E.W.: 1965, J. Chem. Phys. 43, 3123, 3133.
3. Glassman, T.A., Cooper, C., Harrison, L.W., and Swift, T.J.: 1971, Biochemistry 10, 843.
4. Schneider, P.W., Brintzinger, H., and Erlenmeyer, H.: 1964, Hel. Chim. Acta 67, 992
5. Eichhorn, G.L., Clark, P., and Becker, E.D.: 1966, Biochemistry 5, 245.
6. Kainosho, M. and Ajisaka, J.: 1975, J. Amer. Chem. Soc. 97, 6839.
7. Yathindra, N. and Sundaralingam, M.: 1974, Biochem. Biophys. Res. Commun. 56, 119.
8. Saran, A., Berthod, A., and Pullman, B.: 1973, Biochim. Biophys. Acta 331, 154.
9. Lavallee, D.K. and Zeltmann, A.H.: 1974, J. Amer. Chem. Soc. 96, 5552.
10. Barry, C.D., Martin, D.R., Williams, R.J.P., and Xavier, A.H.: 1974, J. Mol. Biol. 84, 491.
11. Bloembergen, N.: 1957, J. Chem. Phys. 27, 572.
12. Solomon, I.: 1955, Phys. Rev. 99, 559.
13. Watenpaugh, K., Dow, J., Jensen, L.H., and Furberg, S.: 1968, Science 159, 206.
14. Coulter, C.L.: 1069, Acta Cryst. B25, 2055.
15. Donohue, J. and Trueblood, K.N.: 1960, J. Mol. Biol. 2, 363.
16. Fazakerley, G.V. and Russell, J.C.: paper in preparation.
17. Esperson, W., Hutton, W., Chow, S., and Martin, R.: 1974, J. Amer. Chem. Soc. 96, 8111.
18. Fazakerley, G.V. and Jackson, G.E.: paper in preparation.

DISCUSSION

Sarkar: Am I correct in understanding that you did your experiments at only one pH and not as a function of pH? In the absence of species distribution in the system, as you have not reported any as a function of pH, I presume that you are making an assumption of having a single species in interpreting your EPR data.

Fazakerley: The experiments with different metal ions on each substrate were carried out at constant pH. With Gd(III and Cu(II) only 1:1 species formed, the remaining sites being blocked.

Rode: The interpretation of your data is based on the assumption, that there would not occur any conformational changes in the DNA molecule upon metal ion bonding. If this bonding occurs at N7, there might exist, however, some influence of this kind. This would lead, however, to a change in the relaxation contribution due to purine- and ribose-proton interactions and, therefore, not allow an anambiguous assignement of the changes in H-8 relaxation as a merely metal induced one. Is there any evidence, that such a conformational change would occur or not?

Fazakerley: We know that binding to phosphate does not significantly change the time averaged conformation. There is no evidence that binding at N7 produces a change, but it cannot be ruled out yet.

Kistenmacher: I may have been somewhat misleading this morning. Most of the known Cu(II)-Purine crystal structures show N(7) binding. However, two solution studies (T$_1$, Martin et al., Jacs **96**, 8111 (1974) and Marzilli et al., Inorg. Chem. **14**, 2568 (1975)) clearly show that N(1) and N(7) are both important in solution.

Fazakerley: This is so. It is interesting that the crystal structures seem to favour N7 binding so strongly and presumable refelct factors not present in solutions.

Simon: I do not see why you are worried because of the differences in correlation times of H$_2$/H$_8$ vs. H$_2$'/H$_3$' etc. Does this not simply mean that there is segmental motion in the molecule?

Fazakerley: We have not been able to measure the correlation times directly but only that they seem to be rather different. It appears that the correlation times of the base are longer than those of H$_2$', H$_3$', etc.

Sletten: Why do you suggest a bidentate binding for Cu(II) and Mn(II) instead of monodentate binding to N(7)?

Fazakerley: With inosines we have found that Mn(II) is bidentate (O-N7) whereas Cu(II) is monodentate. With adenosines Mn(II) binding is weak. Cu(II) on the bulk of the evidence seems to be monodentate N7 and this we have assumed.

Dobson: Have you considered the possibility that your relaxation data could be influenced by outer-sphere (non-specific) relaxation by the metal ions, or by binding to hydroxyl groups of the ribose ring? Such effects have been observed in similar systems (see for example, Barry, C.D., Glasel, J.A., Williams, R.J.P., and Xavier, A.V.:1974, J. Mol. Biol. **S4**, 471).

Fazakerley: We considered it by looking at adenosine itself, and did not find any significant specific effects. Since the ribose conformation in cAMP is more rigid perhaps you are correct in that case.

Laniv: The approximation of the Solomon-Bloembergen equation for

T_{1M}^{-1} presented here is oversimplified. The equation should contain at least the frequency dependent function, and some explanation on the relevant correlation times should be given.

The correlation times for the different protons which are neccessary for the exact calculation of distances between the paramagnetic metal ions and these protons can be obtained from the frequency dependence of the relaxation rates (preferably - T_{1M}). In these systems it is best to concentrate on frequency dependency at low frequencies, since the correlation times are relatively short and are governed mainly by the rotational correlation time, although there is some contribution from the electronic spin correlation time. The alternative method for the determination of TO, mainly the T_{1p} /T_{2p} ratio is not fruitful in the systems presented here since TC is too short and since the hyperfine term contribution is too small.

Fazakerley: We have shown on model systems that the dominant contribution to τ_C is τ_r for all the metal ions employed.

INFLUENCE OF HETEROCYCLIC NITROGEN ATOMS ON THE INTERACTION BETWEEN
PURINE ANALOGUES AND METAL IONS

Wolfgang Lohmann, Anita Pleyer, Max Hillerbrand, Volker Penka
*Institut für Biophysik der Universität, Giessen, and Abteilung
für Biophysikalische Chemie, GSF, Neuherberg*

and

Karl G. Wagner and Hans-Adolf Arfmann
Gesellschaft für Molekularbiologische Forschung, Stöckheim

SUMMARY. The charge-transfer complexes formed between different purine
analogues and Cu^{2+} have been investigated by means of ESR and NMR studies.
It could be shown that the number as well as the position of the heterocyclic nitrogens influence the strength of these complexes.

1. INTRODUCTION

Recently we have shown that the purine bases exhibit a stronger affinity
to cupric ions than pyrimidine bases (Lohmann and Penka, 1973). Since
theoretical results have shown that the electron donor properties of the
purines are more pronounced than that ones of the pyrimidines (Pullman and
Pullman, 1966, it was assumed that the metal ion-nucleobase association is
due, at least partly, to a charge-transfer interaction. According to these
calculations the heterocyclic nitrogen atoms should contribute to this
type of interaction. In this case, the relative strength of such an association ought to be the greater the more nitrogen atoms participate.
In order to elucidate the involvement of the nitrogen atoms purine analogues with different numbers of heterocyclic nitrogens located at different sites were selected for the investigations. Cu^{2+} was used as an
electron acceptor since most of the investigations concerning metal ion-
nucleic acid interactions have been done with this metal ion. Its charge-
transfer interaction with the purine analogues and the contribution of
the heterocyclic nitrogen atoms to such an association have been determined by means of electron spin resonance and nuclear magnetic resonance
techniques.

2. MATERIALS AND METHODS

Indole, 7-aza-indole, benzimidazole, purine as well as $Cu(NO_3)_2$ and di-

*B. Pullman and N. Goldblum (eds.) Metal-Ligand Interactions in Organic
Chemistry and Biochemistry, part 1, 115-126. All Rights Reserved.
Copyright © 1977 by D. Reidel Publishing Company, Dordrecht-Holland.*

methylsulfoxide were purchased from Merck AG., Darmstadt, Germany. 1-deaza-purine was prepared according to a method described elsewhere (Arfmann, 1976). Deuterated dimethylsulfoxide (d_6-DMSO, 99.5%) was obtained from Sharp & Dohme GmbH., Munich, Germany. All substances were of reagent-grade quality and were used without further purification. The test solutions were always prepared shortly before the spectra were recorded. Since in a few instances the spectra changed with time the measurements were repeated at certain times after preparation. All measurements were done at room temperature except in the case of electron spin resonance studies where a few experiments were conducted at 77 K, too.

The proton magnetic resonance (^{1}H NMR) spectra were recorded on an HA 100 Varian Associates spectrometer. Chemical shifts were measured with a frequency counter relative to an internal DMSO standard and are given in parts per million (ppm) relative to tetramethylsilane (TMS); the NMR probe temperature was maintained at ca. 30 °C.

The ^{13}C NMR spectra were recorded on a CFT 20 Varian spectrometer.

The electron spin resonance (ESR) spectra were determined with a Varian E9 100-kc ESR spectrometer using a liquid sample cell accessory. At 77 K, the samples were measured in a Dewar flask containing liquid nitrogen. All spectra presented were obtained at identical experimental conditions except where marked specificly. The relative spin concentration was obtained by double integration of the spectra by means of a planimeter.

The numbering of purine was used throughout the paper despite the fact that the official and usual numbering of indole, e.g., is different. In this way, a certain number designates always the same proton which is very convenient for discussing, for example, the NMR spectra. Thus, the numbering shown in Figure 1 applies to all 5 substances used.

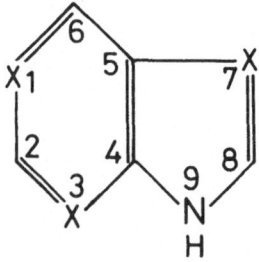

Purine:	$X_1 = X_3 = X_7 = N$
1-Deaza-Purine:	$X_1 = CH, X_3 = X_7 = N$
Benzimidazole:	$X_1 = X_3 = CH, X_7 = N$
7-Aza-Indole:	$X_1 = X_7 = CH, X_3 = N$
Indole:	$X_1 = X_3 = X_7 = CH$

Fig. 1. Structural formula of the substances investigated with the numbering of purine used troughout this paper

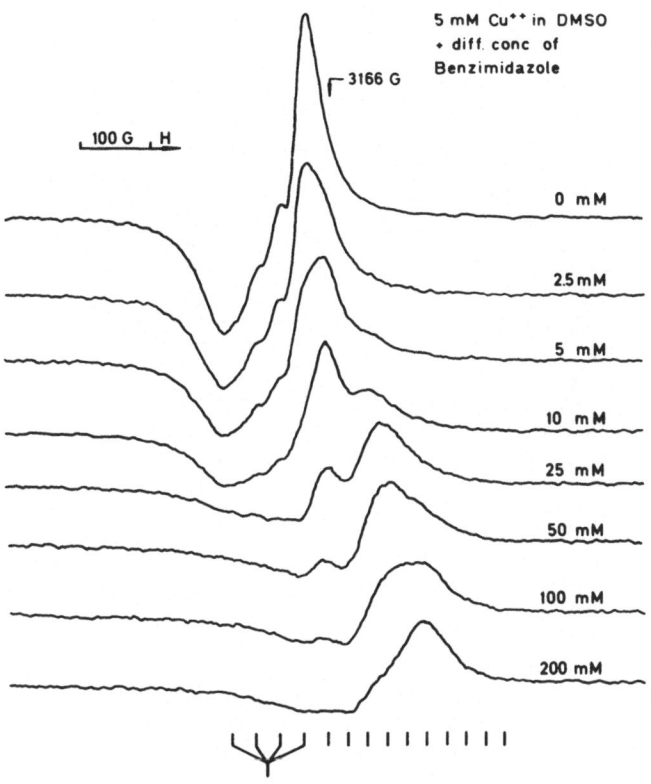

Fig. 2. The effect of varying concentrations of the different compounds on a 5.0 mM Cu^{2+} ESR spectrum in DMSO
   (a) benzimidazole (the first 4 lines at the bottom mark the Cu^{2+} hf structure, the following 10 lines a superhyperfine structure caused probably by nitrogen),
   (b) 7-aza-indole,
   (c) 1-deaza purine,
   (d) purine.

The S.D. of the results obtained is about ± 1 Hz (chemical shift, ^1H NMR and ^{13}C NMR) and ± 10% (spin conc., ESR).

3. RESULTS AND DISCUSSION

In Figure 2a the effect of increasing concentrations of benzimidazole on a Cu^{2+} ESR spectrum in DMSO is shown. As can be seen, already at equimolar concentrations several new peaks appear at a somewhat higher field while the original Cu^{2+} spectrum decreases considerably but still maintains the 4 line hf structure. After passing through different intermediate steps, the final spectrum seems to be established at a con-

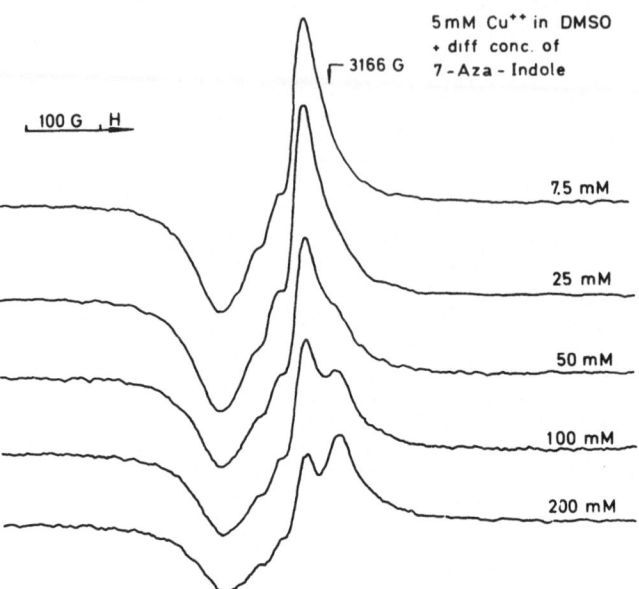

Fig. 2b. 7-aza-indole

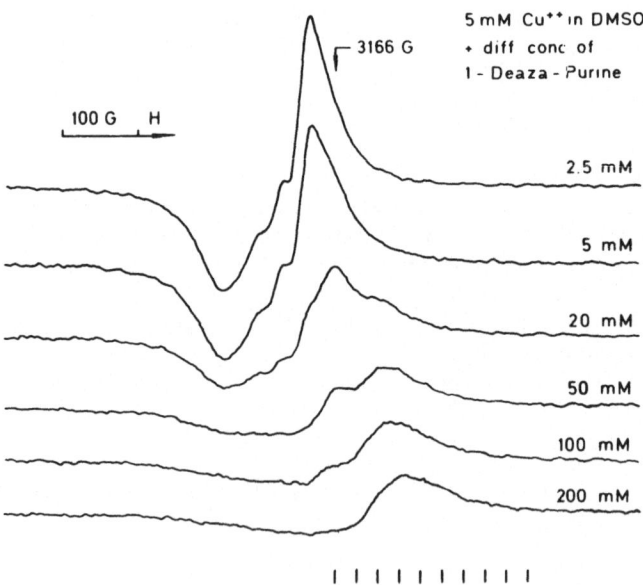

Fig. 2c. 1-deaza-purine

centration of about 200 mM exhibiting a partly resolved superhyperfine
structure with about 10 different peaks spaced equidistantly (marked
by the lines at the bottom of the picture).

   In the case of 7-aza-indole, only at relatively large ligand con-

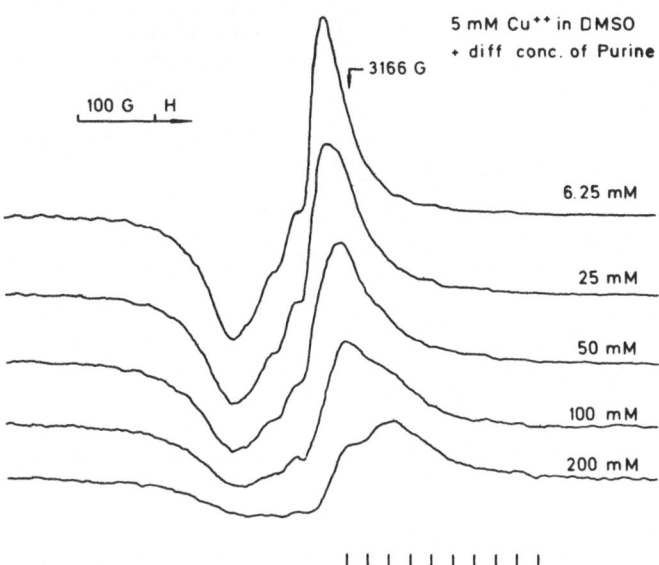

Fig. 2d. purine

centrations a new peak appears at a somewhat higher field while the original Cu^{2+} spectrum is decreased slightly (see Figure 2b). It should be pointed out that the same concentrations of indole don't modify the structure of the spectrum at all and reduce only slighty the spin concentration.

A very similar effect as shown with benzimidazole is obtained with 1-deaza-purine and purine (s. Figures 2c, d) although the changes are less expressed. For comparison, the 10 lines at the bottom mark again the same position as in benzimidazole.

A plot of the rel. spin concentration vs. the ligand concentration is shown in Figure 3. As expected, only indole and 7-aza-indole exhibit a slight decrease. The effect exerted by benzimidazole is more expressed and is even larger for 1-deaza-purine. Surprisingly, purine itself doesn't reduce the spin concentration as much as 1-deaza-purine.

From the ESR results obtained one might conclude, therefore, that heterocyclic nitrogen atoms are necessary for the interaction between Cu^{2+} and the purine analogues. Moreover, more than one nitrogen atom seems to be required for this association, whereby nitrogen positions at 7 — 9 dominate over positions at 3 — 9. These conclusions are also supported by ESR low temperature measurements at 77 K. The concentration of each of the compounds investigated was kept constant. The addition of indole (Figure 4) results only in a better resolution of the hf structure of Cu^{2+}, while 7-aza-indole also modifies slightly the magnetic parameters. It is evident, however, that all the other three substances form a stronger association with Cu^{2+}. According to the spectra, the basic reaction seems to favor N-7 and N-9 which presumably is present in a tautomeric form. The color, which is different for all of these 5 solutions, and the superhyperfine (shf) structure (of the last 3 compounds

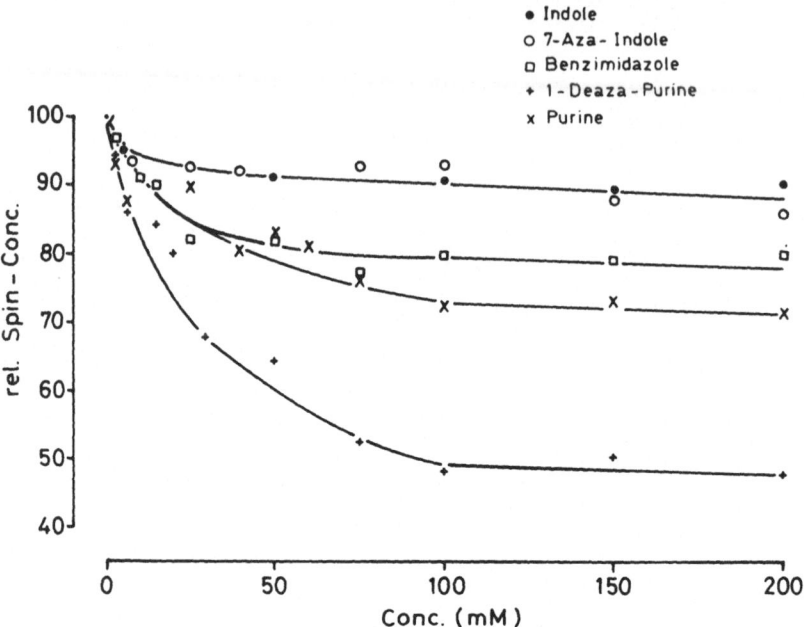

Fig. 3. The relative spin concentration of different Cu^{2+}
-purine analogue complexes vs the purine analogue concentration used

in Figure 4) change gradually with time. After about 5 days the shf
structure is more pronounced and better resolved. This new benzimidazole
spectrum exhibits 10 peaks. Since the two center lines have about the
same height the shf structure seems to consist of 2 sets of 5 peaks.
This splitting is caused very likely by two nitrogens. Due to tautomerism the hyperfine constants of the two nitrogens (N-7, N-9) might be the
same. Even though if they would differ by as much as 20 to 30 per cent
they would still produce a five-line spectrum as has been shown by Lord
and Blinder for solutions of DPPH (Lord and Blinder, 1961). The two sets
of 5 lines might be caused by the two copper isotopes, the splitting
of which is of the same order of magnitude as the nitrogen hyperfine
splitting. It is interesting to note that additional nitrogens (as in
the cases of 1-deaza-purine and purine, resp.) don't seem to change the
hyperfine structure indicating that the basic CT interaction might occur between the metal ion and N-7 and N-9. Furthermore, the poorly resolved high-field hyperfine structure observed at room temperature measurements (Figure 2a) might be due to the Cu^{2+}-nitrogen interaction
described.

The 100 MHz spectra of indole, 7-aza-indole, benzimidazole, 1-deaza
-purine, and purine in DMSO solution are shown in Figure 5, lower 5
traces. The peak of the N-H proton is not shown; it is located further
down-field (between 11 and 13 ppm depending on the compound,see also Table
I). The assignments of the indole peaks have been done by Hiremath and
Hosmane (Hiremath and Hosmane, 1973). Additional heterocyclic nitrogens

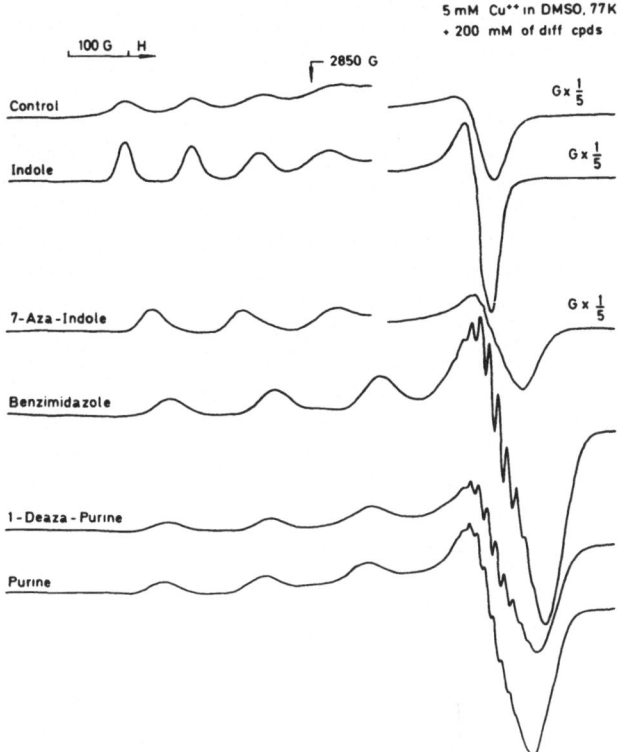

Fig. 4. The effect of different purine analogues (200 mM each) on a 5.0 mM Cu^{2+} ESR spectrum in DMSO at 77 K. *Note:* The experimental conditions were kept constant except for the right portion of the upper 3 spectra (they were recorded at 1/5 sensitivity of the other spectra).

cause, as anticipated, a considerable down-field shift of the adjacent protons. Due to spin-spin interactions the protons could be assigned.

In the case of benzimidazole, H-3/H-6 as well as H-1/H-2 represent an *AA'BB'* spectrum indicating the two tautomeric forms.

From the foregoing discussion, it is apparent that the three protons of purine (with 4 N's) are displaced down-field at the most which has been already reported by Bullock and Jardetzky (Bullock and Jardetzky, 1964).

Addition of Cu^{2+} to the compounds investigated in a ratio of 1:10000 affects the diverse proton peaks differently as can be seen in the upper five spectra of Figure 5: There is no effect on indole at all even up to a concentration ratio of 1:500. In the case of 7-aza-indole, the H-2 peak is broadened predominantly and is hardly measurable at a concentration range of 1:2500 suggesting that Cu^{2+} interacts with N-3. An extensive effect, however, can be observed when Cu^{2+} is added to benzimidazole. The H-8 peak broadens considerably already at minute Cu^{2+} concentrations indicating that Cu^{2+} interacts with the lone-pair

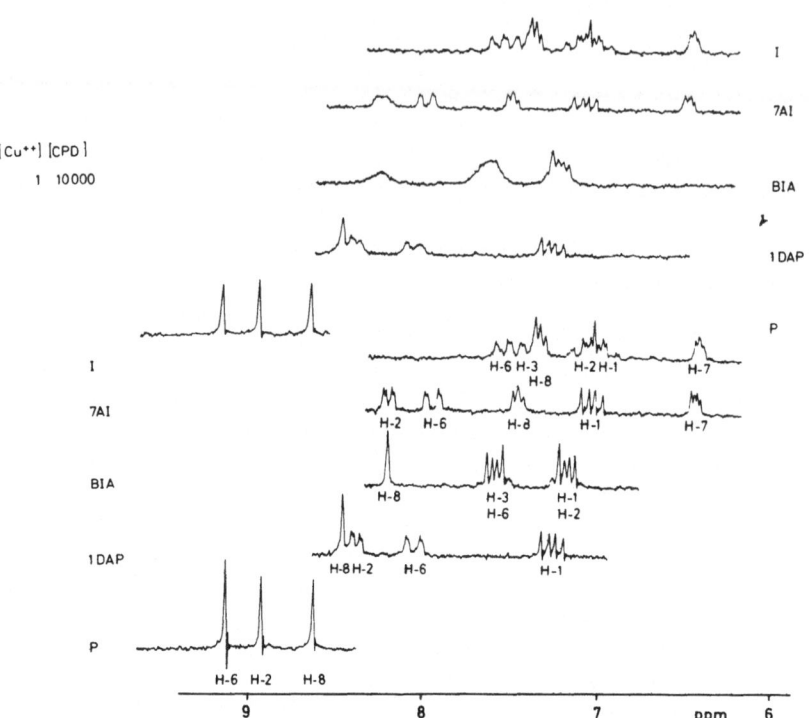

Fig. 5. ^{1}H NMR spectra of different purine analogues with
(upper 5 spectra) and without an 1 : 10000 Cu^{2+}/compound
(CPD) addition. P $\cong$ purine, 1 DAP $\cong$ 1-deaza-purine, BIA $\cong$
benzimidazole, 7 AI $\cong$ 7-aza-indole, I $\cong$ indole.

electrons of N-7 and N-9 present in a tautomeric form. Further proof
for this assumption is obtained with 1-deaza-purine. Addition of Cu^{2+}
to this compound (see also Figure 6) results mainly in a broadening of H-
followed by H-2 and, then, H-6. This, also, suggests very strongly that
the metal ion is associated with the lone-pair electrons of the nitrogens. There might be a possibility that the metal ion is located over
the imidazole ring. In this case, the nitrogen at position 3 will displace Cu^{2+} somewhat towards the benzene ring. This shift or the more
likely weak CT interaction between Cu^{2+} and N-3 results in a more effective broadening of H-2 in the case of 1-deaza-purine. These findings
are supported by the results obtained with ^{13}C NMR measurements. Addition of Cu^{2+} to 1-deaza-purine in a molar ratio of 1:100 (Figure 7)
results in a considerable broadening of C-4 and C-5; one of the two peaks
C-8/C-2, which, unfortunately, cannot be assigned disappears completely.
From the foregoing discussion one might conclude that the C-8 peak disappears. The ^{13}C NMR results also emphasize that the metal ion is interacting almost equally with N-7 and N-9 so that it might be placed over
the imidazole ring with a slight shift (compared to benzimidazole) towards the benzene ring due to the electron withdrawing property of N-3.

This property is still more pronounced in the case of purine since the position 1 is occupied by an N. Compared to benzimidazole, the electron density seems to be distributed more uniformly over the whole ring system which is expressed by the fact that all three purine protons are broadened more or less to the same extent (H-6 slightly more than the others).

The chemical shifts obtained for the protons of the 5 substances investigated are summarized in Table I. It is interesting to note that for each substance the proton peak (except the NH peak) which is located down-field at the most is broadened most (values are framed). The proton peak broadened next is underlined. The relatively low field position of all three purine protons is responsible that the broadening effect is about the same for all three of them.

These experimental NMR results are in agreement with theoretical studies. According to these the down-field shift of the protons is the larger the smaller their shielding, that is the smaller the electron density. This means, in the special case of benzimidazole, that the two nitrogen atoms adjacent to H-8 exhibit the highest electron density. Thus, they are favored for a CT interaction with Cu^{2+}. Such an interaction should be the stronger the more N atoms are present resulting in a displacement of negative charge density towards the metal ion. In an

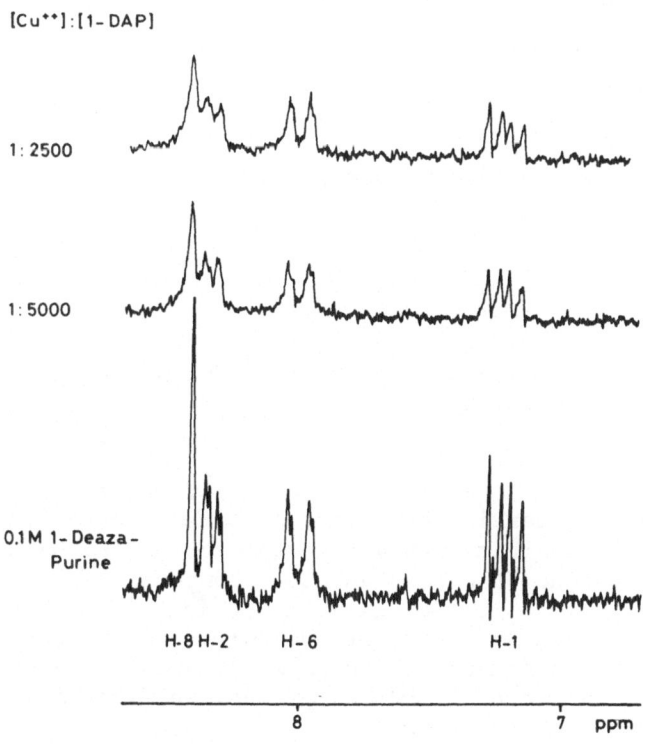

Fig. 6. ^{1}H NMR spectra of 0.1 M 1-deaza-purine with and without Cu^{2+} addition

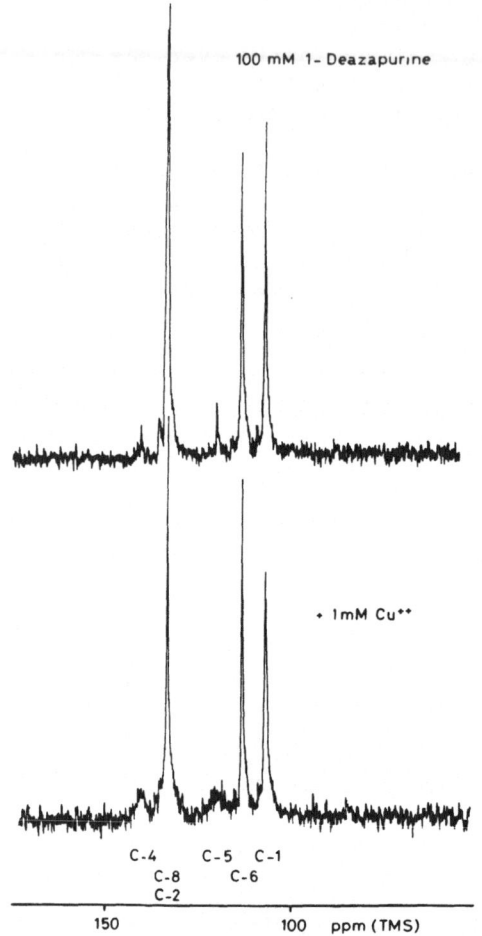

Fig. 7. ^{13}C NMR spectra of 100 mM 1-deaza-purine with and without Cu^{2+} addition.

extreme case this can lead to a reduction of Cu^{2+}. However, the more Cu^{++} is produced the smaller is the effect of paramagnetic line broadening. The NMR results exhibit such a behavior: comparing benzimidazol, 1-deaza-purine, and purine at a constant Cu^{2+}/compound ratio, the broadening effect decreases from benzimidazole to purine. The ESR results discussed above (Figure 3) are in accordance with the NMR findings except for purine, the spin concentration of which should have been decreased more than in the case of 1-deaza-purine. An explanation cannot be offered yet for this observation.

TABLE I

Chemical shifts of the protons of the different purine analogues investigated. framed values: broadened the most, underlined values: broadened second most. NH protons are not considered in this context.

	H-9	H-1	H-2	H-3	H-6	H-7	H-8
Indole	11.03	6.96	7.05	7.38	7.52	6.40	7.30
7-Aza-Indole	11.57	7.04	8.16	—	7.93	7.44	7.42
Benzimidazole	12.38	7.17	7.17	7.58	7.58	—	8.18
1-Deaza-Purine	12.85	7.22	8.33	—	8.0	—	8.4
Purine	13.15	—	8.9	—	9.1	—	8.6

ACKNOWLEDGEMENTS

We are indebted to Dr H.O. Kalinowski, Dept. of Chemistry, for conduc-
ting the ^{13}C NMR measurements. The excellent technical assistance of
Miss E. Christ and Miss H. Pickl is greatly appreciated. This work was
supported in part by the Fraunhofer Gesellschaft.

REFERENCES

Arfmann, H.A.: 1976, to be published.
Bullock, F.J. and Jardetzky, O.: 1964, *J. Org. Chem.* **29**, 1988 — 1990.
Hiremath, S.P. and Hosmane, R.S.: 1973, in A.R. Katritzky and A.J.
   Boulton (eds.), *Advances in Heterocyclic Chemistry*, vol. 15, Academic
   Press, New York, 1973, pp. 277 — 324.
Lohmann, W. and Penka, V.: 1973, *Biophysik* **10**, 61 — 67
Lord, N.W. and Blinder, S.M.: 1961, *J. Chem. Phys.* 34, 1693 — 1708.
Pullman, A. and Pullman, B.: 1966, in P.O. Löwdin (ed.), *Quantum Theory
   of Atoms, Molecules, and the Solid State,* Academic Press, New York,
   pp. 345 — 359.

DISCUSSION

Fazakerley: It has been shown that T_2 is not necessarily dominated by
the dipolar term and that information on metal binding through selective
line broadening can be misleading.
      The interpretation of ^{13}C nmr line broadening is greatly compli-
cated by the high concentrations used at which base stacking is very
significant.

Lohmann: to 1: I agree that the results obtained by paramagnetic line
broadening can lead to misinterpretations. However, whenever applied
carefully it can support results obtained by other methods. Several
papers published during the last few months have shown the validity of
this method.

to 2: Unfortunately, relatively high concentrations have to be used quite often at which a stacking effect has to be encountered. Despite this fact, the line broadening of all C's, especially C-4 and C-5, seems to be caused by Cu^{2+}.

Laniv: Did you attempt to calculate Cu to protons distances from the nmr relaxation times?

Lohmann: No, we have not done it yet.

Eichhorn: Since a statement was made about the controversy over the line broadening technique, it may be useful to summarize the nature of that controversy. It has been proposed by B. Martin, who has pointed out that the scalar factor in broadening may be large enough to prevent accurate estimates of binding sites. He also points out that, since T_1 relaxation measurements can now be made, there are a much more accurate reflection of distances between protons and paramagnetic ions. However Marzilli *et al.* have shown that the results that have been obtained by the line broadening technique are generally confirmed by T_1 relaxation studies. He also recommends T_1 relaxation studies for quantitation but indicates that qualitative indications of binding sites obtained by line broadening are correct.

THE INVESTIGATION OF METAL-AMIDE INTERACTIONS WITH THE AID OF QUANTUM
CHEMICAL MODEL CALCULATIONS AND CORRESPONDING MODEL EXPERIMENTS

BERND M. RODE
*Institut für Anorganische und Analytische Chemie, Universität
Innsbruck, Innrain 52a, A-6020 Innsbruck, Austria*

1. INTRODUCTION

Until the early sixties, the theoretical background of cation binding
to amides was still an almost white spot on the map of chemical knowledge. The interest to know more about such interactions was stimulated
simultaneously by the increasing use of simple amides as nonaqueous
solvents in inorganic and organic chemistry and by new insights into
biological phenomena like ion transport, bioelectrical membrane potentials and the working mechanism of the nerves. New experimental techniques and, recently, the possibility to involve quantum chemical
methods in such investigations, have led to numerous papers of both
experimental [1–19] and theoretical [19–27] nature.
      For this reason, the knowledge about formation and properties of
metal ion-amide complexes extends now already over a pretty wide range
of chemical and biological questions, but several of the intrinsic
problems of this field of research have remained partly or completely
unsolved. To know more about the theoretical foundations of the
chemistry of amides acting as complex ligands seems to be of great
importance, therefore, in order to explain numerous questions, such
as the high specifity of various amide molecules for different cations,
the influence of the metal ions on structure and molecular properties
of peptides and their transport through membranes by carrier proteins
[27, 28].
      We believe, that one of the main problems in this research (as
in many other fields of chemistry) is the more or less wide distance
between the theoretical work, which has already been carried out,
giving us the possibility to obtain a deeper understanding of the
molecular background of such interactions, and the experimental work,
which deals with the reality of solutions, solids and biosystems. The
main scope of our work has been, therefore, to combine both ways of
research in order to test theoretical models, as well as to accomplish
a stepwise approach from the model calculation to the experimental
situation. This intention led us to several crucial problems, owing
to the fact that the accuracy of the theoretical calculations must
decrease successively, the larger (and the more real) the system becomes

On the other hand, there is the difficulty to choose suitable experiments
reflecting physical quantities which are directly accessible from the
calculations and allow thus to obtain some reliable information about
the capabilities of the simplified models in the prediction of the behaviour and the properties of the complexes under the conditions *in
vitro* and *in vivo*.

2. SIZE AND ACCURACY OF THE QUANTUM CHEMICAL MODELS

Biological substrates and even simple solvate complexes of metal cations
with substituted amides are extraordinarily large systems from the
viewpoint of the quantum chemist. For example, an *ab initio* calculation
of the $Li^+(HCONH_2)_4$ complex in a moderate (e.g. $8/4)_0$GLO basis set
implicates already the evaluation of more than $4x10^{10}$ integrals. This
system, however, belongs still to the very small ones for the experimental chemist and represents a rather unsophisticated model for a
solvated cation. On the other hand, this kind of calculations is still
far from the limits of accuracy, which we reach today by the use of
larger basis sets and the inclusion of calculations describing also
the influence of electron correlation. If we use, however, more simplified
chemical models (i.e. 1:1 complexes), we can lose more information
about the chemical reality than we have gained by the improvement of
the calculation method. This statement can be founded on experimental
and theoretical results concerning the well known chelate effect [24],
which often occurs only upon binding of more than one ligand to the
cation.
      From these first considerations we can derive one important con-
clusion about the chance of quantum chemistry in the research of ion-
amide interactions: We have to restrict the sophistication of our
calculations, to reduce the basis sets or even to use semiempirical
methods. Only within this methodical framework are we able to calculate
also chemically larger systems or simple biologically important systems.
A careful control of the loss of information due to simplification of
the calculations will be necessary, however, and should be carried out
in two ways: first, by a comparison of the results of model calculations
on small systems with varying degree of accuracy, second, by means of
experiments, in wich we can examine the quality of the model in the
description of the observed chemical behaviour. The general idea of
using small and minimal basis sets and/or molecular fragments for the
calculation of large compounds has been proposed and applied successfully already by numerous quantum chemists [29—33], and we believe this,
at the time, to be the most promising chance of quantum chemistry to be
useful in the investigation of metal-amide complexes, which are of more
than academic interest.

3. A METHODICAL COMPARISON OF SIMPLIFIED MODEL CALCULATIONS AND THE
   RELATION OF THE MODEL CALCULATION TO REALITY

In this section we will not consider any of the principal limitations
of quantum chemical calculations like the neglect of entropy factors

or the loss of information due to the use of nonrelativistic one deter-
minantal wave functions. It has been argued already, that we should
restrict our model calculations to rather simplified *ab initio* or to
semiempirical MO SCF methods, in order to keep the computational problem
within reasonable and practicable limits. This means, that the main
questions will now be: How small can we chose the basis set in order to
reflect still correctly most of the molecular properties? Which effect
has the introduction of empirical parameters, and how many ligand
molecules have to be considered in order to obtain a correct picture
of the complexes realized in the experiment and in nature, respectively?

At this point, one may ask, whether such systems as being calculated
usually by the quantum chemist, can be realized experimentally, i.e. can
we produce an isolated metal ion bonded to one or two ligand molecules,
unaffected by any other chemical systems in the environment? In order
to answer this question, we have performed the following experiments:
Some metals (Li, Na, K, Zn, Cd, Hg,) were evaporated at 200—250 ^{o}C in
vacuo and ionized by an electron beam of the lowest possible energy.
The ionized metal was allowed to react with dimethylformamide (DMF)
vapour (60 ^{o}C), the products being observed in the mass spectrometer.
The separate mass spectra of the reactants do not show any peak higher
than the atomic one for the metals and the molecular one for DMF, respectively.

Upon reaction, however, we observe molecular peaks for the $M^{+}(DMF)_n$
complexes, n being 1 and 2 in the case of Li and Na, and 1 for K, Zn
and Cd. Mercury does not seem to form any complex. Higher fragments in
the Li/DMF and K/DMF spectra prove also the formation of 1:4 (for Li)
and 1:2 (for K) complexes with the amide.

Since the pressure during these experiments was kept constant at
5.10^{-6} T, corresponding to a mean free path of 10—30 m, we can state,
that it is well possible to realize the quantum chemical models in the
laboratory.

After having obtained now this first link from theoretical to
experimental chemistry, we should pay attention to the first two questions mentioned before: the size of the basis set and the influence of
empirical parameters in the calculations.

Calculations on cation complexes with water [34], aldehydes [35]
and carbonic acids [36, 37] have already shown the deficiency of
semiempirical methods like CNDO or INDO in the prediction of molecular
geometries. We have investigated, therefore, the energy surface for
the formation of the $Li^{+}/HCONH_2$ complex by the CNDO method and by means
of *ab initio* calculations with an optimized 4/2 [32] and an optimized
2/1 (i.e. minimal) basis set. The results of this investigation is
illustrated in Figure 1. There is no significant difference between
both *ab initio* surfaces, whereas the semiepirical one shows two incorrect
minima. This effect is mainly due to artificial cation bonding to
hydrogen atoms (NH and CH) and clearly an artefact of the CNDO method
[25]. In analogy with similar complexes [37], this artefact is to be
expected also for other semi-empirical methods.

We can conclude, therefore, that *ab initio* calculations, even with
a minimal basis set, have to be prefered for any geometry optimization
of such complexes. Only if we fix the correct geometry, can we make some

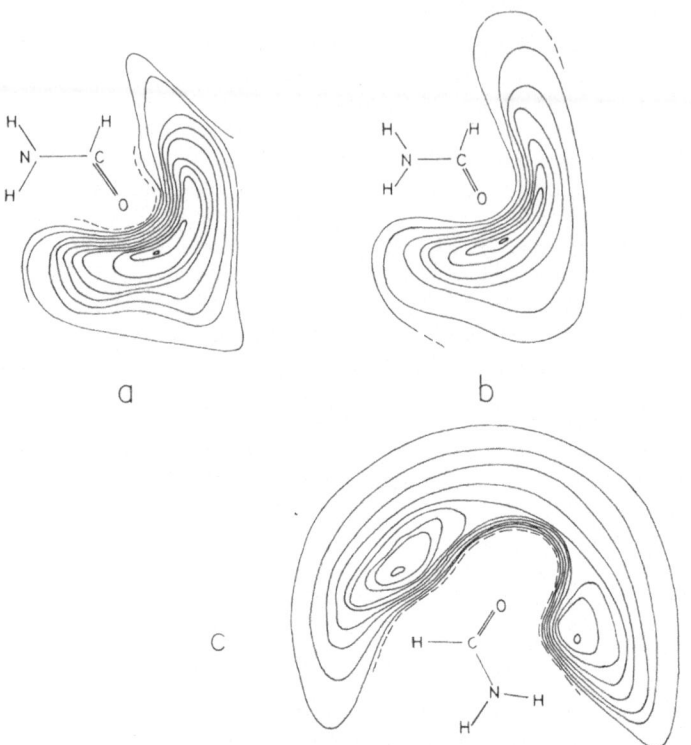

Fig. 1.   Energy surfaces for the Li$^+$(HCONH$_2$) complex
(a) calculated with 4/2 GLO basis set,
(b) calculated with minimal (2/1) GLO basis set,
(c) calculated with semiempirical (CNDO/2) method.

use of semiempirical methods, since then most of the other results, like
changes in bond strength and atomic net charges are obtained in a more
or less satisfactory qualitative agreement with the *ab initio* calculation.
    The above mentioned failure, together with the artificial preference
for cyclic structures resulting from semiempirical calculations, prevent,
however, the prediction of chelating properties. This seems to be a very
serious deficiency of the semiempirical methods, since such properties
play an extremely important part in the chemistry of amide-metal complexes. We can illustrate the influence of the chelate effect by the
result of an *ab initio* calculation on the Li$^+$(HCONH$_2$)$_2$ complex in two
different geometrical arrangements (Figure 2). In the special case of
Li$^+$(HCONH$_2$)$_2$, configuration A remains the absolutely more stable, but
the stabilization energy per ligand decreases by 7% (similar to cation-
water complexes), compared to the corresponding 1:1 complex. In configuration B, however, we find an increase of this quantity of 63%. This
energy gain consists of hydrogen bond energy (~ 90%) and a further
stabilization effect, which we can call chelate energy gain [24].
    A more detailed analysis of this phenomenon shows a preference for
B due to the electronic energy reaching the phantastic amount of 18.6
megacalories (cf. Table I), which, however, is nearly compensated by the

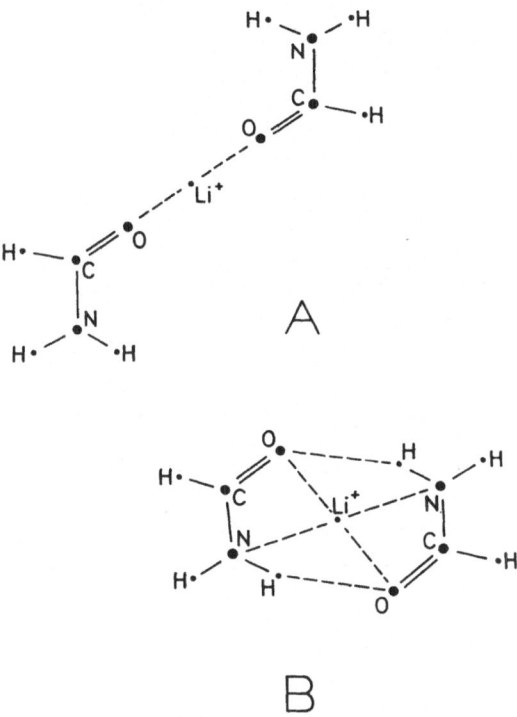

Fig. 2. Geometrical arrangements in the calculations on the $Li^+(HCONH_2)_2$ complex
A....energy minimum configuration,
B....chelate configuration.

strong nuclear repulsion in this geometry, at least for $Li^+(HCONH_2)_2$. We can expect, therefore, a more dominating chelate effect for various amide complexes of metal cations of higher atomic number. This means, that such cations may be preferably bonded simultaneously to nitrogen *and* oxygen.

Summarizing the considerations of this chapter we can conclude, that we should favour nonempirical methods with minimal or small basis sets over the semiempirical methods for the calculation of any larger metal-amide system. From this results we will try to move now to the next step of our investigation, namely that from the gaseous phase to solution chemistry.

4. QUANTUM CHEMICAL MODELS FOR METAL-AMIDE INTERACTIONS IN SOLUTION

Proceeding from the isolated complex to the dilute solution, we have to consider two important differences: the abundance of ligand molecules (being the solvent), leading to higher coordination numbers also

TABLE I
Energies for complexes of Li^+ with formamide from LCGO-MO-SCF computations with an uncontracted 4/2 GLO basis set.
I....Li^+ in the C=O axis
II...Li^+ bonded to N and O (chelate position)

(a) 1:1 complexes $Li^+(HCONH_2)$

	I	II
Total energy(a.e.u.)	−172.05089	−171.96946
$\Delta E/n$ (kcal)	55.9	13.2

(b) 1:2 complexes $Li^+(HCONH_2)_2$

	I	II
Total energy (a.e.u.)	−336.86632	−336.77066
$\Delta E/n$ (kcal)	51.5	21.5
Electronic energy (a.e.u.) electronic	−563.69397	−593.33149
$\Delta E/n$ (a.e.u.)	− 3.37962	− 18.19838

fo_ weakly interacting metal ions, and the influence of the amide molecules which are not directly involved in the complex formation, i.e. the environment of the bulk solvent.

How can the quantum chemical model take into account these effects? The straightforward answer is the enlargement of the model up to a size, where the addition of further amide molecules would not cause any more effects. Such a procedure can be performed, if at all, only within the framework of the above discussed minimal basis set calculations, since the description of just ten of the smallest amide molecules, $HCONH_2$, in this basis means already to handle Hartree-Fock-matrices of the dimension 200 x 200 and more.

Another way would be the consideration of the further amide molecules as an environmental perturbation by means of an additional term in the Hamiltonian. It has been proposed already [38] to describe solvation effects by such procedures, but in the case of the rather large and well structurated amide molecules, the success of this method seems to be questionable. In order to obtain a somewhat consistent model, we should consider explicitly at least all amide molecules in the first coordination sphere, the outer spheres being considered as the averaged perturbation. With the exception of hydrogen bond forming amides, where this method will necessarily fail [38], the outer spheres cannot be expected to influence considerably the properties of the amide complexes. For this reason, such procedures will not really facilitate the model calculations.

One of the fundamental questions will be, therefore: how small can we choose our models for the liquid phase, in order to obtain still a correct and at least semiquantitative description of the ion-amide interaction? The question has to be extended further, if we want to include also solutions other than very dilute ones, where ion pair formation and the increasing statistical probability for the interaction of one amide molecule with more than one metal ion will require the use of alternative models. For the answer to this problem, namely, which are the smallest possible quantum chemical representations of specific conditions realized in the solution, an experimental control of these theoretical models seems to be inevitable.

What kind of experiments will be suitable now, to perform such a control? High pressure mass spectroscopy, which has been applied extensively to solvation investigations [39—43], can be regarded as one of the most powerful and illustrative means for this purpose. In these experiments we can observe directly the increase of coordination numbers for a cation by increasing pressure (= concentration) of the solvent (= ligands). Unfortunately, this experimental technique has not been used so far for the investigation of metal-amide complexes, but it promises to lead to highly interesting and valuable results, which could be useful to detect successively any changes in complex formation due to increasing ligand concentration.

The final examination of any model, however, must be performed by experiments in the liquid phase itself. Thus we have to choose suitable experiments in solution, allowing a sensitive comparison of model and reality, in other words, to measure physical quantities, which are easily accessible from the calculations. We could choose, for example, chemical shifts or the change of force constants of the amide molecule due to metal ion bonding. Both methods have been used already for such purposes in similar complexes [44]. Both chemical shifts and force constants implicate, however, a rather complex physical background and, consequently, quite rough simplifications are being made in their evaluation from the calculations. For this reason, we have made use of an outstanding property of the amides, which allows a much more unambiguous comparison of calculated and experimental results: the partial double bond character of the C-N bond in the NH-CO-group, leading to a barrier to internal rotation about this bond. Apart from the methodical advantages, it is just this peptide group, which deserves most interest in the metal ion-amide interaction, since this group plays the most important part in the cation bonding to simple amides as well as to biological substrates like depsipeptides or proteins [21, 26].

The barrier to internal rotation can be calculated by the quantum chemist as a simple energy difference of different geometrical complex configurations. Experimentally, this energy is usually evaluated from velocity constants, obtained by NMR line shape analysis at several temperatures, using the Arrhenius equation. Any accurate calculations of such activation energies have to involve large basis sets containing several polarization functions and should take into account even electron correlation. As we have stated already earlier, such quantum chemical methods are rather impracticable for the treatment of ion amide complexes. We should examine, therefore, to which extent the predictions of small, unsophisticated model calculations will be correct, at least for the estimation of relative differences in the influence of various metal ions on the ligand molecule's rotational barrier. In the corresponding

TABLE II
Temperatures of coalescence for CH_3-proton signals of metal
chloride solutions in DMF (molar ratio 1:20) and percentual
raise of activation energy for internal rotation of the amide
molecule under the influence of the cation, obtained from
NMR line shape analysis.

Metal ion	T_{coal} °C	ΔE_a %
Be^{2+}	124.0	+ 45.2
Al^{3+}	117.2	+ 31.6
Li^+	124.1	+ 22.0
Mg^{2+}	128.4	+ 20.8
Zn^{2+}	123.4	+ 18.0
Na^+	132.5	+ 12.8
Ca^{2+}	127.6	+ 12.0
K^+	132.9	+ 6.8
Rb^+	132.6	+ 4.0
Cs^+	128.2	+ 2.0

experiments we have a sensitive instrument — reflecting changes in electronic structure and bonding — for the desired control of the reliability
of the calculations.

An increase for the rotational barrier in amides upon ion bonding
has been predicted already for some examples [20, 22] by means of SCF
methods. For the Li^+/DMF system, an increased temperature of coalescence
for the CH_3 proton signals has been regarded as an experimental indication for this increase [20]. We should point out, however, that this
elevated temperature of coalescence does not prove anything in this
case, since the chemical shifts are also changed strongly by ion influence. This can be demonstrated by the results collected in Table II.
These results have been obtained for dilute solutions (molar ion/amide
ratio 1:20) of the metal chlorides in DMF, which seemed to be the most
suitable small amide for such investigations, since its anion solvating
properties are almost negligible [45, 46].

The changes in the coalescence temperature are in clear contrast
to the change in the activation energy obtained by the more sophisticated
method of NMR line shape analysis. The temperatures would predict the
strongest effects for sodium, only a moderate one for the most strongly
interacting beryllium, and an inverse influence for aluminium, which,
however, raises the barrier by more than 30%. The results of the line
shape analysis, however, agree quite well also with intuitive chemical
expectations, as can be demonstrated by a diagram of the barrier increase versus the ionic radii, since these radii give us an average
measure for the effective net charge experienced by the amide molecule
(cf. Figure 3).

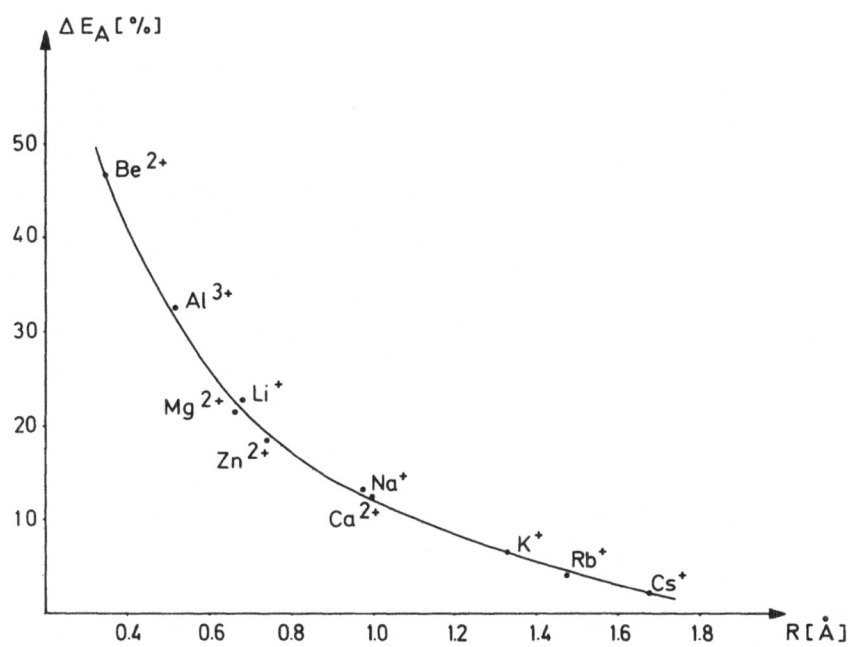

Fig. 3. Percentual change of rotational barrier of dimenthyl-
formamide due to metal ion bonding versus ionic cadii.

How strongly concentration can influence all the molecular parameters,
can be demonstrated also by the rotational barrier. If we increase
sucessively the salt concentration in LiCl/DMF solutions, we observe
first a further barrier increase until a molar ratio metal ion to
amide of 1:10, followed by a rather striking decrease, which, at a
ratio of 1:4, leads to a value for E_a even lower than for the pure DMF!
Consequently, quite different theoretical models are required to describe the respective interactions between the cations and the amide
molecule.

These experimental results obtained for the DMF solutions could
be regarded as a suitable basis for the test of the required theoretical
models. First of all, we wanted to exame the previously discussed
LCGO-MO-SCF calculation methods, which in the case of a satisfactory
description of the small complexes with DMF, would allow an extension
of the application to relatively large chemical systems. For some of
the model complexes, we extended the minimal basis also to double size
in order to have some methodical control. The (energy optimized) Gaussian
exponents for the atoms in the minimal basis set are give in Table III,
those of the 4/2 basis were taken from the literature [3]. As amide
molecules, both formamide and DMF were used in their experimental geometries [47, 48], optimizing only the cation position(s) with respect
to total energy. This means, of course, a further restriction of the
model, since a complete geometry optimization will sometimes improve

TABLE III
Optimized GLO exponents for minimal basis set

	H	C	N	O	Li^+	Be^{2+}	Na^+
S_1	0.395	9.407	13.080	17.360	2.511	6.671	56.750
S_2	—	0.304	0.440	0.606	0.432	1.350	12.170
S_3	—	—	—	—	—	—	1.530
S_4	—	—	—	—	—	—	0.448
P_1	—	0.253	0.367	0.505	—	—	1.918
P_2	—	—	—	—	—	—	0.400

the results [22]. We decided to avoid this procedure, because its application to large systems wouls lead again to unreasonable computing times.
      Which specific complexes should serve now as the desired model
compounds? Dilute solutions, containing mainly $M^+(DMF)_n$ species, were
described first of all by the 1:1 complexes of Li^+, Be^{2+} and Na^+
with formamide, and of Li^+ and Be^{2+} with DMF. The results of these
calculations are presented in Table IV. They prove, that even this
strongly simplified quantum chemical model enables us to make some
semiquantitative predictions about the behaviour of the amide molecule
under ion influence in the liquid phase, as long as we refer to dilute
solutions. It is to be expected, that the consideration of the further
amide molecules in the first solvation layer will improve the quantitive
agreement of calculations and experiment more than the calculation of
the 1:1 pair potential with large and even largest (near HF-limit)
basis sets, similar to the description of other ion-molecule-interac-
tions [49]. Such investigations are being carried out presently, so that
the results of this model improvement will be available soon.
      In this context we should mention also briefly the influence of
a moderate basis extension in the calculations on such metal complexes.
It has been found, that such a moderate increase of the basis size
(4/2, 6/3) does not improve significantly the characteristic deviations
from experimental values, such as the overrating of stabilization and
activation energies and cation-ligand distances [22, 23, 25, 37]. On the
other hand, the use of the smallest basis did not change seriously the
direction of the calculated dipole moment, which is very important for
a correct description of the interaction of ligands and cations. Such
an agreement we obtain also for the qualitative trends of polarization
effects, i.e. the changes in atomic net charges upon ion bonding (Table
V). We believe, therefore, that an only moderate expansion of the basis
set will not bring about advantages, wich would compensate for the much
higher computing times and the greater problems in the treatment of
larger molecules. The use of larger chemical models in the calculations
with the smaller basis set seems to be a much more promising way for
an improved description of the chemical properties of the metal-amide
complexes in the liquid phase, similar to other solvate complexes [50].

TABLE IV

Comparison of calculated and experimentally determined ion influence on rotational barriers in amides (1...4/2 GLO basis, II...minimal GLO basis)

(a) $HCONH_2/M^{n+}$ - models

Ion	Basis	exptl.system	$\Delta E_a^{exptl.}$	model	$\Delta E_a^{calc.}$
Li^+	I	LiCl/DMF(1:20)	+ 22%	$Li^+/HCONH_2$	+ 31%
Li^+	II	LiCl/DMF(1:20)	+ 22%	$Li^+/HCONH_2$	+ 17%
Be^{2+}	II	$BeCl_2$/DMF(1:20)	+ 45%	$Be^{2+}/HCONH_2$	+ 44%
Na^+	II	NaCl/DMF(1:20)	+ 13%	$Na^+/HCONH_2$	+ 12%

(b) $HCON(CH_3)_2/M^{n+}$ - models

Ion	Basis	exptl.system	$\Delta E_a^{exptl.}$	model	$\Delta E_a^{calc.}$
Li^+	II	LiCl/DMF(1:20)	+ 22%	Li^+/DMF	+ 16%
Be^{2+}	II	$BeCl_2$/DMF(1:20)	+ 45%	Be^{2+}/DMF	+ 59%

TABLE V

Changes of atomic net charges of ligand molecules upon metal complex formation, calculated with different GLO basis sets.

Complexation reaction	Basis	H_{CH}	C	O	N	H_{NH}^{trans}	H_{NH}^{cis}
Li^+ + $HCONH_2$	4/2	+0.103	+0.115	−0.215	+0.044	+0.054	+0.028
	2/1	+0.063	+0.117	−0.278	+0.032	+0.053	+0.024
Li^+ + HCOOH(37)	6/3	+0.072	+0.143	−0.278	—	—	—
	2/1	+0.058	+0.137	−0.239	—	—	—
Li^+ + $(CH_3NHCO)^-$	8/4	—	+0.128	−0.201	+0.022	+0.022	+0.121
	2/1	—	+0.111	−0.332	+0.332	+0.028	+0.082

It has been pointed out by some authors [22], that the cation may occupy also another position in amide complexes, bonded to both nitrogen and oxygen. This was indicated also by the results of our gas phase experiments reported previously [18]. Such a position could favour then the rotated form of the amide [22]. The results of all calculations show, however, that the stabilization energy for such complexes is much to low in the case of small cations, where other stabilizing effects, like chelating, would not predominate [22, 24].

How we can describe, therefore, the lowering of the rotational barrier in the concentrated solutions containing lithium ions? The considerations for this case are based on the statistical distribution of the cations among the amide molecules. The average solvation numbers

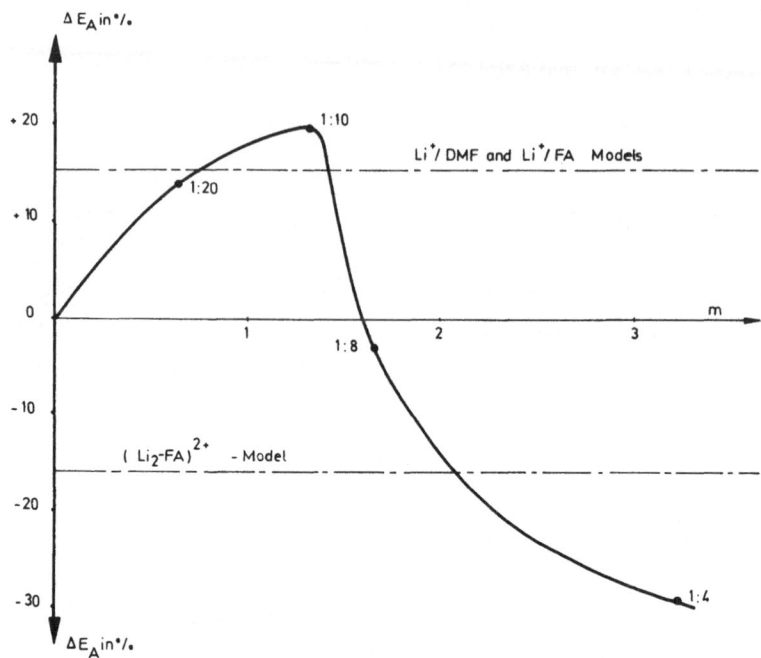

Fig. 4. Percentual change of rotational barrier of dimethylformamide with increasing salt concentration (molar ion/amide
ratio 1:20 to 1:4) in comparison with the predictions of the
simplest possible quantum chemical model calculations.

of the ions [45, 46, 51, 52], especially the low solvation number of the
anions ensure the presence of fully solvated and sufficiently separated
cations up to a molar ratio of 1:10. For a cation/amide ratio of 1:6 or
even 1:4, however, the probability for one amide molecule to interact
with two cations at the same time, has increased rather drastically. The
second interaction will occur most probably at the nitrogen atom. Geometry optimization of a formamide molecule with one Li^+ bonded to the
oxygen and a second one to nitrogen (leading to a pseudotetrahedral
geometry at the latter) leads, within the MO SCF formalism, to a
stabilized complex, in which rotation around the C–N bond is strongly
facilitated. Figure 4 shows the comparison of the actual changes of
E_a over a wide concentration range of LiCl in DMF with the two model
complexes for both dilute and concentrated solutions.

     The results indicate, that it should be possible, even for rather
complicated systems such as solutions, to find, with the aid of corresponding model experiments, quantum chemical models, which are not too
large and still represent various conditions *in vitro* or *in vivo*. The
molecular orbital calculations will allow then to obtain some basic
molecular properties, which may be of interest and importance for the
understanding of the chemical behaviour of such complexes. This seems
to be an encouraging confirmation, that even the — necessarily — simple
quantum chemical methods have a good chance to be a helpful tool in the

investigation of the various kinds of interactions between metal ions
and amides or peptides.

Our results cover only one single measurable quantity until now,
the rotational barrier. We should consider, therefore, whether these
results justify already a general optimism about the model calculations.
We think they do, since the change in the activation energy for inner
rotation involves several simultaneously occuring molecular proecesses,
which will determine also other chemical properties. From these molecular
changes we want to mention especially the charge transfer to the cation,
the polarization of the ligand molecule, the change of the bond characters
within this molecule and the necessary redistribution of atomic net
charges. If we can describe a physical quantity depending on all of
them, in a qualitatively or even semiquantitatively correct way, we
have good reason to hope, that we can obtain also other valuable information about the chemical properties of the system from such calculations.

At this point we should remember, that the cation influence on the
rotational barrier of amides does not only represent a useful quantity
for the examination of the theoretical models. It deserves also special
interest for peptide chemistry, since it has been discussed [53, 54]
in connection with conformational changes in polypeptides [55] under
the influence of inorganic salts. Employing the minimal basis set calculations, which have been successful in the case of the small amide
complexes, we can extend this approach to medium size peptide molecules.
This procedure promises to supply some more information about the mechanism of this salt influence on the peptide structure.

5. SOLID CATION-AMIDE COMPLEXES

One of the characteristic properties of amides is the formation of
numerous well defined stable adducts with alkali and alkaline earth
halides [2, 56]. This tendency shows up already in the liquid phase,
where sometimes an unexpected behaviour is found for concentrated
solutions (e.g. melting points, electrical conductivity and vapour
pressure [46]).

Similarily as from the gaseous to the liquid phase, we can perform
therefore, a more or less continuous transition in our model experiments
from the solution to the solid adducts. For the quantum chemical model,
this means a further expansion, since a representative part of the
crystal contains a large number of amide molecules. Even employing
minimal basis sets we have to face serious computational problems,
although the availability of very reliable structural data from X-ray
studies will facilitate such investigations to some extent. We should
consider, however, that the limited accuracy of these calculations will
make it difficult, to determine significant differences in the interactions in the liquid and the solid phase. It seems to be more promising,
therefore, to collect first all available infirmations about the theoretical foundations of the metal-amide interaction in the gas phase and
the solution, and to use this information in the interpretation of
carefully chosen experiments on solid state complexes. It is not to
be expected, that the fundamental laws of interaction, valid in the

liquid phase, will change dramatically, so that a general understanding
of the solid amide complexes should be obtained from the model calculations on the liquid phase and the chemical knowledge about the solid
state, without explicit calculations on these complex species.

Another way could be the application of HFS-SW-SCF or typical solid
state calculations to these solid complexes, which, however, means again
to deal with the general problems of more or less semi-empirical methods
containing several simplifying assumptions. For principal reasons, we
should not expect too much information about the changes in the amide
molecule properties from these methods, but it is these changes which
are most interesting and important for the chemist working in this field
of research.

6. CONCLUSION

Concluding our attempt to show, how quantum chemistry and experiment can
be combined in the study of complex formation between metal ions and
amide group containing ligands, we want to illustrate by a short example,
that sometimes quantum chemistry can be the only possibility to answer
questions in this field of research, even if we do not intend to discuss
details of the electronic structure and bonding, which surely belong to
the domain of this branch of chemistry.

In addition to the alkali ion complexes $M^+(DMF)$, the corresponding
complexes $M^+(DMF-CH_3)$ can be observed in the gas phase. Zn and Cd do not
form such complexes, indicating thereby a possible different structure
of their gas phase complexes [18]. There does not exist any experimental
method, which would allow to decide between the two possible structures
of these complexes, namely the isocyanate (A) and the amide (B) type
(Figure 5) under the conditions of their formation (cf. section 3). For
quantum chemistry, this problem is accessable, especially for these
specific experimental conditions. Calculations employing an 8/4 (5111/31)
GLO basis set [57] and the previously discussed minimal one proved (58),
both in excellent agreement, that the amide structure (B) is favoured
over (A) by more than 100 kcal. This result inicates, that there is
almost no reason to assume a migration of the proton within the DMF
fragment, even if complexation with alkali metal ions occurs.

This example, in connection with the discussion given in the previous chapter allows us to summarize briefly:

(1) Quantum chemical model calculations can be regarded as a most
helpful, sometimes even as the most favourable tool in the investigation
of metal-amide interactions.

(2) The relative reliability of even strongly simplified model
calculations allows the extension of such investigations to quite large
ligand molecules or to higher coordination numbers (i.e. systems with
30 and more heavy atoms C, N and O), including thereby the possibility
of calculating effects occuring in the liquid phase or chelating properties.

(3) If we want to improve the theoretical models, the treatment of
larger chemical systems using small basis sets promises to give more
reliable chemical information than the accurate evaluation of pair po-

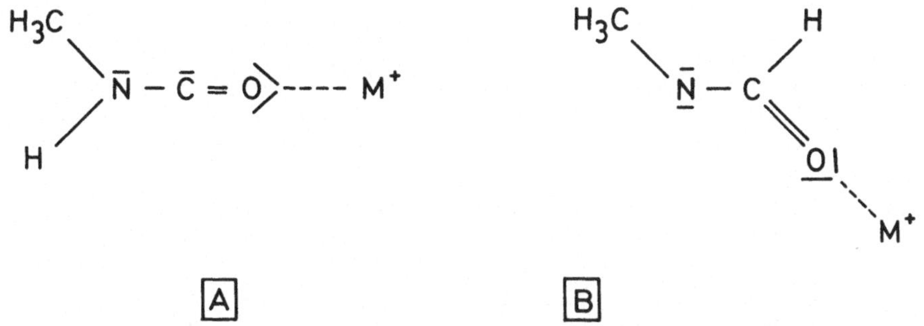

Fig. 5. Possible structures of the gas phase complexes of DMF-CH$_3$ with Li$^+$, Na$^+$ and K$^+$.
A...isocyanate structure, B...amide structure.

tentials resulting from calculations on 1:1 complexes.

(4) Carefully chosen model experiments represent a valuable and sometimes necessary tool for the development and the control of the simplified theoretical models. They will indicate any inevitable necessary improvement of the calculations, but they will give also further inspirations for the quantum theoretical investigations.

(5) It is to be hoped, that the combined experimental and quantum chemical research on ion-amide interactions, of wich we have just reported some of the first tentative steps, will prove soon as a successful way in the treatment of larger chemically and biologically important complexes of this kind.

ACKNOWLEDGEMENT

The author is very indebted to the Fonds zur Förderung der wissenschaftlichen Forschung, Vienna, Austria, for financial support (Project No. 2432).

REFERENCES

1. Craig, R.A. and Richards, R.E.: Trans Farady Soc. 59, 1972, (1963).
2. Gentile, P.S. and Shankoff, T.A.: J. Inorg. Nucl. Chem. 27, 2301, (1965).
3. Ram Chand Paul et al.: J. Phys. Chem. 73, 741, (1969).
4. Matwiyoff, N.A.: Inorg. Chem. 5, 788, (1966).
5. Movius, W.G. and Marwiyoff, N.A.: Inorg. Chem. 6, 847, (1967).
6. Fratiello, A. et al.: Mol. Phys. 12, 111, (1967).
7. Matwiyoff, N.A. and Movius, W.G.: J. Am. Chem. Soc. 89, 6077, (1967).
8. Movius, W.G. and Matwiyoff, N.A.: Inorg. Chem. 8, 925, (1969).
9. Erlich, R.H. et al.: J. Am. Chem. Soc. 92, 4989, (1970).
10. Erlich, R.H. and Popov, A.I.: J. Am. Chem. Soc. 93, 5620, (1971).

11. Adams, M.J. et al.: J. Chem. Soc. Faraday II 70, 1114, (1974).
12. Dechter, J.J. and Zink, J.I.: J. Chem. Soc. Chem. Comm. 96, (1974).
13. Gudlin, D. and Schneider, H.: J. Magn. Resonance 16, 362, (1974).
14. Nakamura, T.: Bull. Chem. Soc. Japan 48, 1447, (1975).
15. Good, A.: Chem. Rev. 75, 561, (1975).
16. Zhitomirskii, A.N. et al.: Ukr. Khim. Zh. 41, 346, (1975).
17. Rusin, G.G. et al.: Nauk Pr., Ukr. Silskogospod. Akad. 119, 142,
    (1975).
18. Rode, B.M.: Chem. Phys. Letters 35, 517, (1975).
19. Rode, B.M. and Fussenegger, R.: J. Chem. Soc. Faraday II 71, 1958,
    (1975).
20. Balasubranian, D. et al.: Chem. Phys. Letters 17, 482, (1972).
21. Perricaudet, M. and Pullman, A.: Internat. J. Peptide Protein Res.
    5, 99, (1972).
22. Armbruster, A.M. and Pullman, A.: FEBS Letters 49, 18, (1974).
23. Rode, B.M. and Preuss, H.: Theoret. Chim. Acta 35, 369, (1974).
24. Rode, B.M.: Chem. Phys. Letters 26, 350, (1974).
25. Rode, B.M.: Mh. Chemie 106, 339, (1975).
26. Pullman, A.: Int. J. Quantum Biology Symp. 1, 33, (1974).
27. Kostetcky, P.V. et al.: FEBS Letters 30, 205, (1973).
28. Noble, D.: Biological Memranes, Oxford Univ. Press, 1975.
29. Preuss, H. and Janoschek, R.: J. Mol. Struct. 3, 423, (1969).
30. Clarke, P.A. and Preuss, H.: Z. Naturforsch. 27a, 1294, (1972).
31. Frost, A.A.: J. Chem. Phys. 27, 3707, 3714, (1967).
32. Mely, B. and Pullman, A.: Theoret. Chim., Acta 13, 278, (1969).
33. Dreyfus M et al.: Theoret. Chim. Acta 17, 109, (1970).
34. Russegger, P. et al.: Theoret. Chim. Acta 24, 191, (1972).
35. Russegger, P. and Schuster, P.: Chem. Phys. Letters 19, 254, (1973).
36. Rode, B.M.: Chem. Phys. Letters 20, 366, (1973); 25, 369, (1974).
37. Rode, B.M. et al.: Chem. Phys. Letters 32, 34, (1975).
38. Germer, H.A.: Theoret. Chim. Acta 34, 145, (1974).
39. Dzi, I. and Kebarle, P.: J. Phys. Chem. 74, 1466, (1970).
40. Kebarle, P. et al.: J. Am. Chem. Soc. 89, 6393, (1967).
41. Searles, S.K. and Kebarle, P.: J. Phys. Chem. 72, 742, (1968).
42. Hogg, A.M. and Kebarle, P.: J. Chem. Phys. 43, 449, (1965).
43. Tang, I.N. and Castleman, A.W.: J. Chem. Phys. 62, 4576, (1975).
44. Rode, B.M.: Chem. Phys. Letters 32, 38, (1975); Mh. Chemie 105, 308,
    (1974).
45. Rastogi, P.P.: Z. Phys. Chem. NF. 73, 163, (1970).
46. Fussenegger, R. and Pontani, T. : unpublished results.
47. Ladell, J. and Pust, B.: Acta Cryst. 7, 559, (1954).
48. Sutton, L.E.: Tables of Interatomic Distances, The Chemical Society,
    London, 1958.
49. Schuster, P. et al.: in Chemical and Biochemical Reactivity, The
    Jerusalem Symposia on Quantum Chemistry and Biochemistry, vol. VI,
    (1974).
50. Pullman, A. and Armbruster, A.: Chem. Phys. Letters 36, 558, (1975).
51. Pontani, T.: Thesis, Univ. Innsbruck/Austria (1976).
52. Pontani, T. and Rode, B.M.: in preparation.
53. Harrington, W.F. and Sela, M.: Biochem. Biophys. Acta 27, 24, (1958).
54. Mandelkern et al.: J. Am. Chem. Soc. 84, 1383, (1962).

55. Von Hippel, P.H. and Schleich, T.: Structure and Stability of Biological Macromolecules, M. Dekker, N.Y., 1969.
56. Lagowski, J.J.: The Chemistry of Nonaqueous Solvents, Part II. Academic Press, N.Y./London, 1966.
57. Huzinaga, S.: Techn. Report, Univ. of Alberta, 1971.
58. Rode, B.M. and Ahlrichs, R.: Z. Naturforsch. <u>30a</u>, 1792, (1975).

DISCUSSION

A. PULLMAN: Concerning the influence of cations on the barrier to rotation, it would be interesting to evaluate separately the polarization, electrostatic and repulsion contributions to the phenomenon, in order to obtain a clearer view of the features underlying the fact that the cations of the pairs Li^+ - Mg^{++} and Na^+ - Ca^{++} behave in a similar way in their influence on the barrier.

RODE: I agree to that the separate evaluation of the energy components of the ion-amide interaction would be of high interest and would like to add also the respective charge transfer to be one of the components which should be considered. In order to perform such an analysis of the various components determining the interaction one should carry out, as I believe, calculations with more sophisticated basis sets than we have done. For a rather large complex, as being realized in the solution, our experiments refer to, this means a quite big computational problem. In the case of DMF it might be possible, however to obtain some quite reliable information from the calculation of 1:1 complexes. This would not necessarily hold for other amides, for reason mentioned in the course of this lecture.

EISENMAN: Do sufficient data or calculations exist to compare the one to one energies of the group Ia cations with dimethyl formamide as compared to H_2O? This is a direct test for the molecular asymmetry (Eisenman and Krasne, 1975)* postulated for amide carbonyls as H_2O ligands to underlie the selectivity of carriers such as valinomycin. Restating my point, it is possible to decide whether a ligand is symmetrical to water or asymmetrical by asking a plot of the energy of Ion-Ligand *vs* Ion-H_2O is a straight line (in which case it is symmetrical in the sense of Eisenman + Krasne) or is curved (in which case it is asymmetrical). e.g.

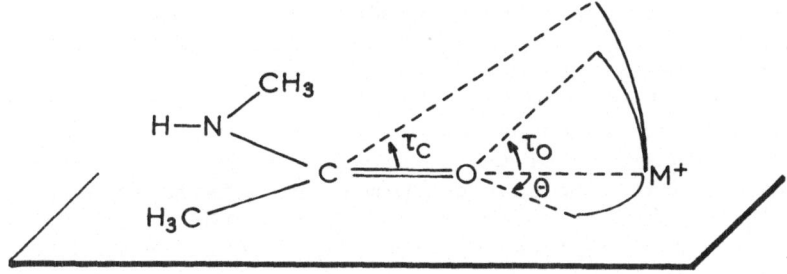

---

* MTP, *Int. Review of Science Biochemistry Series,* cf. Fox.

RODE: There exists quite a lot of calculations for the interaction energies of H_2O with Ia metal ions, and also some concerning the Ia ion/DMF system. Unfortunately, they use quite different basis sets; a comparison as that mentioned by you can be made, however, only for calculations using the same basis set, which should also be a quite large one, since the differences may be only small and, therefore, should be and can be evaluated only by quite accurate calculations. As far as we have already obtained comparable data, I would prefer to say that the plot your question refers to, leads to a curved line.

A. PULLMAN: The comparison of the intrinsic affinity of group Ia cations for amide carbonyls, ester carbonyls and water has been made by computations at the STO3G level only, and only for Li^+, Na^+, K^+ (M. Perricaudet and A. Pullman, FEBS Letters 34, 222, (1973). The differences obtained are large enough for the couple amide-water to indicate that the intrinsic affinity for the amide carbonyl is larger thans for water for the three cations and this conclusion is not likely to be changed by a refined computation. For the ester carbonyl the STO 3G result places the affinities slightly below those for water but the difference is small and our present better experience with the STO 3G basis makes me reluctant to conclude without more computations.

Concerning the relationship between $\Delta E_{amide-ion}$ and ΔE_{H_2O-ion} it could be done but has not been yet.

SIMON: Is it not very dangerous to talk about intrinsic selectivities of an ester carbonyl group vs. an amide carbonyl group without specifying clearly the conformations involved?

RODE: I think it is obvious, that the stereochemistry as well as the substituents in the neighbourhood of an ion binding site influence strongly the activity and affinity of this centre for the ion binding process. Thus, talking about the intrinsic affinity of a binding site I would think we should extend the definition of the binding site to a rather large part of the molecule containing the carbonyl group which will bind the metal ion, e.g. to a part including of least all the atoms within a radius of $5 - 7$ Å around the > C=O group.

A. PULLMAN: When I speak about intrinsic affinities, I simply wish to emphasize the properties to be expected of the best possible conditions can be realized. When trying to interpret experimental data in the area of cation complexation one constantly uses some concept that one has of this intrinsic binding properties of the ions, and more often than not these concepts are based on very approximate notions like cation radius for instance. Theory nowadays allows to go farther and bring step by step more accurate knowledge of various features. As to the role of conformation for instance, the binding energy as a function of the fundamental changes of conformation may be computed. In the case of the carbonyl groups of methylacetate and methylacetamide we have studied two such rotations of the carbonyl group away from the cation out of the most stable coplanar conformation (details in A. Pullman, 7th Jerusalem Symposium on Molecular and Quantum Pharmacology, p. 401). A comparison with the in-plane rotation away from the most stable position has permitted to find various characteristcs which would be difficult to guess otherwise: for example the rotation of the cation in the plane of the amide or ester linkage is very easy for both Na^+

and K^+ : very little energy is lost upon in-plane displacement of the cation in a relatively large angular region around the carbonyl oxygen. On the other hand the rotation of the cation out of the plane has a much stronger effect: for a $45°$ rotation at a constant distance from the carbon atom of the carbonyl (τ_c) (see figure) about half of the binding energy is lost for Na^+ and K^+, at $90°$ nearly all the binding energy is lost. For a rotation at a constant distance of O (τ_o) the loss in energy is much slower and half of the binding is still present at $90°$.

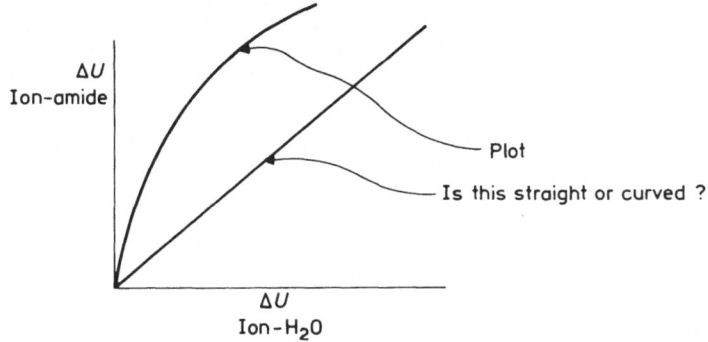

These are examples of the kind of knowledge that theoretical studies may provide in this area. I call them intrinsic binding properties and they represent building blocks necessary to construct our understanding of the mechanism of functionning of the ion-carriers.

B. PULLMAN: I would like to strongly support Dr. Rode's assertion about the inadequacy of the CNDO method for studies of cation binding to biological substrates. We have come recently in our laboratory accross three examples which show that the results of this method are in contradiction with those of *ab initio* computations and represent artefacts: (1) in computations concerning the binding of Na^+ and K^+ to amides and esters (M. Perricaudet and A. Pullman, FEBS Letters 34, 222, (1973) in which CNDO predicts an incorrect out-of-plane minimum; (2) in computations on the influence of cation binding upon the conformation of the phosphodiester linkage (B. Pullman, N. Gresh, and H. Berthod, Theoret. Chim. Acta 40, 71, (1975) in wich CNDO erroneously predicts that the binding of the B_{13} site should produce a preference for a trans-trans conformation of the phosphodiester unit; and (3) in computations on cation binding to uracil (D. Perahia, A. Pullman and B. Pullman, Theoret. Chim. Acta, in press) in which CNDO erroneously predicts the $C_5 = C_6$ double bond to be the preferential site for such a binding. A detailed analysis of this last case shows that this erronous results is an artefact due to an exaggeration by the CNDO method of the charge transfer between the ligand and the cation.

INTERACTION OF ALKALI AND ALKALINE EARTH CATIONS AND PROTONS WITH
AMIDES AND RELATED SYSTEMS

C.N.R. RAO
*Dept. of Chemistry, Indian Institute of Technology, Kanpur
208016, India*

1. INTRODUCTION

Interaction of alkali and alkaline earth metal ions is of significance
to the biophysical chemistry of proteins and polypeptides. Effect of
salts are now known to involve binding with specific sites rather than
general effects or alteration of water (solvent) structure [1 — 4]. Such
specific interactions would be important to consider in conformational
transitions of polypeptides, denaturation of proteins, selective ion
complexation by macrocyclic anti-biotic ionophores etc. Lithium salts
are also known to affect brain chemistry [5]. Interaction of protons
with the peptide bond (protonation) is of considerable importance in
understanding helix-coil transitions of proteins and related phenomena
[1, 6]. In this paper, we present the results of theoretical and spectroscopic investigations carried out in this laboratory on the interaction
of alkali and alkaline earth ions as well as protons on model amides and
related systems. These studies have provided the relative strengths of
interaction of these cations and the consequences of such interaction
on the structure and spectra of the donor molecules. Vibrational, electronic and nuclear magnetic resonance spectra of the metal ion adducts
have been examined.

2. SEMI-EMPIRICAL CALCULATIONS

While semi-empirical molecular orbital calculations would not yield good
values of M^{z+}-donor interaction energies or bond distances, they do provide the right trends in energies and distances in a related series of
complexes of the type $M^{z+}(donor)_n$ as verified recently by studies of vapour phase complexes in some instances [7]. Typical results of our CNDO/2
calculations on the interaction of various group IA and IIA cations with
amides and related compounds are shown in Table I. We see that the major
changes are in the C=O and C=N distances. There is an increase in C=O
distance and a decrease in the C-N distance. In amides, such a variation
in distances would imply an increase in the barrier height to rotation
about the C-N bond. Consequences of these variations in distances are

*B. Pullman and N. Goldblum (eds.), Metal-Ligand Interactions in Organic
Chemistry and Biochemistry, Part 1, 147-157. All Rights Reserved.
Copyright © 1977 by D. Reidel Publishing Company, Dordrecht-Holland.*

TABLE I
CNDO/2 Calculations on complexes of oxygen donors [a]

	Li$^+$			Na$^+$			Mg^{2+}			H$^+$		
	r, (Å)	ΔE	Δr(CO)	r, (Å)	ΔE	Δr(CO)	r, (Å)	ΔE	Δr(CO)	r, (Å)	ΔE	Δr(CO)
1:1 Complexes												
HCHO	2.30	85	0.01	2.9	59	<0.01	2.6	100	0.01	1.0	227	0.03
H$_2$NCHO [b]	2.30	116	0.02	2.8	76	<0.01	2.6	144	0.01	1.0	296	0.05
NMA [b]	2.30	152	0.02									
1:2 Complexes												
HCHO	2.40	80	0.01	3.0	56	<0.01	2.7	94	0.01	1.1	230	0.03
H$_2$NCHO [b]	2.40	112	0.02	2.9	72	<0.01	2.7	140	0.01	1.1	297	0.04
1:4 Complexes												
HCHO	2.40	50	0.01	3.0	30	<0.01	2.7	88	0.01	1.1	220	0.02
H$_2$NCHO [b]	2.40	81	0.01	3.0	43	<0.01	2.7	110	0.01	1.1	245	0.03

[a] Results are given for the most stable configuration with <COLi = 60°. In the table, r is the cation-oxygen distance, ΔE the binding energy in kcal m^{-1} per M–O bond and Δr(CO) is the increase in C–O bond length.

[b] In the case of amides C–N bond distance decreases; Δr(CN) is -0.22 ± 0.01 Å for Li$^+$ and -0.05 ± 0.01 Å for H$^+$; Na$^+$ has negligible effect and effect of Mg^{2+} is in between that of Li$^+$ and H$^+$. The barrier height to rotation about the C–N bond is much higher in the metal ion complexes (see ref. [7]).

also seen in vibrational, electronic and NMR spectra as will be detailed later.

The order of binding energies is $H^+ > Li^+ > Mg^{2+} > Na^+$. This trend is consistent with the observed spectral changes. Perricaudet and Pullman [8] find the order of binding energies to be $Li^+ > Na^+ > K^+$. Another interesting finding from the calculations (Table II), is the essential constancy of the cation-oxygen distances in the complexes with different oxygen donors. This has some bearing on the vibrational spectra of the complexes as will be shown in the next section.

Based on the results of these calculations, one could represent the interaction of cations with amides as follows:

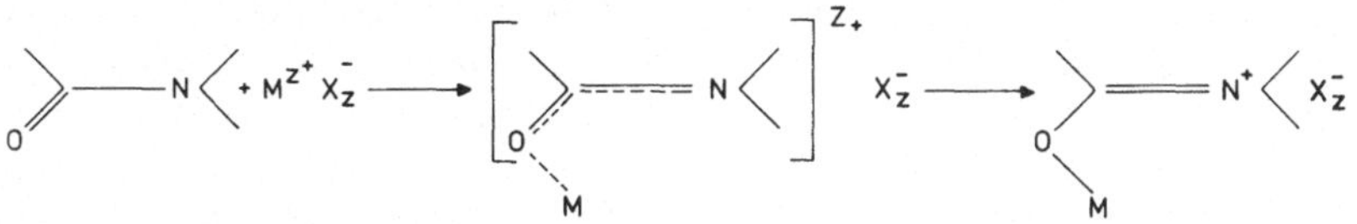

The carbonyl oxygen is shown to be the preferred site for interaction in the above scheme. In a 1:1 complex, the cation-oxygen-carbon angle is around 60° in the peptide plane so that the cation is sufficiently close to the nitrogen atom as well.

CNDO/2 calculations on the interaction of cations with nitrogen donors like CH_3CN, NH_3 and pyridine (Table II) show that Li^+ and Ng^{2+} interact strongly. Interaction of nitrogen donors with Mg^{2+} is particularly strong. Calculations show that ammonium ions also bind to oxygen and nitrogen donors [9].

TABLE II
CNDO/2 Calculations on Complexes of Nitrogen Donors[a]

	Li^+		Na^+		Mg^{2+}	
	r, Å	ΔE	r, Å	ΔE	r, Å	ΔE
1:1 Complexes						
NH_3	2.18	95	2.80	62	2.40	144
CH_3CN	2.20	193	2.60	116	2.40	181
Pyridine	2.20	164	2.80	113	2.40	202
1:4 Complexes						
NH_3	2.30	81	2.90	50	2.50	128

[a] Here, r is the M-N distance and ΔE the binding energy in kcal m^{-1} per M-N bond. Metal ion is along lone pair direction in the case of NH_3. Metal ion is perpendicular to the C≡N bond in CH_3CN. In the case of pyridine $\emptyset = 75^o$ as in ref. [7].

3. VIBRATIONAL SPECTRA

When salts of alkali and alkaline earth ions are dissolved in solvents
like ethers, ketones, amines and amides, infrared bands due to
quantised vibrations characteristic of cations are seen in
the low-frequency region [10]. The bands are around 420, 200, 150, 120
and 100 cm^{-1} for Li$^+$, Na$^+$, K$^+$, Rb$^+$ and Cs$^+$ respectively and 400 cm^{-1} for
Mg^{2+}. These cation vibration bands are also exhibited by alkali and alkaline earth metal oxide systems containing metal-oxygen polyhedra indicating that similar coordination may also be present in the solutions
of oxygen donor solvents. Thus, the alkali metal-ligand ratio in solution is 1:4, while in glasses the alkali metal ion has tetrahedral coordination (MO$_4$). The cation vibration bands seem to be useful probes to
study the primary solvation of cations and structures of electrolyte solutions. Preliminary studies [3, 10, 11] of the infrared spectra of crystalline complexes of amides with Li$^+$ indicate the presence of the low
frequency bands due to vibrations involving Li$^+$. We have carried out a
detailed investigation of the infrared and Raman spectra of amide complexes of alkali and alkaline earth metals and compared the results with
those of crown-ether complexes and other systems containing similar MO$_4$
or MO$_6$ metal-oxygen polyhedra [12].
     The major bands in the infrared and Raman spectra of the alkali
metal complexes of N-methylacetamide (NMA) of the general composition
M$^+$(NMA)$_4$ are listed in Table III along with the bands of Mg^{2+}(NMA)$_6$. All
the metal complexes generally show a lower frequency for the amide I band
indicating that the metal ion is coordinated to the carbonyl oxygen. The
amide II bands appear at higher frequencies in the complexes compared
to free NMA (assuming that there is negligible hydrogen bonding in the
complex); the amide III band which also has considerable contribution
from C-N stretching is at higher frequencies in the complexes. These band
shifts in the complexes are consistent with a decreased C=O order and
an increased C-N bond order. This is what was predicted by molecular orbital calculations on the interactions of these cations with amides
(Table I).
     As mentioned earlier, an important feature in the infrared spectra
of the alkali metal complexes of NMA is the appearance of a band in the
low frequency region characterictic of the metal ion. Such bands have
been found in solutions of alkali metal salts in oxygen donor solvents
like ethers, ketones and amides and have been ascribed to the metal-
oxygen polyhedra of the primary coordination sphere. By comparison with
the infrared spectra of oxide glasses and oxyanion salts of alkali metals
it has been shown that these bands found around 400, 200 and 150 cm^{-1}
are characterictic of the MO$_4$ tetrahedra of Li$^+$, Na$^+$ and K$^+$ respectively
[10, 11]. It is indeed interesting that the crystalline 1:4 amide com
plexes also show these bands in the infrared spectra. We, therefore, conclude that the M(O)$_4$ frame in these complexes are essentially tetrahedral
although the splitting of the Amide I band (Table III) in some complexes
may indicate lower overall symmetry.
     Assuming a tetrahedral structure for the M(O)$_4$ frame, we have car-
ried out the normal coordinate analysis for the in-plane modes of the
alkali metal-NMA complexes employing the Urey-Bradley force field. The

TABLE III
Major band assignments in alkali and alkaline earth metal complexes of NMA [a]

NMA		Li$^+$(NMA)$_4$		Na$^+$(NMA)$_4$		K$^+$(NMA)$_4$		Mg^{2+}(NMA)$_6$		Assignment in complex
IR	R	IR	R	IR	R	IR	R	IR	R	
1660(s)	1657(s)	1646(s)	1645(s)[b]	1654(s)	1653(s)[b]	1636(s)	1634(s)[b]	1649(s)	1648(s)[b]	ν(C=O), Amide I
1700			(1642)		(1647)		(1647)			
(free)										
1569(s)	1569(s)	1569(s)		1569(s)		1569(s)		1570(s)		δ(NH), Amide II
1500										
(free)										
1301(s)	1298(s)	1304(s)	1305(s)	1305(s)	1305(s)	1304(s)	1304(s)	1304(s)	1306(s)	ν(CN)+ν(CC), Amide III
1260			(1314)		(1315)		(1318)			
(free)										
883(w)	879(m)	882(w)	885(m)	880(w)	880(m)	883(w)	890(m)	885(w)	887(m)	ν(CC)+ν(CN)
			(863)		(876)		(875)			
	725(m)		715(m)		700(m)		700(w)		700(w)	NH out-of-plane bend, Amide V
	648(free)									
628(m)	628(m)	629(m)	625(m)	630(m)	629(m)	630(m)	630(b)	633	630(b)	δ(OCN)+ν(CC), Amide IV
			(640)		(650)		(640)			
600(m)	600(w)	590(m)	605(w)	570(m)	590(w)	595(m)	600(w)	593(m)	600(w)	CO out-of-plane bend, Amide VI
585(free)										
289(m)	290(m)	290(m)	290(m)	290(m)	290(m)	280(m)	270(m)	290(m)	300?	δ(CNC)+δ(OCN)
192		202		200?		210		210		τ(CN); Amide VII
136		145(b)		140(b)		155?		146		τ(CH$_3$)
		388	(390)	190(b)	(189)	146(b)	(137)	393	(390)	M-O asym. str.
		180	(206)	125?	(117)	122?	(108)	155?		δ(OMO) asym. bend + M-O asym. str.

[a] All values in cm^{-1}. Values in brackets are calculated from normal coordinate analysis. Values with question marks are doubtful. The values of some of the bands of NMA in the free non-hydrogen bonded states are given.
[b] Appear as doublets.

force constants and structural parameters of NMA and related molecules
are available in the literature. By taking the $M^+ - O$ stretching constants
in the range $0.6 - 0.3$ mdyne $\mathring{A}^{-1}$ (Table IV), and reasonable values of
M-O and other bond distances in the complexes [12], we could obtain satisfactory agreement with the observed frequencies (Table III). The calculations establish that the cation-dependent infrared frequencies at
388, 190 and 146 cm^{-1} in Li^+, Na^+ and K^+ complexes are due to the M-O
asymmetric stretching vibration in the MO_4 tetrahedra. The decreasing
M-O stretching force constant (Table IV) from Li^+ to K^+ shows that the
lowering of the characteristic M-O frequency in the series is not due
to mass effect alone. This trend in force constants is also consistent
with the binding energies found for Li^+ and Na^+ complexes by M.O. calculations (Table I). From Table IV we see that the only stretching force
constants of NMA which show any significant changes are for the C=O and
C-N stretching vibrations as excepted from M.O. calculations.

Normal vibration calculations also predict vibration frequencies
of the MO_4 tetrahedra other than the highest frequency due to the M-O
asymmetric stretching vibration. We seem to find bands in the infrared
spectra of complexes which can be ascribed to the asymmetric OMO bending
mode (Table III). However, we do not see distinct bands in the infrared
or the Raman spectra due to the symmetric stretching and bending modes
of the MO_4 tetrahedra. In the case of Li^+, we see a weak band around
190 cm^{-1} in the Raman (calc 200 cm^{-1}) which could be due to the symmetric
stretching mode. It is interesting that we do not see the cation-charac-
teristic M-O asymmetric stretching mode in the Raman spectra. This could
be possibly due to the predominance of electrostatic forces between the
metal ion and the amide donor.

From Table III we see that the C-N torsional frequency is somewhat
higher in the amide complexes compared to the parent amide. This is consistent with the increased C-N bond order and barrier to rotation in the
complexes (see Table I). The amide IV band involving OCN bending is
slightly higher in the complexes. In the case of the potassium complex
this band appears as a doublet possibly indicating a lower symmetry.

We have examined the spectra of the complex of Li^+ with N,N-dimethyl
formamide of the composition $Li^+(DMF)_4$. In this case, we see a lower C=O
stretching frequency, higher C-N frequency (band at 1395 cm^{-1} in DMF)
and higher OCN bending frequency. The torsional mode at 350 cm^{-1} also
appears at a higher frequency. In addition, a band around 400 cm^{-1} due
to the Li-O asymmetric stretching mode is seen. These data are consistent
with an essentially tetrahedral MO_4 unit in this complex.

The $Mg^{2+}(NMA)_6$ complex shows features similar to the alkali metal
complexes (Table III). The coordination here is octahedral and the MgO_6
octahedron gives rise to the asymmetric stretching mode frequency around
390 cm^{-1} in the infrared spectrum. A weak band found in the Raman around
200 cm^{-1} may be due to the symmetric stretching mode. A preliminary calculation assuming an Mg-O distance of 2.6 $\mathring{A}$ and a stretching force constant of 0.5 mdyne $\mathring{A}^{-1}$ gave the Mg-O asymmetric stretching frequency
around 390cm^{-1}. The assignment of the 390 cm^{-1} band to the asymmetric
stretching vibration of the MgO_6 octahedron is justified by the spectra
of oxide glasses as well as oxyanion salts of Mg^{2+} which contain such
octahedra.

TABLE IV
Force constants (in mdynes Å^{-1}) alkali metal-NMA complexes

Force constants	NMA	$M^+(NMA)_4$
K (Li^+...O)		0.60
K (Na^+...O)		0.45
K (K^+...O)		0.35
K (C=O)	7.137	6.937
K (C...N)	6.428	6.528
K (C–C)	3.026	3.026
H (O=C–N)	0.955	0.80
F (O...N)	1.154	1.00
H (O–M^+...O)		0.20
F (O···O)		0.40
H (C'–C=O)	0.347	0.347
F (C'...O)	0.653	0.653
H (N–C–C')	0.786	0.786
F (N...C')	1.122	1.122
H (M^+–O–C)		0.30
F (M^+...C)		0.40
F' = F/10		

4. ELECTRONIC SPECTROSCOPY

Alkali and alkaline Earth metal ions cause marked shifts of the $n-\pi^*$
and $\pi-\pi^*$ transitions of ligands such as carbonyl and thiocarbonyl compounds, the magnitude of the shift depending on the cation and solvent.
Among the group IA cations, Li^+ causes the maximum spectral perturbation,
while among group 2A cations, Ca^{2+} shows the maximum effect [13]. Spectra of ligands like acetone or cyclopentatone recorded with varying concentrations of cations show progressive band shifts with increasing cation concentration and the presence of ososbestic points. Acids (protons)
show effects similar to IIA cations, but the perturbations are more marked. In the light of these results, we have examined the electronic spectra of a few amide [14] and thioamide derivatives in the presence of various salts.

The first set of experiments was concerned with the effect of group
IA and IIA cations on the far ultraviolet spectral bands of dimethyl formamide and lactams. The $\pi-\pi^*$ bands of these compounds occur around 195
nm in aqueous solutions. Addition of alkali and alkaline earth metal
salts causes a blue shift of this band maximum along with a reduction
in band intensity. Among the Group IA cations, Li^+ causes the maximum
spectral perturbations, while it is Ca^{2+} among Group IIA. These spectral
effects are analogous to the shifts seen in the spectra of other carbonyl compounds [13] and are consistent with the scheme given in section
2 for the mode of interaction between cations and amides.

We have observed the presence of a solvation equilibrium between
the cations and amides in solution. Figure 1 shows the progressive red
shifts of the 225 nm band of a model amide, benzamide, brought about
by these metal cations, and also by hydrogen ions (protons). An important
result in all these cases is the presence of isosbestic points near 230
nm which are indicative of the presence of equilibrating species. The mag
tudes of the shifts, when plotted against the molarity of added electroly
(see inset in Figure 1), point to a competition between the amide ligand
(L) and solvent (S) water for solvating the cation, and also to some
degree of cooperativity involved in this ligand substitution processes.
The following type of step-wise process appears likely:

$$M^{z+}(S)_m + L \rightarrow M^{z+}S_{m-1}L + S \tag{1}$$

$$M^{z+}S_{m-1}L + L \rightarrow M^{z+}S_{m-2}L_2 + S \text{ etc} \tag{2}$$

$$M^{z+}SL_{n-1} + L \rightarrow M^{z+}1_n + S \tag{3}$$

$$M^{z+} + nL \rightarrow M^{z+}L_n \tag{4}$$

Equilibria (1) and (2) would be predominant at concentrations at which
the ions are nearly fully solvated (up to 8M in Li^+). Then it appears
that further ligand binding occurs causing more pronounced shifts of the

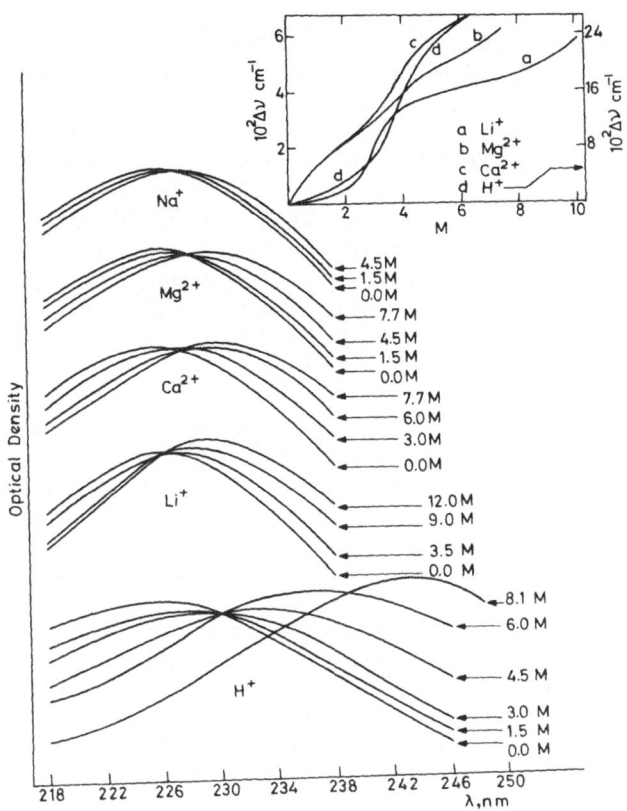

Fig. 1. Effect of addition of Group IA and IIA cations and acid on the spectrum of benzamide. In the inset the spectral shifts are plotted against concentration of cation.

ligand spectrum. This would still retain the isosbestic point provided the spectrum of the ligand bound to the cation remains nearly the same in the presence or absence of a second ligand (or the solvent molecule) on the ion. At high concentrations of cations, perturbations become more marked since all the solvent molecules would be insolved in solvation and equilibria (3) and (4) become predominant. This happens at 8M in Li^+ or at 4M in Ca^{2+}; above these concentrations, heat effects and perturbations of ligand vibrations also become large. The cation-induced spectral shifts vary with the solvent in the order $H_2O<CH_3OH<THF$; this is probably related to the relative strengths of binding of the solvent and the ligand molecules to the cation.

Interaction of protons with amides can be directly compared with those of the metal cations. As shown in Figure 1, the proton-amide interaction is also characterized by an isosbestic point and co-operativity. The mode and general features of the interaction in this case are quite similar to the alkali cations, except with a greater strength. Based on the spectral shifts produced, the strength of interaction of cations with amides is found to be in the order $H^+ > Li^+ \approx Ca^{2+} > Mg^{2+} > Na^+ > K^+$. This trend is consistent with the results of molecular orbital calculations discussed earlier.

It is interesting that the isosbestic point near 230 nm seen in low to moderate acid strengths is lost in stronger acids (see Figure 1, curve for 8M H_2SO_4), suggesting the presence of a third species which is most likely to be the protonated amide. Thus, it would appear that the interaction of protons with amides involves two stages: (i) at low acid strengths, equilibria similar to (1) – (4) discussed earlier in the case of IA and IIA cations and (ii) in stronger acids,

$$H^+(amide)_n = [amide-H]^+ + (n-1) \text{ amide} \qquad (5)$$

where the species in the square brackets is the protonated amide.

It would be pertinent to make a few comments regarding the above equilibria. It has been repeatedly suggested that amides are protonated in acid media, at the carbonyl oxygen site. Benderly and Rosenheck [15] have recently provided far ultraviolet spectral evidence for this. However, the suggestion has been made by Liler [16] that, at low to moderate acid concentrations, the amide is protonated at the nitrogen atom, with the site changing to the oxygen at high acid media. This seems unlikely to us on two grounds: (i) Rosenheck's data suggest the amount of N-protonated species to be negligible at all acid concentrations, and (ii) we have compared the effect of varying concentrations (0-6 *M*) of protons and of alkali metal ions on the far ultraviolet spectra of DMF and lactams. In both cases, the effects (blue-shifts and hypochromism of the peptide $\pi-\pi^*$ band) are very similair, indicating that both these cations interact with the amide group in an identical manner, i.e., at the carbonyl oxygen site. Thus, the 230 nm isosbestic point observed in benzamide in low acid media seems to owe its origin to aquilibria (1) – (4) (with $M^+ = H^+$) rather than the N-protonation. Distinction between equilibria (4) and (5) is rather subtle. Equilibria of the kind suggested here may account for the reversible acid-induced structural changes in polypeptides and proteins.

5. NMR SPECTROSCOPY

Based on M.O. calculations, we find an increased C—N bond order on interaction of amides with IA and IIA group cations. Shifts in amide II bands confirmed such bond order changes. An increase in C—N bond order in amide, implies an increase in barrier height to rotation about the C—N bond. While preliminary results from temperature-dependent NMR studies [2 – 4] showed some evidence for such an increase in the barrier height, the method employed for the study was not sufficiently reliable. Dr K.G. Rao of this laboratory has just completed a detailed analysis (including computer fitting of profile) of dimethyl formamide-d_1 + IA and IIA cation systems There appaers to be no doubt that there is an increase in coalescence temperature; there is also an increase in barrier height, finite although small, on addition of the cations, lithium ion causing the maximum change.

ACKNOWLEDGEMENT

The author is thankful to the National Institutes of Healt for support of this research (01-078-1).

REFERENCES

1. Rao, C.N.R.: in Conformation of Biological Molecules and Polymers, *The Jerusalem Symposia on Quantum Chemistry and Biochemistry*, Vol. 5 1973.
2. Balasubramanian, D., Goel, A., and Rao, C.N.R.: *Chem. Phys. Letters* 17, 482 (1972).
3. Balasubramanian, D. and Shaikh, R.: *Biopolymers* 12, 1639 (1973).
4. Balasubramanian, D. and Misra, B.C.: *Biopolymers* 14, 1019 (1975).
5. Samuel, D. and Gottesfeld, Z.: *Endeavour* 32, 122 (1973).
6. Rao, C.N.R., Rao, K.G., Goel, A., and Balasubramanian, D.: *J. Chem. Soc.,A.* 3077 (1971).
7. Gupta, A. and Rao, C.N.R.: *J. Phys. Chem.* 77, 2888 (1973).
8. Perricaudet, M. and Pullman, A.: *FEBS Letters* 34, 222 (1973).
9. Pullman, A. and Armbruster, A.M.: *Chem. Phus. Letters* 36, 558 (1975).
10. Rao, C.N.R. and Mol, J.: *Struc.* 19, 493 (1973) and references cited therein.
11. Rao, C.N.R., Bhujle, V.V., Goel, A., Bhat, U.R., and Paul, A.: *J. Chem. Soc. Chem. Comm.* 161 (1973).
12. Rao, C.N.R., Randhawa, H.S., Reddy, N.V.R. and Chakravorty, D.: *Spectrochim. Acta.* 31A, 1283 (1975).
13. Rao, C.N.R., Rao, K.G., and Reddy, N.V.R.: *J. Amer. Chem. Soc.* 97, 2918 (1975).
14. Rao, C.N.R., Rao, K.G., and Balasubramanian, D.: *FEBS Letters* 46, 192 (1974).
15. Benderly, H and Rosenheck, K.: *J. Chem. Soc. Chem. Comm.* 179 (1972).
16. Liler, M.: *J. Chem. Soc. D., Chem. Comm.* 527 (1972).

ALKALI ION BINDING TO POLYPEPTIDES AND POLYAMIDES

'D. BALASUBRAMANIAN and B.C. MISRA
*Dept. of Chemistry, Indian Institute of Technology, Kanpur
208016, India*

ABSTRACT. Lithium ion binding to two polymers, poly vinyl pyrrolidone
(PVP) and poly α,β-hydroxyethyl dl-aspartamide (PHEA) has been studied
using equilibrium dialysis and hydrodynamic methods. Preferential binding of Li^+ ions to these polymers in aqueous solutions has been established. Upon cation binding, these two neutral polymers expand in a
pseudopolyelectrolyte fashion. These results offer support to the predictions made from studies of cation binding to model monomeric amides.

1. INTRODUCTION

An area where the interaction of group IA and IIA salts with the amide
group is of relevance is the study of the conformation of proteins and
polypeptides. Polyamides such as of the Nylon type have been shown [26]
to interact with lithium salts and as a consequence the melting temperature of the polyamide is altered. The addition of li^+ and Ca^{+2} to synthetic poly α-amino acids such as poly L-glutamic acid [5, 25], poly
L-lysine [25], poly L-serine [16], and poly L-proline [10] alters their
spectroscopic properties in such a way as to suggest cation binding to
the peptide groups. Even poly ω-peptides, e.g., the optically active
polyamides studied by Overberger et al. [15] are influenced by LiBr.
Lotan [9] has recently shown that polyhydroxyalkyl glutamines (PHEG,
PHBG) bind Li^+ strongly and upon such binding, exhibit pseudopoluelec-
trolyte behaviour. And finally, proteins such as Bovine Serum Albumin
[13, 23], and Collagen and elastin [22] have been shown to bind Li^+
and Ca^{2+} efficiently. The effect of salt binding in these cases is to
alter the conformation of the polypeptides or the proteins. In several,
if not all, of the cases cited above, there appears the suspicion of
specific ion effects on the properties of the polymer. It is in this
perspective that we have studied the binding of alkali cations to two
polymers: the polypeptide poly α,β-hydroxyethyl dl-aspartamide (PHEA),
and the vinyl polymer polyvinyl pyrrolidone (PVP) where each monomer
carries an amide moiety (pyrrolidone group) in the side chain.

*B. Pullman and N. Goldblum (eds.), Metal-Ligand Interactions in Organic
Chemistry and Biochemistry, part 1, 159-169. All Rights Reserved.*

2. MODEL AMIDE STUDIES

There has been considerable interest in recent years on the problem of
the interaction of group IA and IIA cations with model monomeric amides.
Semiempirical calculations and spectroscopic studies (1, 17) on cation
binding to model amides suggest the following mode of interaction:

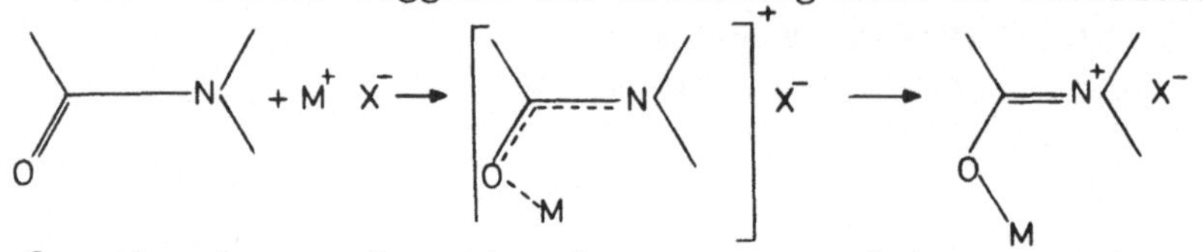

Evidence for the above scheme has been presented in several papers from
our own group [14, 18] and those of others [19]. For a given cation,
the strength of interaction with ligands varies in the order amide
>ester>water>ether, while with a given ligand the cation sequence is
as $Li^+ \approx Ca^{2+} > Mg^{2+} > Na^+ > K^+$. The effect of such a binding is to alter the
electronic distribution in the ligand atoms and consequently to alter
the spectroscopic and conformational properties of the ligand. We have
shown in the case of amides that alkali (and alkaline earth) cation
binding leads to an increase in the C-N bond order at the expense of
the C-O bond order, and a change in the electronic, vibrational, nuclear
magnetic resonance, and circular dichroic spectra of the ligand amides
[2, 18]. In the alkali series, Li^+ interacts strongest with amides,
while it is Ca^{2+} in the group IIA. Recently, Rode and Fussenegger [19]
have shown that Be^{2+} interacts with dimethyl formamide even stronger
than Li^+, but its solubility is low. It is for these reasons that we
have chosen to work wit Li^+ as the probe cation for binding studies
with polyamides.

3. STUDIES WITH TWO TEST POLYMERS

3.1. *Polyvinyl pyrrolidone (PVP)*

We have chosen, as a severe test of the model, to work with polyvinyl
pyrrolidone (PVP), a polymer that has the structure:

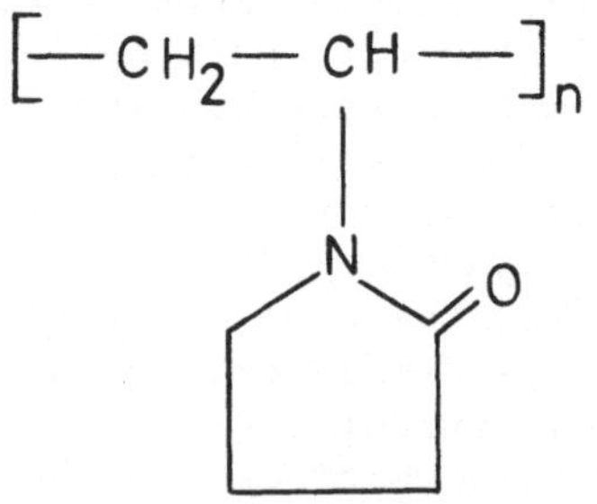

(PVP)

The only groups capable of interacting with added ions are the sidechain
pyrrolidone amide functions. PVP is a water soluble polymer with a
randomly coiled conformation in solution. Its chain conformation and

binding properties (to mainly nonpolar compounds) have been studied
in detail by Frank and coworkers [8, 11]. Hydrodynamic studies indicate
that binding of Na carboxylate salts to PVP occurs and causes a slight
expansion of the polymer coil as monitored by measurements of the radius
of gyration of polymer. We have studied the interaction of LiCl with
PVP in aqueous and in methanolic solutions by equilibrium dialysis and
viscosity experiments.

3.2. *Equilibrium dialysis:*

This technique has been used by Noelken and Timasheff [13,14] successfully to study the preferential binding of solvent components to a protein dissolved in a mixed solvent, e.g. aqueous salt solutions. The
metod involves dialysing a polymer solution against the chosen mixed
solvent across a membrane impermeable to the polymer, in *a closed system.*
When osmotic equilibrium is attained, the solvent components (water and
salt) are redistributed across the membrane. The component that interacts stronger with the polymer will be found to be at a slightly larger
concentration in the vicinity of the polymer (bound to the polymer),
than in the other compartment across the membrane. This preferential
concentration difference may be readily monitored by several methods.
We have chosen to use the technique of differential refractometry to
monitor preferential binding of LiCl to PVP dissolved in aqueous LiCl
solutions.
    If components 1, 2 and 3 designate water, PVP, and LiCl respective-
ly, then preferential interaction of LiCl with PVP can be expressed as

$$\left[\frac{\partial C_3}{\partial C_2}\right]^0_{T,\mu,\mu_3} = \frac{\left[\frac{\partial n}{\partial C_2}\right]^0_{T,\mu_1,\mu_3} - \left[\frac{\partial n}{\partial C_2}\right]^0_{T,P,C_3}}{\left[\frac{\partial n}{\partial C_3}\right]^0_{T,P,C_2}}$$

at a given temperature T and pressure P. μ_i is the chemical potential
and C_i the concentration (molarity) of component i, n the refractive
index, and the superscript 0 refers to infinite dilution. The refrac-
tive increment $(\partial n/\partial C_2)^0_{T,P,C_3}$ is measured by comparing PVP solution in
a given molarity of LiCl with the same molarity of salt as reference in
a differential refractometer. This measurement is done before dialysis.
$(\partial n/\partial C_2)^0_{T,\mu_1,\mu_3}$ is the refractive increment obtained for PVP dissolved
a given molarity of aqueous LiCl compared to its dialysate after equi-
librium is achieved (usually 48 h).
$\left[\frac{\partial C_3}{\partial C_2}\right]^0_{T,\mu_1,\mu_3}$ >0 represent preferential salt binding to PVP while a negative
value indicates preferential hydration. In order to eliminate the polymer

volume, one can write the corrected binding parameter as

$$\left[\frac{\partial C_3}{\partial C_2}\right]^{corr.}_{T,\mu_1,\mu_3} = \left[\frac{\partial C_3}{\partial C_2}\right]^{0}_{T,\mu_1,\mu_3} + C_3\bar{v}_2$$

where $\bar{v}$ is the partial specific volume of PVP. Or alternatively

$$\left[\frac{\partial g_3}{\partial g_2}\right]^{0}_{T,\mu_1,\mu_3} = \frac{g_3}{\bar{v},C_3}\left\{\left[\frac{\partial C_3}{\partial C_2}\right]^{0}_{T,\mu_1,\mu_3} + C_3\bar{v}_2\right\}$$

where g is the concentration in g/g water and $\bar{v}$ the partial specific volume. And finally, the number of moles of salt preferentially bound per mole of PVP can be written as the binding parameter.

$$\left[\frac{\partial m_3}{\partial m_3}\right]_{T,\mu_1,\mu_3} = \left[\frac{\partial g_3}{\partial g_2}\right]^{0}_{T,\mu_1,\mu_3} \frac{\text{Mol.wt. of polymer}}{\text{Mol.wt. of salt}}$$

The experiment involves equilibrating a solution PVP (concentration C_2) in aqueous LiCl (concentration C_3) in a closed system equilibrium dialysis assembly using a dialysis membrane, and measuring the refractive increments of the solution before and after dialysis. The preferential binding data are expressed in terms of either

$$\left[\frac{\partial C_3}{\partial C_2}\right]^{corr.}_{T,\mu_1,\mu_3} \quad \text{or} \quad \left[\frac{\partial m_3}{\partial m_2}\right]_{T,\mu_1,\mu_3}$$

for various molarities of salt.

The results are shown in Table I. It is clear that at higher molarities of LiCl, i.e., 4 M and 6 M LiCl, preferential salt binding occurs. The binding parameters $(\partial m_3/\partial m_2)$ values are 113 and 192 at 4 M and

TABLE I
Preferential binding parameters of polyvinyl pyrrolidone (PVP) to LiCl

Salt concentration	$\left[\frac{\partial n}{\partial C_2}\right]_{T,\mu_1,\mu_3}$ $\times 10^{-1}$	$\left[\frac{\partial n}{\partial C_2}\right]^{0}_{T,P,C_3}$ $\times 10^{-1}$	$\left[\frac{\partial C_3}{\partial C_2}\right]^{0}_{T,\mu_1,\mu_3}$	$\left[\frac{\partial C_3}{\partial C_2}\right]^{corr.}_{T,\mu_1,\mu_3}$	$\left[\frac{\partial g_3}{\partial g_2}\right]^{0}_{T,\mu_1,\mu_3}$	$\left[\frac{\partial m_3}{\partial m_2}\right]_{T,\mu_1,\mu_3}$
2 M	1.20	1.66	−0.197	−0.135	−0.134	−127
4 M	1.30	1.31	−0.004	+0.120	+0.119	+113
6 M	1.21	1.24	0.017	+0.204	+0.203	+192

6 M LiCl respectively. The variation of the binding parameter with salt
molarity is illustrated in Figure 1. The molecular weight of PVP (a
Sigma chemicals product) was found by light scattering to be 44 000,
which means a degree of polymerization of around 350. The stoichiometry
Li^+: monomer is thus 192:350 or roughly 1:1.8, which is in excess of the
1:4 ratio seen for Li^+:amide complexes in anhydrous media, and emphasizes
the role of the solvent water. It is also noteworthy that at low molarity
(2 M LiCl), there is preferential hydration. Of relevance to this are
the earlier calorimetric studies on LiCl binding to aqueous DMF where
we had noticed a sigmoidal increase in the heat of mixing near 4 M LiCl
[4].

EQUILIBRIUM DIALYSIS OF PVP in LiCl/H_2O

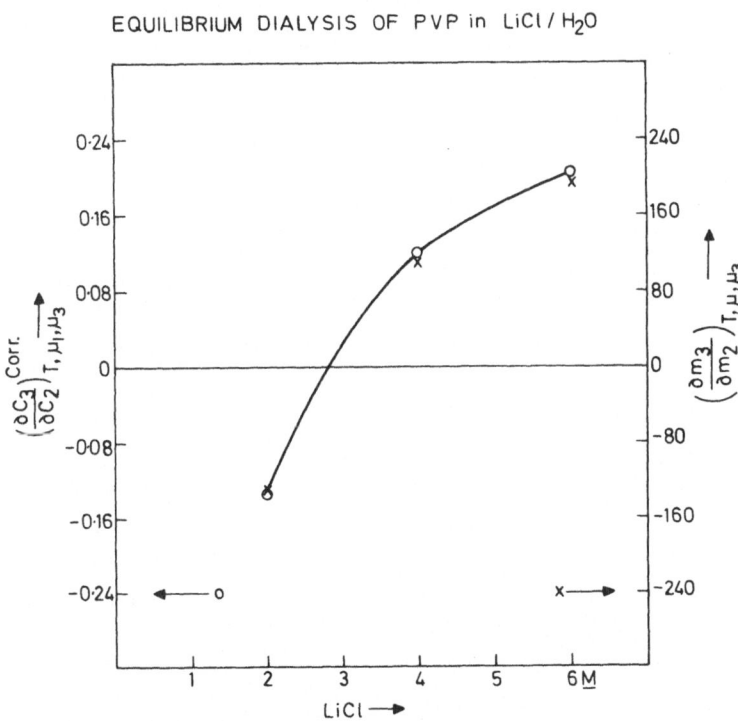

Fig. 1. Preferential salt binding to PVP in aqueous LiCl
solutions, at various molarities of salt. Negative values
denote preferential hydration.

Hydrodynamic studies on PVP have been done by Scholtan [21] and by
Frank et al. [8, 11]. In aqueous solutions, and in methanol, PVP adopts
a randomly coiled conformation. We undertook a study of the hydrodynamic properties of PVP in high salt media, but light scattering studies
had to be abandoned at high LiCl concentrations because of turbidity
problems caused by a slight opalescence. Hence we resorted to viscosity
measurements.

The Huggins plot, i.e., a plot of $\eta_{SP/c}$ $(=(\eta-\eta_o)/\eta_o c)$ against c (mg/ml)
of PVP in water and in methanol yielded straight lines and a limiting
viscosity number, $[\eta]$, of 1700 and 1790 ml g^{-1} respectively. Frank's
analysis [8] of the viscosities of PVP samples in methanol have yielded
the parameters K and a of the Mark-Houwink equation:

$$[\eta] = KM^a$$

as 2.3×10^{-1} and 0.65 respectively. Using this, we obtained a value of
40 000 as the viscosity average molecular weight of our sample of PVP
in methanol, in reasonable agreement with the light scattering result.
The Huggins plots of PVP in methanolic LiCl solutions are nonlinear, as
shown in Figure 2. The increase in the value of the reduced specific
viscosity at low concentrations of PVP in salt-containing media is typical of polyelectrolytes [24]. Curves of this type can be fitted by the

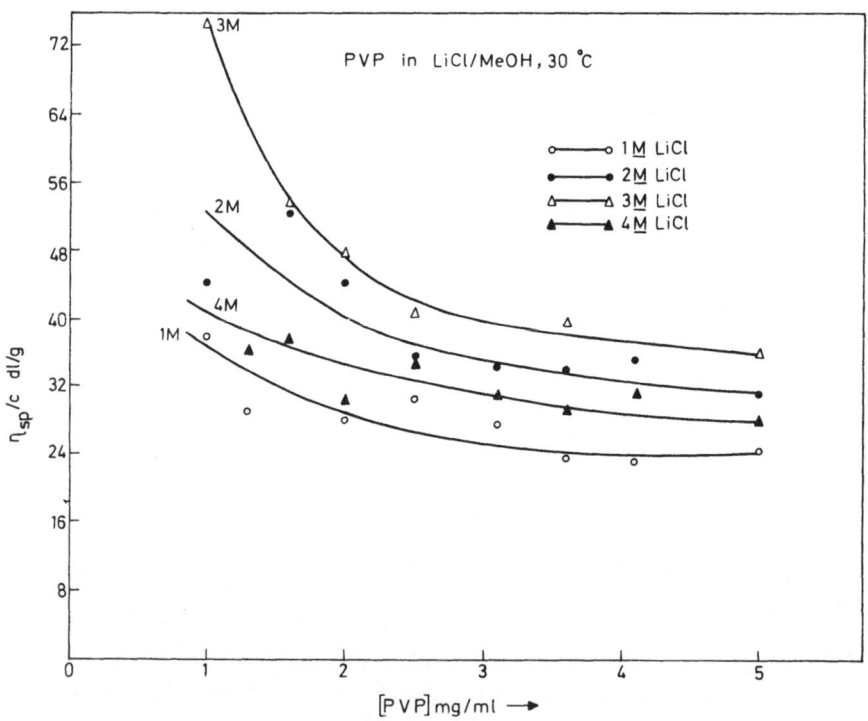

Fig. 2. Huggins plots of the variation of the reduced specific viscosity of PVP in methanolic LiCl solutions.

Fuoss-Strauss equation for polyelectrolytes [6]:

$$\eta_{SP/C} = \frac{A}{1 + B\sqrt{C}}$$

and from a plot using the above equation, one can obtain the [η] values of PVP in methanolic LiCl solutions. According to polymer theory [24], the intrinsic viscosity of a flexible coil polymer chain is related to its hydrodynamic radius R_e and radius of gyration R_G ($R_G = \xi\, R_G$, where $\xi = 0.775$ for a flexible chain) by the equation:

$$[\eta] = \frac{10\,\pi}{3\,M}\, \xi^3 R_G^{\,3}$$

where N is Avogadro number, and M the molecular weight. Using this expression, we have computed the radii of gyration of PVP in several of the media mentioned above. Notice from Table II that the presence of LiCl in the solvent causes an expansion of the PVP chain. Strong binding of Li^+ to the polymer, at the pyrrolidone amide groups, apparently causes inter-segment electrostatic repulsions, making the polymer behave as a polyelectrolyte, as seen from the viscosity curves. Indications of such a chain expansion have come from the early work on sodium carboxylates binding to PVP [11]. Lotan [9] has shown convincingly that such pseudo-polyelectrolyte effects occur in polyhydroxyalkyl glutamines in (LiCl + CH_3OH). In each case, the effect of the salt is to increase the chain dimensions. (In the extreme case of saturation binding, the chain ought to expand to a long rodlike shape; under these circumstances, use of the above equation to calculate R_G will be unjustified since it applies to a coil.)

TABLE II
Chain dimensions of PVP in LiCl/MeOH

Solvent	$[\eta]^{30°}$, ml/g	R_G, Å
CH_3OH	1790	280
CH_3OH + 1 M LiCl	6150	433
CH_3OH + 2 M LiCl	6900	450
CH_3OH + 3 M LiCl	8300	480
CH_3OH + 4 M LiCl	9100	494

3.2. *Polyhydroxyethyl DL-aspartamide (PHEA)*

This synthetic polypeptide is an analogue of the Lotan polymers, and has both α - and β-linked peptide units as shown below:

α, β-Poly(2-hydroxyethyl) DL-aspartamide
(PHEA)

Interest in this compound, first synthesized by Neri et al. [12], has come from the fact that it contains both α, and β peptide bonds, it is water soluble, and is used satisfactorily as a blood plasma volume expander. We have studied the cation binding properties of this polypeptide by equilibrium dialysis and viscosity methods, and find it to be an efficient Li$^+$ binder.

TABLE III
Preferential binding parameters for PHEA/LiCl

Salt molarity (M)	$\left[\dfrac{\partial n}{\partial C_2}\right]^0_{T,\mu_1,\mu_3}$ $\times 10^{-1}$	$\left[\dfrac{\partial n}{\partial C_2}\right]^0_{T,P,C_3}$ $\times 10^{-1}$	$\left[\dfrac{\partial C_3}{\partial C_2}\right]^0_{T,\mu_1,\mu_3}$	$\left[\dfrac{\partial C_3}{\partial C_2}\right]^{corr.}_{T,\mu_1,\mu_3}$	$\left[\dfrac{\partial g_3}{\partial g_2}\right]^0_{T,\mu_1,\mu_3}$	$\left[\dfrac{\partial m_3}{\partial m_2}\right]_{T,\mu_1,\mu_3}$
1 M	1.575	1.500	+0.0322	+0.0630	+0.0627	+74
2 M	1.410	1.600	−0.0810	−0.0190	−0.0189	−22
3 M	1.147	1.090	+0.0245	+0.1170	+0.1165	+138
4 M	1.280	1.200	+0.0340	+0.1580	+0.1573	+186
6 M	1.400	1.300	+0.0429	+0.2295	+0.2286	+270

Table III summarizes the results of equilibrium dialysis measurements of PHEA in aqueous LiCl solutions. It may be noticed that binding of Li$^+$ in this case is as efficient as in PVP. But the values of the binding parameter in PHEA are to be viewed in light of the fact that in PHEA, each monomer residue contains two peptide bonds (plus an OH group), one in the backbone and the other in the sidechain. The molecular weight of the PHEA sample used here (a kind gift of Dr Neri) is reported to be 50 000, or a degree of polymerization of 320. At 6 M LiCl 270 molecules of LiCl are bound per molecule (320 monomers and 640 peptide groups) of PHEA, which may be compared to the 192:350 Li$^+$: monomer value in PVP at the same molarity. The concentration dependence of the binding of Li$^+$ to PHEA is shown in Figure 3. After an initial drop in binding, prefer-

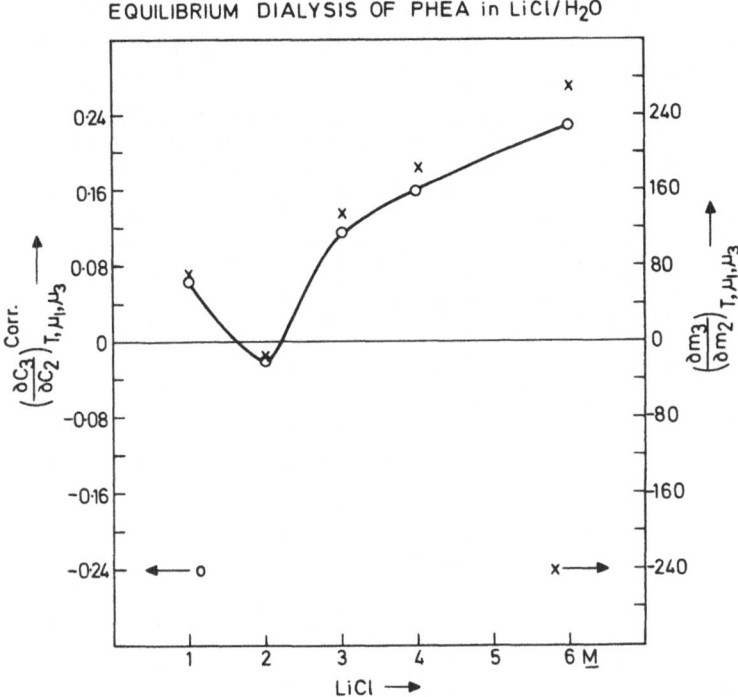

EQUILIBRIUM DIALYSIS OF PHEA in LiCl/H₂O

Fig. 3.   Preferential salt binding to PHEA in aqueous LiCl
solutions, at various molarities of salt. Negative values
denote preferential hydration.

ential binding of LiCl increases steadily and the shape of the curve
resembles the one reported by Noelken for LiCl binding to BSA [13].
    PHEA is virtually insoluble in methanol, but is solubilized in
this solvent upon the addition of LiCl. In Figure 4 we show the Huggins
plot of PHEA dissolved in methanolic LiCl, where again one notices the
induction of pseudopolyelectrolyte behaviour on Li^+ binding to the neutral polymer. More intensive hydrodynamic studies, currently in progress,
are expected to yield information about the details of the molecular
shape and chain dimensions of this polypeptide in various media.
    It is evident from the studies on PVP and PHEA that cation inter-
action with neutral dipolar sites in polymers such as polyamides occurs,
and causes noticeable changes in the polymer properties. This is a factor to be kept in mind when one interprets spectral and other properties
of polypeptides and proteins in media containing salts such as LiCl or
$CaCl_2$ and so on. These studies also corroborate the expectations of the
model compound study with regard to ligand-cation interactions and the
associated canges in the physical properties of the interacting substrate.

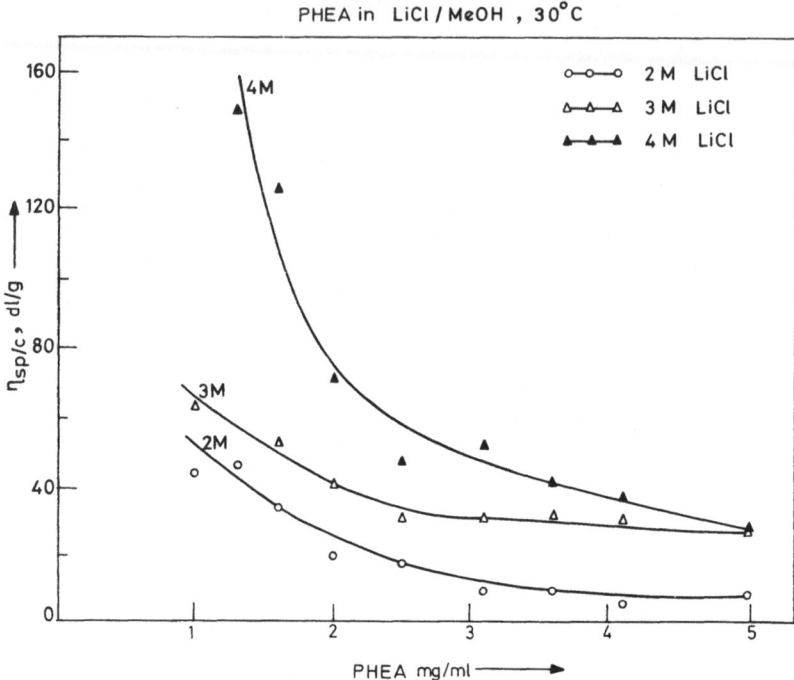

Fig. 4. Huggins plots of the variation of the reduced specific viscosity of PHEA in methanolic LiCl solutions.

4. POSSIBLE MEDICAL RELEVANCE

An interesting outcome of this work with PVP and PHEA has been in the field of blood plasma volume expanders. These expanders are water sol-uble polymers administered to an organism that has lost blood, and help alleviate pathological shock and to restore normal blood level recovery. Commonly used plasma expander substances are [7]: purified plasma frac-tions, hemoglobin, albumin, gelatin, starch, dextran, PVP, or synthetic polypeptides such as PHEA, PHEG, PLGA [12]. The diversity in the chemi-cal structures but the common requirements demanded of these plasma expander substances point out to a physicochemical basis for their ac-tion. We have shown above that PHEA and PVP are both efficient binders of alkali cations, to the extent that Li^+ is a representative example. All the other expander substances mentioned above have also been shown by others to be able to bind alkali cations. We have discussed this point in detail elsewhere [3] and have suggested that efficient alkali cation binding is a feature common to all plasma volume expanders. This feature may be relevant to the physicochemical basis of expander action. Again, the importance of Li^+ in brain chemistry has been highlighted recently [20]. This property of Li^+ to efficiently interact with model amides, polypeptides and some proteins in a feature that may be well worth exploring in this connection.

ACKNOWLEDGEMENTS

We are most grateful to Dr P. Neri of I.S.V.T. Sclavo, Siena, Italy for his gift of a sample of PHEA. This research was supported financially by the Council of Scientific and Industrial Research of India.

REFERENCES

1. Balasubramanian, D., Goel, A., and Rao, C.N.R.: Chem. Phys. Letters 17, 482 (1972).
2. Balasubramanian, D. and Misra, B.C.: Biopolymers 14, 1019 (1975).
3. Balasubramanian, D. and Misra, B.C.: 1976, manuscript submitted.
4. Balasubramanian, D. and Shaikh, R.: Biopolymers 12, 1639 (1973).
5. Discherl, W. and Dohr, H.: Hoppe-Seyler's Z. Physiol. Chem. 355, 1135, (1974).
6. Fuoss, R.M. and Strauss, U.P.: J. Poly. Sci. 3, 246, 602 (1948).
7. Gruber, U.F.: Blood Replacement, Springer-Verlag, Berlin (1969).
8. Levy, G.B. and Frank, H.P.: J. Poly. Sci. 17, 247 (1955).
9. Lotan, N.: J. Phys. Chem. 77, 242; also in E.R. Blout et al. (eds.), Peptides, Polypeptides & Proteins, Wiley-Interscience, New York, 1974, p. 157
10. Mattice, W.L. and Mandelkern, L.: Biochemistry 8, 1049 (1969).
11. Molyneux, P. and Frank, H.P.: J. Amer. Chem. Soc. 83, 3169, 3175 (1961).
12. Neri, P., Antoni, G., Benevenuti, F., Cocola, F., and Gazzei, G.: J. Med. Chem. 16, 893 (1973).
13. Noelken, M.E.: Biochemistry, 9, 4122 (1970).
14. Noelken, M.E. and Timasheff, S.N.: J. Biol. Chem. 242, 5080 (1969).
15. Overberger, C.G., Montaudo, G., Nishimura, Y., Sebenda, J., and Veneski, R.A.: J. Poly. Sci., B. 7, 219 (1969); also Overberger, C.G. and Shimokawa, Y.: ibid. 9, 161 (1971).
16. Quadrifoglio, F. and Urry, D.W.: J. Amer. Chem. Soc. 90, 2760 (1968).
17. Rao, C.N.R.: Paper presented in this symposium (1976).
18. Rao, C.N.R., Rao, K.G., and Balasubramanian, D.: F.E.B.S. Letters 46, 192 (1974).
19. See, e.g., Rode, B. and Fussenegger, R.: J. Chem. Soc. Faraday Trans. II 71, 1958. (1975).
20. Samuel, D. and Gottesfeld, Z.: Endeavour 32, 122 (1973).
21. Scholtan, W.: Makromol. Chem. 7, 209 (1952).
22. Starcher, B.C. and Urry, D.W.: Bioinorg. Chem. 3, 107 (1974); also Urry, D.W.: Proc. Natl. Acad. Sci. U.S. 68, 810 (1971).
23. Sun, S.F., Del Rosario, N.O., and Goldstein, L.A.: Int. J. Peptide Protein Res. 5, 337 (1973).
24. Tanford, C.: Physical Chemistry of Macromolecules, John Wiley, New York (1961).
25. Tiffany, M.L. and Krimm, S.: Biopolymers 8, 347 (1969).
26. Valenti, B., Bianchi, E., Greppi, G., Tealdi, A., and Ciferri, A.: J. Phys. Chem. 77, 389 (1973).

INFRARED AND RAMAN STUDIES OF INTERACTIONS BETWEEN SALTS AND AMIDES OR ESTERS

M.H. BARON, H. JAESCHKE, R.M. MORAVIE, C. DE LOZÉ,
and J. CORSET
Laboratoire de Spectroscopie Infrarouge et Raman C.N.R.S.
2 Rue Henri Dunant, 94320 – Thiais – France

The surrounding medium may play an important role in the activity of biological macromolecules. The action of salts, among others, may modify the macromolecular conformation, such as for gelatin, lithium bromide induced mutarotation of polyproline (Ramachandran, 1967; Von Hippel, 1969), or conformational changes of polyglycine (Baron, 1973). The ion transfer through membranes takes place through antibiotic-salts interactions (Mumoz et al., 1972; Ovchinnikov and Ivanov, 1975). These biological molecules may be either polypeptide chains, as in proteins, or chains with alternate ester and peptide groups, as in depsipeptides, or cyclic ethers and methyl esters. We therefore thought it of interest to start an investigation, through infrared and Raman spectroscopy, of the interactions between some esters and amide molecules and alkaline Earths or alkaline salts. Infrared and Raman spectroscopy are well adapted for studying short life interactions, and Raman laser spectroscopy makes it possible to examine aqueous solutions. On certain points the present work cross-checks the results obtained from quantum chemistry calculations (Pullman, 1974; Perricaudet, 1973). We shall first show how the organization of carbonyl groups around the cations may be determined through infrared and Raman spectroscopy. We shall then discuss the nature of interactions between cations and these molecules.

1. COMPLEXATION SITE AND STRUCTURE OF THE SOLVATION SHELL

Upon solution of a salt in an organic solvent $\underline{S}$ the M^{z+} cations of this salt are usually solvated; they form $M^{z+}S_n$ species when these salts are dissociated, which is generally the case for perchlorates in dilute solution. Through examining the vibration spectra of the S solvent molecules located in the cation first solvation shell, information can be gained on the interaction sites of the solvent molecule with ions, on the nature of the binding forces and on the organisation of the solvent molecules in this first solvation shell. This last point may often be reached only for relatively concentrated solutions, so that the spectrum of the cation perturbed solvent molecules may be distinguished from that of other solvent molecules. Moreover, as we shall see, the spectrum of the molecules in the first solvation shell may be perturbed through couplings between the vibrations of neighbour molecules, as was observed in organic crystals of molecules with strongly polar vibration modes (Hexter, 1960).

1.1. *Complexation Site*

An easy way of examining interactions between a cation and the molecules of an S_2 solvent is to dissolve a salt of this cation in a diluting S_1 solvent, less basic than S_2 and to use the preferential solvation of the cation by S_2. We were thus able (Regis A and Corset, 1973; Baron and de Lozé C., 1972; Moravie, 1975) to specify the nature of the interactions occurring between alkaline and alkaline Earth cations and basic molecules by dissolving these molecules and the perchlorates of these cations in moderately basic solvents such as nitromethane, acetonnitrile or tetrahydrofurane.

To illustrate this point Figure 1 presents the infrared and Raman spectra of the νCO vibration of N,N-dimethylacetamide (DMA) in acetonitril solution in the absence of salts and in the presence of barium perchlorate For amide concentrations of 0.5 M and 1 M, the infrared and the Raman bands have the same frequency: this shows the disappearance of the dipolar interactions which exist in the pure liquid solvent and are characterized by an infrared band at 1648 cm^{-1} and a Raman band at 1631 cm^{-1} (Garrigou-

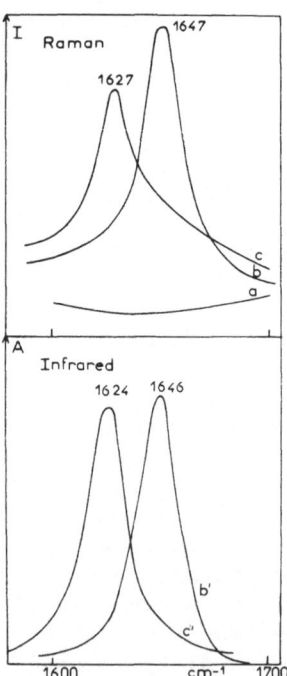

Fig. 1.   Infrared and Raman spectra of N,N-dimethylacetamide dituted in CH$_3$CN with or without salt added - a, pure CH$_3$CN ; b and b', DMA 1M ; c and c', DMA 0.5 M, [Ba(ClO$_4$)$_2$] = 0.25 M.

Lagrange et al. 1970). In spite of the high amide concentration[*] the addition of barium perchlorate with an amide/salt ratios of 2 yields a new νCO band, which is by about 20 cm^{-1} lower. Although an average of two amide molecules per Ba^{2+} cation are complexed, the infrared frequency is very close to the Raman one. Besides the infrared spectrum of this solution is very similar to that of much more diluted solution where the band is at 1624 cm^{-1} and must be assigned to the $[Ba-DMA-(CH_3CN)_n]^{2+}$ species. This band is therefore related to the νCO vibration of an amide molecule interacting with a Ba^{2+} cation through the carbonyl oxygen.

Depending on the S_1 diluting solvent transparency, various vibrations of the S_2 molecule may be observed. For DMA as for the esters studied (methyl acetate, propionate, pivalate and isobutyrate) the interactions between the alkaline or alkaline Earth cations and the carbonyl oxygen have been demonstrated through analyzing the frequency perturbations and comparing them with those of analogous molecules. Such conclusions have been reached through comparing N,N-dimethylacetamide and N-methylacetamide (Baron and de Lozé, 1972) and through comparing the esters with ethers or acetone which have only one interaction site for cations (Moravie, 1975). For instance the complexation of Mg^{2+} on the oxygen of methyl ether (Kress and Guillermet, 1973) decreases the frequency of the two C—O bands; in the case of Ba^{2+} interaction with methylisobutyrate (Table I) the C—O bond in α to the carbonyl has an increased

TABLE I
Infrared frequencies of some vibrations of methylisobutyrate dissolved in acetonitrile with or without salt added

Vibrations	Ba(ClO$_4$)$_2$			LiClO$_4$		NaClO$_4$	
	ν_F	ν_a	$\Delta\nu$	ν_a	$\Delta\nu$	ν_a	$\Delta\nu$
ν (C$_1$=O)	1735	1704	−31	1719	−16	1722	−13
ν (C$_1$—O)	1207	1232	+25	1231	+24	1222	+15
ν (C$_0$—O)	993	982	−11				
ν (C$_2$—C$_1$)	829	839	+10				

frequency, while the O—CH$_3$ bond is lowered in frequency, which shows a complexation on the carbonyl oxygen: the double bond character of the C—O bond in α to the carbonyl is reinforced upon complexation of this oxygen.

[*] The amide concentration is chosen high so that the same mixtures may be examined in infrared and in Raman.

1.2. *Structure of the First Solvation Shell of the Ions*

In order to observe the frequency perturbations of all the vibrations of the solvent molecule interacting with the cations, it is necessary to examine the spectra of binary mixtures with relatively high salt concentrations. The precise analysis of the spectrum makes it necessary to know the composition of the solvation shell for the considered cation.

1.2.1. Composition of the cations solvation shell

Lithium and barium perchlorates, in which we are mainly interested, are very soluble in amides and esters and lead to liquid or crystalline solvates. In methyl acetate, propionate and isobutyrate it is possible to dissolve one mole of $Ba(ClO_4)_2$ for six moles of ester or two moles of $Li(ClO_4)$ for three moles of ester. Barium perchlorate is poorly soluble in methyl pivalate but a needle-shaped solvate crystallizes from saturated solutions with the stoichiometry $Ba(ClO_4)_2 \cdot 4(CH_3)_3CCOOCH_3$. Liquid mixtures of N,N-dimethylacetamide and $Ba(ClO_4)_2$ may be prepared with composition up to $Ba(ClO_4)_2 \cdot 6$ DMA, at this stoichiometry the mixture crystallizes. Similarly the dehydratation of $Cu(ClO_4)_2$–DMA mixtures lead to isolating a cryslalline solvate $Cu(ClO_4)_2 \cdot 4$ DMA.

For liquid mixtures, the number of solvent molecules surrounding the cations may be measured through infrared spectroscopy. If D_0 is the optical density of the band for the pure solvent, and D the optical density for the solvent bound to the cation, $(D_0-D)/D_0 = n\rho$ where ρ is the ratio of the number of salt moles over the number of solvent moles in the solution (Roche, 1971). The number of solvent moles must be high enough to consider the salt as completely dissociated. In Figure 2 are

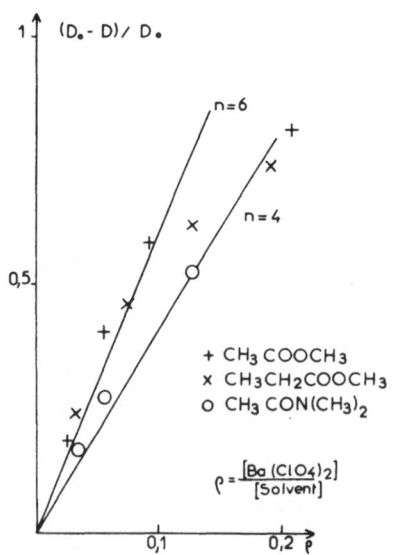

Fig. 2. Determination of the solvation number of Ba^{++} from ATR spectra of the $\nu(C=O)$ vibration band.

presented the points corresponding to solutions of barium perchlorate
in methyl acetate and isobutyrate and in DMA. For the esters, the points
corresponding to dilute solutions fall close to the straight line with
slope n = 6. The points corresponding to DMA are located between the
lines with slopes n = 4 and n = 6, which will lead us into considering
both hypothesis n = 4 and n = 6 to interpret the spectra.

1.2.2. Structure of the cations first solvation shells

A detailed study of ester-salts mixtures has shown that the comparison
of infrared and Raman spectra gives information on the organisation of
the carbonyl groups in the solvation shell of the considered cations
(Moravie and Corset, 1976; Moravie, 1975). Through these results we were
led into revising some interpretations of the spectra of amides in interaction with cations (Baron and de Lozé, 1972; Baron et al., 1972).
Figure 7 and Table II summarize our results and the method will be illustrated by example concerning DMA. On Figure 3 are presented the infrared
and Raman Spectra of the liquid and crystalline solvate Li(ClO$_4$).4 DMA.

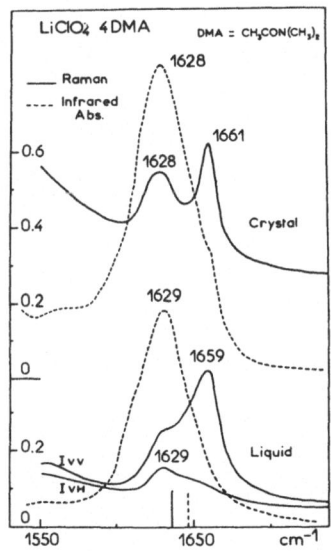

Fig. 3. Infrared (-----) and Raman (————) spectra of LiClO$_4$.4
DMA liquid and crystal.

The spectra of the solid and the liquid include two bands in Raman and
only one band in infrared. Raman polarization (I_{VV} and I_{VH} spectra) has
been performed on the molten solid: the Raman band corresponding to the
one observed in infrared is depolarized, while the high frequency band
is polarized. The presence of only one band in infrared, shifted from
the band of the non-complexed amide (represented by the vertical lines

TABLE II

Different types of environment of Li$^+$, Ba^{++} and Cu^{++} cations in mixtures of perchlorates with N-N-DMA and esters

Compound and phys. state	Sym. a)	Obs. frequencies cm^{-1}		ν_0 frequency cm^{-1} obs.	calc.		Sym.ClO$_4^-$ b)
LiClO$_4$.4 (CH$_3$)CON(CH$_3$)$_2$ solid	T$_d$	ν_1 (A$_1$)	1661			**D. (ion-dipole)Å c)** calc. $(r_c + r_w + \frac{r(C=0)}{2})$	T$_d$ ou C$_{3v}$
		ν_2 (F$_2$)	1628	1635	1635,5	3.0 2.68	
2 LiClO$_4$.3 CH$_3$COOCH$_3$ liquid	D$_{3h}$ or D$_3$	ν_1 (A$_1$)	1739			**d (dipole-dipole) Å** calc.d) $(2r_w + r(C=0)\sqrt{3})$ e)	C$_{3v}$
		ν_2 (E')	1710	1720		3.74	
2 LiClO$_4$.3 (CH$_3$)$_2$CHCOOCH$_3$ liquid	D$_{3h}$ or D$_3$	ν_1 (A'$_1$)	1731			3.84	C$_{3v}$
		ν_2 (E')	1707	1719	1715	3.90	
Ba(ClO$_4$)$_2$.6 CH$_3$CON(CH$_3$)$_2$ solid	C$_i$	A$_g$	1661 1628			**D (ion-dipole) Å c)** calc. $(r_c + r_w + \frac{r(C=0)}{2})$	≤C$_{3v}$
		A$_u$	1646 1629 1602	1624	(1632)		
liquid	O$_h$	ν_1 (A$_{1g}$)	1657			3.3 3.34	T$_d$
		ν_2 (F$_{1u}$)	1627	1624	1627		
		ν_3 (E^1_g)	1618				

Salt (state)	Symmetry[a]	Mode	Observed freq.					ClO$_4^-$ symmetry
Ba(ClO$_4$)$_2$·4 CH$_3$CON(CH$_3$)$_2$ liquid	D$_{4h}$	ν_1 (A$_{1g}$) 1657 ν_2 (E$_u$) 1627 ν_3 (B$_{1g}$) 1618	1624	1631	3.0	3.34		T$_d$
Ba(ClO$_4$)$_2$·6 CH$_3$CH$_2$COOCH$_3$ liquid	O$_h$	ν_1 (A$_{1g}$) 1721 ν_2 (F$_{1g}$) 1702 ν_3 (E$_u$) 1698	1704	1704	3.4	3.34		≤ C$_{2v}$
Ba(ClO$_4$)$_2$·6 (CH$_3$)$_2$CHCOOCH$_3$ liquid	O$_h$	ν_1 (A$_g$) 1728 ν_2 (F$_{1g}$) 1706 ν_3 (E$_u$) 1702	1708	1704	3.2	3.34		≤ C$_{2v}$
Ba(ClO$_4$)$_2$·4 (CH$_3$)$_3$CCOOCH$_3$ solid	C$_i$	A$_g$ 1722 A$_g$ 1700 A$_u$ 1708 A$_u$ 1698	1706.5 1698					≤ C$_s$
Cu(ClO$_4$)$_2$·4 CH$_3$CON(CH$_3$)$_2$ solid	D$_{4h}$	ν_1 (A$_{1g}$) 1628 ν_2 (E$_u$) 1602.5 ν_3 (B$_{1g}$) 1594	1607		3.2	2.72		≪ C$_{2v}$

a) Symmetry : the arrangement of the carbonyl groups around the cations.

b) $\nu_0 = (f_{rr}\,g)^{1/2}$ calculated through combinations of observed frequencies (Table III); frequencies between brackets are average frequencies corresponding to different f force constant of the carbonyl group.

c) Ion-dipole distance calculated through observed frequencies and through f_{rr}, f'_{rr} expressions related to dipole-dipole interactions.

d) Dipole-dipole distance calculated through observed frequencies and through f_{rr} expressions related to dipole-dipole interaction.

e) Dipole-dipole estimated assuming : (a) O-O distance = 2 r_w (r_w = Van der Waals radius) ; (b) the carbonal dipole is located in the middle of the C=O band.

f) ClO$_4^-$ ion symmetry according to the observed frequencies splittings of its fundamental frequencies.

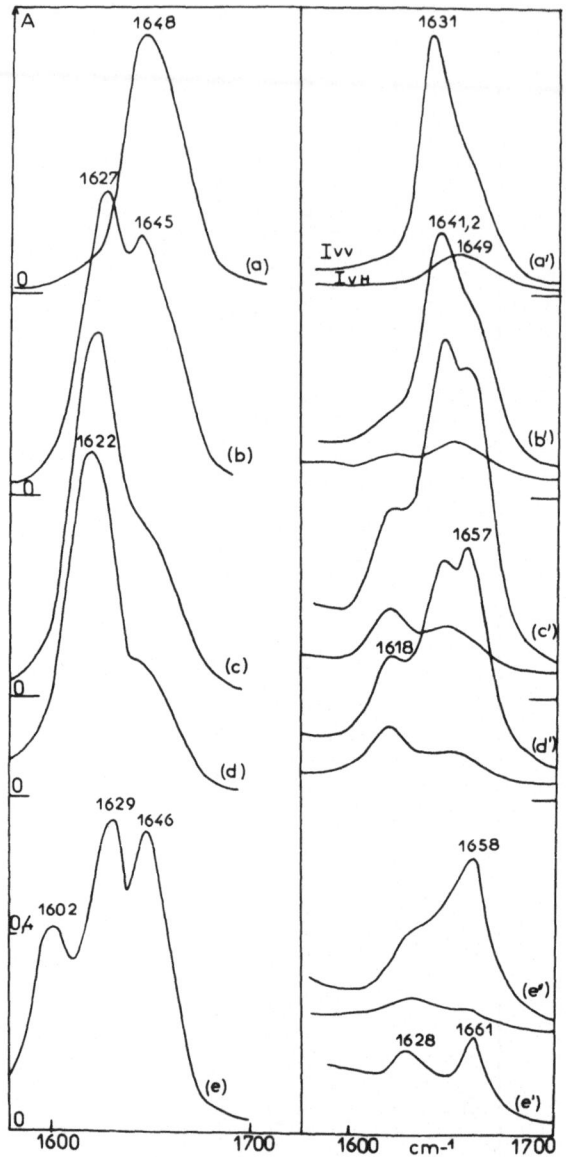

Fig. 4. Infrared and Raman spectra in the ν(C=O) region of
pure DMA (a, a') and of liquid or solid mixtures with
Ba(ClO₄)₂ in the ν(C=O) region. a, a' to d, d' and e" liquid,
e, e' solid [DMA]/[Ba(ClO₄)₂] : b, b' 18.6; c, c' 7.9; d, d'
6.7; e, e', e" 6.

at the bottom of the figure) shows that the whole amide is complexed
with the Li⁺ cation; this cation is therefore surrounded by four CO
groups. If these groups were not coupled only one band would be observed

at the same frequency in infrared and Raman. The presence of couplings between the $\nu(CO)$ vibrations, indicated by two Raman bands 30 cm^{-1} apart, is not surprising in view of the strong polarity of the carbonyl groups. Through the spectrum it is possible to specify the arrangement of the CO groups around the Li$^+$ cation, which has the T$_d$ symmetry of the tetrahedron (Figure 7). The band at 1628 cm^{-1} corresponds to the triply degenerate F$_2$ vibration and the polarized Raman band at 1661 cm^{-1} to the A$_1$ vibration of this tetrahedron.

In Figure 4 are presented the Raman and ATR infrared spectra of Ba(ClO$_4$)$_2$ solutions in DMA and of the solid compound Ba(ClO$_4$)$_2$.6 DMA. The addition of salt in DMA yields an intense infrared band at 1627 cm^{-1} and two Raman bands at 1618 and 1657 cm^{-1}; their intensities increase with salt concentration at the expense of the absorption due to the uncomplexed amide. For the solutions with a high salt concentration the band at 1618 cm^{-1} is depolarized while the more intense band at 1657 cm^{-1} is polarized. The frequency of the infrared band shifts from 1627 to 1623 cm^{-1}, probably due to medium effect when the salt concentration increases. To sum up the spectrum of the DMA carbonyl groups in the first solvation shell of the Ba^{2+} cations is made of two Raman bands (1618, 1657 cm^{-1}) and one infrared band (1623 − 1627 cm^{-1}). As we have seen above, the experimental incertitude lead to a solvation number which can be n = 4 or n = 6*. In the hypothesis n = 4 the spectrum may be interpreted by an environment with the D$_{4h}$ symmetry, where the four CO groups are oriented around the cation along the diagonals of a square plane (Table II and Figure 7). In the hypothesis n = 6 the spectrum is consistent with an octahedral environment of the Ba^{2+} cation such as can be seen (Figure 7) for the solvate Ba(ClO$_4$)$_2$.6 (CH$_3$)$_2$CHCCOOCH$_3$. For a spectrum including several components active either in infrared or in Raman, the normal mode vibrations may be calculated through the Wilson method (Wilson et al., 1955) for the various arrangements of the carbonyl groups with an f$_r$ force constant, and leads to the equations given (Table III). The f$_{rr}$ and f'$_{rr}$ force constants are respectively the interactions constants between neighbour and opposite carbonyl groups. As we shall see later, a dipolar interaction model leads to a relationship between f$_{rr}$ and f'$_{rr}$, which makes it possible to calculate the ratio $(\nu_2{}^2 - \nu_3{}^2)/(\nu_1{}^2 - \nu_2{}^2)$. The ratio is equal to 0.21 for the octahedron (O$_h$) and to 0.36 for the square plane (D$_{4h}$). The experimental value found for the Ba^{2+} cation surrounded by six ester molecules is 0.21, which confirms the octahedral structure. For the DMA, for which the ν_2 frequency varies from 1623 to 1627 cm^{-1}, this ratio is found between 0.15 and 0.30. This is in favor of the hypothesis n = 6, which besides corresponds to the more frequent solvation number of the Ba^{2+} cation. Through canceling the interaction constants between the normal mode

---

* The solvation number n = 5 is not frequent, it would moreover lead, even in the most symmetrical surrounding of the tetragonal pyramids (C$_{4v}$) to three $\nu(CO)$ vibrations modes, two of which (A$_1$ and E) are infrared and Raman active and one (B$_1$) Raman active. This contradicts the experimental data, where only one infrared and two Raman bands are observed.

TABLE III
Expression for the carbonyl vibration frequencies in some M^{Z+} (C=O)$_n$
models.

n	Sym.		λ	Activity	Relation between frequencies
4	T_d	A_1	$\lambda_1 = (f_r + 3 f_{rr})g$	R.	$\nu_0^2 = \frac{1}{4} (\nu_1^2 + 3 \nu_2^2)$
		F_2	$\lambda_2 = (f_r - f_{rr})g$	R.IR.	
	D_{4h}	A_{1g}	$\lambda_1 = (f_r + 2 f_{rr} + f'_{rr})g$	R.	$(\nu_2^2 - \nu_3^2)/(\nu_1^2 - \nu_2^2) = 0.36$
		E_u	$\lambda_2 = (f_r - f'_{rr})g$	IR.	
		B_{1g}	$\lambda_3 = (f_r - 2 f_{rr} + f'_{rr})g$	R	$\nu_0^2 = \frac{1}{4} (\nu_1^2 + 2\nu_2^2 + \nu_3^2)$
6	O_h	A_{1g}	$\lambda_1 = (f_r + 4 f_{rr} + f'_{rr})g$	R.	$(\nu_2^2 - \nu_3^2)/(\nu_1^2 - \nu_2^2) = 0.21$
		F_{1u}	$\lambda_2 = (f_r - f'_{rr})g$	IR.	
		E_g	$\lambda_3 = (f_r - 2 f_{rr} + f'_{rr})g \cdot$	R.	$\nu_0^2 = \frac{1}{6} (\nu_1^2 + 2 \nu_2^2 + 3 \nu_3^2)$

$\lambda = (4 \nu^2 c^2 \pi^2)/N \qquad g = \frac{1}{m_c} + \frac{1}{m_0}$

frequencies the uncoupled frequency $\nu_0^2 = f_r gN/4\pi^2 c^2$ may be calculated
(Table III). The obtained value may then be compared to the one observed
in ternary mixtures (Table III). There is usually a good agreement which
confirms our interpretation. However the calculated value is slightly
higher than the observed one in ternary mixtures: this would correspond
to a decrease of the cation-ligand interaction when the number of basic
interacting molecules increases in the solvation shell (Régis and Corset,
1973).

For the solid solvate Ba(ClO$_4$)$_2$.6 DMA (Figure 4 curves e and e'),
the spectra include three infrared and two Raman components. The magnitude of the observed splittings between the components seems to indicate
a coupling between equivalent groups in the unit cell, so that a notice-
tably distorted octahedral environment must be considered with only the
C_i symmetry left for the six DMA carbonyl surrounding the Ba^{2+} cations.

The spectra of the crystal Cu(ClO$_4$)$_2$.4 DMA include two Raman components at 1594 and 1628 cm^{-1} and one infrared component at 1602.5 cm^{-1}.
The structure of this surrounding seems more symmetrical than that of
the crystal Ba(ClO$_4$)$_2$.4(CH$_3$)$_3$CCOOCH$_3$ for which two infrared components
are observed close to each other. The environment of the Cu^{2+} cation
in the oxygenated complexes has usually the shape of a tetragonally
distorted octahedron (Smith, 1972), and the spectrum is consistent with

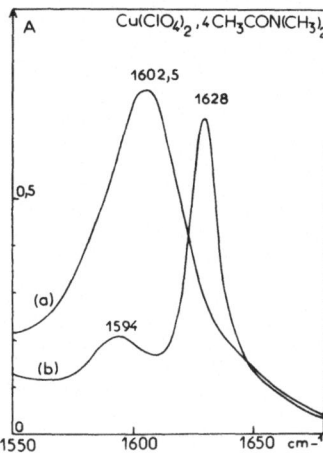

Fig. 5. Infrared (a) and Raman (b) spectra of $Cu(ClO_4)_2 \cdot 4$
DMA crystal in the $\nu(C=O)$ region.

the structure of D_{4h} symmetry presented Figure 7 for the CO groups. The
ratio $(\nu_2^2 - \nu_3^2)/(\nu_1^2 - \nu_2^2)$ is measured equal to 0.33 close to the
theoretical value of 0.36 in the dipol r interaction model. The two
ClO_4^- anions participate in the surrounding of the Cu^{2+} cation and would
have the C_{2v} symmetry (Jaeschke, 1976).
        Complementary information may indeed derive from the symmetry of
the perchlorate ion, which can be deduced from the observation of its
internal vibrations in Raman and infrared. In dilute solutions this
ion keeps the T_d symmetry, characterized by two triply degenerate ν_3
and ν_4 vibrations (symmetry F_2) and one doubly degenerate Raman active
ν_2 vibration (symmetry E). The first two have been assigned at 1100 cm^{-1}
and 628 cm^{-1} for the $Ba(ClO_4)_2$ dilute solutions in DMA, and are very
close to the frequencies observed in dilute aqueous solutions (Ross,
1962). This shows that the ion does not participate in the cation solvation and that the perchlorate is dissociated. The symmetry is lowered
and certain degeneracies are released as soon as the ClO_4^- ion is included in the solvation shell of the cation, or even if the anion stays
in the vicinity of the cation, from which it is separated by a solvent
molecules layer, as for instance in the hydrate $Ba(ClO_4)_2 \cdot 3 H_2O$ (Mani
and Ramaseshan, 1960) Thus the ν_2 vibration, only Raman active in the
T_d symmetry, becomes infrared active in the C_{3v} symmetry, and splits into
two Raman active components when the symmetry is lower than or equal
to C_{2v}. This is illustrated in Figure 6 where it can be seen that Raman
polarization is in favor of these depolarized bands while the bands which
would be due to the esters in this region would be strongly polarized.
In the mixture $Ba(ClO_4)_2 \cdot 6(CH_3)_2 CHOOCH_3$ and in the crystals $Ba(ClO_4)_2$
$Ba(ClO_4)_2 \cdot 4 (CH_3)_3CCOOCH_3$ two bands may be assigned to the ν_2 vibration
at 451 and 476 cm^{-1}, 446 and 474 cm^{-1}, and indicate for the anion a
symmetry lower than C_{2v}. In the mixture 2 $LiClO_4 \cdot 3(CH_3)_2CHCOOCH_3$ the
presence of only one band at 462 cm^{-1} prove the C_{3v} symmetry for this
anion which may be explained by its local symmetry in the solvation shell
of Li^+, Figure 7.

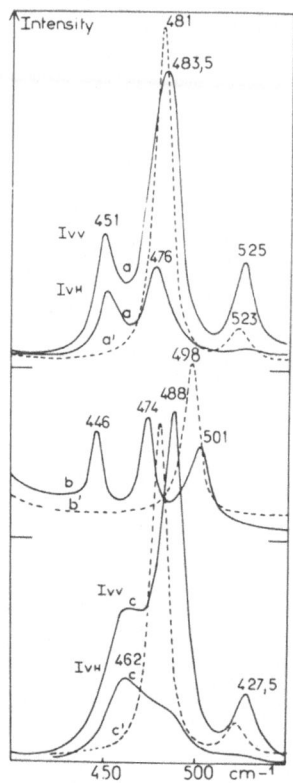

Fig. 6. Raman spectra of the ν_2 vibration of the ClO_4^- anion
for $Ba(ClO_4)_2$ and $LiClO_4$ solvates. (a) $CH_3CH_2COOCH_3$ /
$Ba(ClO_4)_2 \approx 6$ liquid, (b) $Ba(ClO_4)_2.4$ $(CH_3)_3CCOOCH_3$ crystal,
(c) $(CH_3)_2CHCOOCH_3$ / $LiClO_4 \approx 1.5$ liquid. Dashed curves a',
b', c' are the spectra of the corresponding esters.

2. NATURE OF THE ION-MOLECULE AND INTERMOLECULAR INTERACTIONS IN THE IONS SOLVATION SHELLS

2.1. *Nature of the Ion-Molecule Interactions*

The spectrum of the esters bound to cations show (Moravie, 1975) that
all the vibrations of the molecule are perturbed, not only the skeletal
vibrations (Table I), but also the vibrations of the O—CH$_3$ group. This
confirms the theoretical calculations about the interactions of these
compounds with Na$^+$ and K$^+$ cations showing that the charge distribution
over all the atoms is perturbed when an ester or amide carbonyl group
interacts with a cation (Perricaudet, 1973; Perricaudet and Pullman,
1973; Pullman, 1974). Moreover these calculations show that in this ion-
molecule interaction the molecule polarization through the cation is
largely predominant over the charge transfer. In the ester and amide
molecules interacting with cations in ternary mixtures, the smaller and

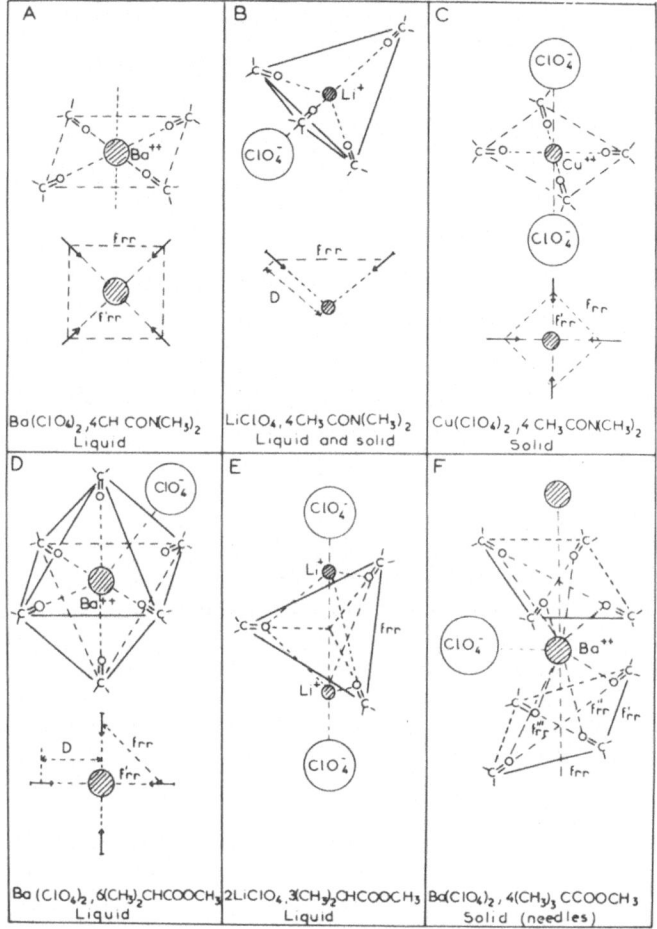

Fig. 7. Models for the different environments by esters and N,N–DMA carbonyl groups and ClO_4^- anions of Li^+, Ba^{++}, Cu^{++}.

the more charged the cation is the more the νCO frequency decreases (Table I and Figure 8). According to a simple model of ion-dipole interaction, where the unpolarizable dipole is assumed located at the center of the $\sub{>}C=O$ bond, the $\nu(CO)$ frequency decreases linearly with the Z/D^2 ratio, Z being the charge of the cation, D the ion-dipole distance (Figure 8). In this model the cation-oxygen distance has been taken equal to the sum of the cation crystalline radius and the oxygen Van der Waals radius (1.40 Å); Table IV shows that there is a good agreement with the X-ray data for cation-oxygen distances in crystalline solvents. The obtained straight lines intercept the ordinate axis at points which are intermediate between the frequencies measured for the gas and for dilute solution in carbontetrachloride. The slopes are very similar for DMA and methyl isobutyrate; the slightly more important

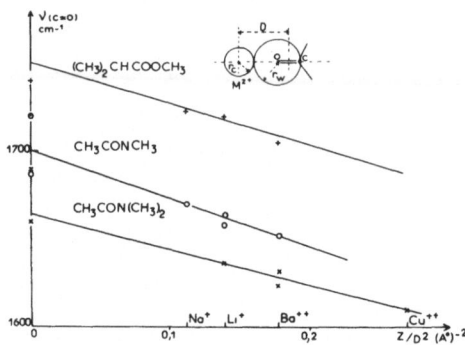

Fig. 8.   Frequencies of the $\nu(C=O)$ vibrations of $(CH_3)_2CHCOOCH_3$
$CH_3CONHCH_3$ and $CH_3CON(CH_3)_2$ as a function of Z/D^2.

slope observed for NMA may be due to a cooperative effect between the
$>C=O...M^{z+}$ interaction and the N—H...S hydrogen bonding (Baron and de
Lozé, 1972). Due to the large difference of molecular dipole moments
between the ester and DMA a more important frequency perturbation would
be expected for the amide than for the ester. A contribution of the
charge transfer would give the same result since according to the donor
Gutmann number (Table V), the amide is more basic than the ester. It
seems therefore that in this approximation the permanent dipole moments
μ_0 of the $>C=O$ bond must be considered very similar for the amide and
for the ester. The integrated intensity A of the $\nu(CO)$ band measured
in dilute CCl_4 solutions (Table V), is directly proportional to the
square of the variation of the dipole moment during the bond stretching
: Δr.

$$A = \frac{N^2\pi g}{3c^2}\left(\frac{\partial \mu}{\partial r}\right)^2 \text{ with } g = \frac{1}{m_0} + \frac{1}{m_C} \text{ and } \mu = \mu_0 + \left(\frac{\partial \mu}{\partial r}\right)_0 \Delta r$$

TABLE IV
Compared properties of the carbonyl group of acetone, methylacetate and
N,N dimenthylacetamide

	μ.D.	A^2 a) cm M^{-1}	B b) Arbitrary units	Donor number (Guttmann, 1968)
CH_3COCH_3	2.77	1.75×10^7	3.5	17
CH_3COOCH_3	1.75	2.49×10^7	3.5	16.5
$CH_3CON(CH_3)_2$	3.81	4.20×10^7	6	27.8

a) Infrared integrated intensity of the ν (C=O) band of the solvent
   diluted in CCl_4, $A = 1/Cl \int \log_c I_0/I \, d_\nu$
b) Raman integrated intensity $B = 1/C \int Id\nu$ measured relatively to an
   internal standard ($\nu = 200$ cm^{-1} band of CCl_4 at the concentration
   of 5% v. by v. in the solvent)

TABLE V
X-ray measured $M^{Z+} \ldots O$ distances

Compound	$M^{Z+} \ldots O$ Å	Ref.	$r_c + r_w$
LiClO$_4$, 3 H$_2$O	2.15		
LiBr, L Alanylglycine, 3;H$_2$O	1.943 O amide	Declercq et al. (1971)	2.08
	1.916 O water		
LiCl, 4 CH$_3$CONHCH$_3$	1.85	Bello, and Haas (1964)	
	2.01		
NaBr, 2 CH$_3$CONH$_2$	2.32 to 2.37	Piret et al. (1966)	
NaI, 3 HCON(CH$_3$)$_2$	2.40	Gobillon et al. (1962)	2.37
NaI, 3 CH$_3$COCH$_3$	2.47	Piret et al. (1962)	
NaI, 3 MeOH	2.47	Piret and Mesureur (1965)	
BaClO$_4$, 3 H$_2$O	2.96 to 3.18	Mani and Ramaseshan (1960)	
Ba(SNC)$_2$, H$_2$OC$_{18}$H$_{40}$N$_2$O$_6$	2.74 to 2.88 O ether		2.74
	2.86 O amide	Metz et al. (1971)	
Cu(gly)$_3$Cl 1 1/2 H$_2$O	1.987 O peptide	Freeman (1966)	2.12

The $\partial\mu/\partial r$ variation is only 1.3 times greater for acetamide than for methylacetate, while the dipole moments ratio is 2.17 for these molecules. A similar comparison between methylacetate and acetone shows that $\partial\mu/\partial r$ is 1.2 times greater for methylacetate than for acetone, while the latter molecule is more polar.

The average frequency ν_0 calculated from the bands observed for the complex $Cu(ClO_4)_2.4$ DMA falls on the line corresponding to the alkaline or alkaline Earth cations, which shows that for this cation the charge transfer contribution is still weak compared to the ion-dipole interaction.

The comparison between integrated intensities in infrared and in Raman shows a modification in the electron distribution of the CO bond upon complexation with the Ba^{2+} cation. The qualitative data of Figure 9 indicate that when going from acetone to methylacetate and to DMA $\partial\mu/\partial r$

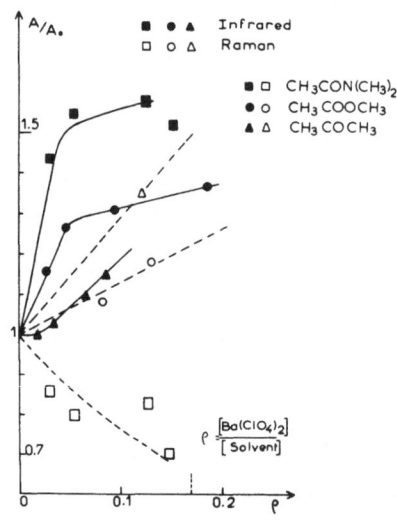

Fig. 9. Relationship between infrared and Raman molar integrated intensity and molar concentration ratio. Comparison of acetone, methylacetate and N,N-dimethylacetamide.
B : ν CO band in $Ba(ClO_4)_2$ solutions
B_0 : ν CO band for pure solvents.
Points ▲ are taken from (Boruka and Kecki, 1972).

increases while $\partial\alpha/\partial r$ decreases. The relative Raman intensity measurement given in Table IV show that $\partial\alpha/\partial r$ is 1.2 times greater for the CO bond of DMA than for methylacetate and acetone, which could explain the Raman intensity decrease of the carbonyl bond upon complexation of DMA with $Ba(ClO_4)_2$.

2.2. *Nature of the Interactions between Molecules in the Ions Solvation Shell*

We have seen in the first part that the spectra of the solvent molecules in the ions first solvation shell could be interpreted using a model of oscillators coupled by interaction constants f_{rr}, f'_{rr} ... Assuming that the dipole are not polarizable the interaction energy between two dipoles is:

$$w_{12} = - \frac{\mu_1 \mu_2}{R^3} \left[2 \cos \theta_1, \cos \theta_2 - \sin \theta_1 \sin \theta_2 \cos(\Psi_2 - \Psi_1) \right]$$

where R is the distance between the dipoles, θ and Ψ define their relative orientation.

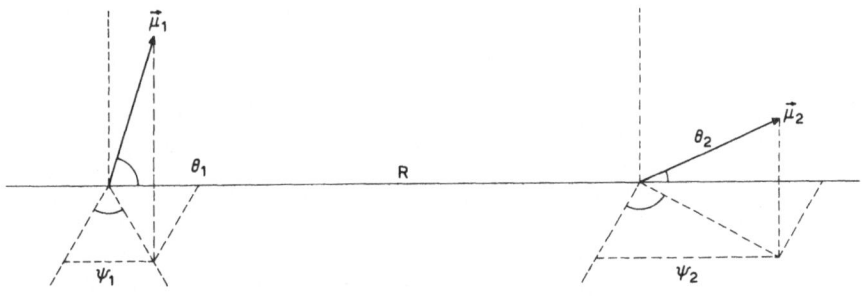

The coupling constant f_{12} between these two dipoles is

$$f_{12} = \frac{\partial^2 w_{12}}{\partial r_1 \partial r_2}$$

When the solvation shell is symmetrical enough it is possible to precise the dipolar orientations and distance in relation to the distance D between the ion and the dipole Figure 7. A simple expression of these interaction constants is thus obtained for instance:

Tetrahedron T_d $\qquad f_{rr} = \dfrac{0.383}{D^3} (\dfrac{\partial \mu}{\partial r})^2$

Octahedron O_h $\qquad f_{rr} = \dfrac{3}{4\sqrt{2}} \dfrac{1}{D^3} (\dfrac{\partial \mu}{\partial r})^2 \quad f'_{rr} = \dfrac{1}{4 D^3} (\dfrac{\partial \mu}{\partial r})^2$

The splittings between the observed frequencies depends only on these interaction constants, Table III, and it is therefore possible to calculate the distance D, since $(\dfrac{\partial \mu}{\partial r})^2$ is related to the integrated intensity A of the carbonyl vibration measured in infrared.

The calculated values of D are given (Table III) and compared with the distances obtained assuming that the cation-oxygen distance is the sum of the ion crystallographic radius and of the oxygen Van der Waals radius and that the dipole is located half way of the C=O bond. These

Fig. 10.   Steric effects of molecules in the solvation shell of cations.

two sets of values are in general fairly close, which confirms the essentially dipolar nature of the forces bonding and orienting the solvent molecules in the ions solvation shell. For the particular case of $LiClO_4$.4 DMA the calculated value is clearly higher than the $(r_c + r_w + \frac{r\ C=O}{2})$ estimated one, which seems to indicate a strong contribution of the repulsion energy of the dipoles in the solvation shell. As shown in Figure 10, where the molecules are figured with the Van der Waals radius of atom or groups $(r_O = 1.40\ \text{Å}, r_N = 1.50\ \text{Å}, r_{CH_3} = 2.0\ \text{Å})$, there is indeed no steric effect which could explain this discrepancy. For the solvates $Ba(ClO_4)_2$.6$(CH_3)_2CHCOOCH_3$ and $Ba(ClO_4)_2$.6 DMA in the liquid state the calculated values are similar and slightly higher for the ester than for the amide. This is consistent with the above analysis of the νCO frequency perturbations which shows a similar polarity of the two carbonyls. It must however be noticed, as shown in Figure 10, that steric bulkiness, which does not seem to exist for the Ba^{2+} octahedral surrounding by six molecules of methyl acetate, propionate or isobutyrate, may occur in DMA and explain the distortion of the octahedron in the crystalline solvate $Ba(ClO_4)_2$.6 DMA where the symmetry is only C_i. Steric bulkiness may also explain that the Cu^{2+}-dipole distance measured from the frequencies is higher than the estimated one (r_c + r_w + (r(CO)/2)), while the Cu^{2+} crystallographic radius is much smaller than that of Ba^{2+} (Figure 10). This abnormally large distance due to dipolar repulsion may explain the essentially dipolar nature of the Cu^{2+}-amide interaction in this crystal, in agreement with the νCO frequency decrease. Finally these steric effects may contribute to the weak solubility of barium perchlorate in methylpivalate, since the octahedral surrounding of the Ba^{2+} cation seems no more possible. In the solvate 2 $LiClO_4$.3 esters we could calculate only the distance d between dipoles. This distance is close to that obtained assuming the O—O distance equal to the sum of the Van der Waals radii (as it is observed in the solvates with analogous structure NaI. 3 CH_3COCH_3 and NaI. 3 $HCON(CH_3)_2$ and the dipole located in the middle of the CO bond).

It is a pity that the spectroscopic measurements do not reach the Li^+-O distance in this model, since the carbonyl perturbation indicated by the ν_0 frequency is close to that observed in solution where only one Li^+ interacts with the carbonyl ($>C=O...Li^+$), while in this solvate two cations interact ($>C=O \cdots Li^+ \atop \cdots Li^+$).

In conclusion we have shown that infrared and Raman spectroscopy give important information on the structure of the cation surrounding in the solid or liquid state. The surrounding of Li^+ and Ba^{2+} cation by esters or DMA molecules is well interpreted through a predominating influence of electrostatic forces of the ion–dipole and dipole–dipole types and through steric effects due to the substituants. The difference between the behavior of DMA and that of esters such as methylacetate seems less due to the difference of their polarities than to a difference of polarizability of their carbonyl groups. Comparing the dipole moments of the molecules may in certain cases be an erroneous way of comparing the dipole moments of the bonds.

REFERENCES

Baron, M.H. et al.: 1972, C.R. Acad. Sci. Paris 274 C, 1321.
Baron, M.H. and Lozé, C.: 1972, J. Chim. Phys. 69, 1084
Baron, M.H.: 1973, Thèse de Doctorat, Université Paris VI - CNRS no AO 7308.
Bello, J. and Haas, D.J.: 1974, Nature 201, 64; and Bello, J.: Personal communication.
Boruka, J. and Kecki, A.: 1972, Roczn. Chem. Ann. Soc. Chim. Polonorum 46, 1829.
Dalley, N.K. et al.: 1972, J. Chem. Soc. Chem. Comm. 90.
Declercq, J.P. et al.: 1971, Acta Cryst. B 27, 539.
Freeman, H.C.: 1966, in The Biochemistry of Copper p. 86.Persach et al. (Eds.), Academic Press.
Garrigou-Lagrange, C. et al.: 1970, J. Chim. Phys. 67, 1936
Gobillon, Y. et al.: 1962, Bull. Soc. Chim. Franc. 85, 282.
Gutmann, V.: 1968, Coordination Chemistry in non-aqueous solutions, Springer-Verlag, N.Y.
Hexter, R.M.: 1960, J. Chim. Phys. 33, 1833.

Jaeschke, H.: 1976, Thèse de Doctorat de 3ème Cycle - Université
   Pierre et Marie Curie, Paris.
Kress, J. and Guillermet, J.: 1973, Spectrochim. Acta 29, 1717.
Mani, N.V. and Ramaseshan, S.: 1960, Z. Krist. 114, 200.
Metz, B. et al.: 1971, J. Am. Chem. Soc. 93, 1806.
Moravie, R.M.: 1975, Thèse de Doctorat - Université Pierre et Marie
   Curie, Paris - CNRS no AO 9437.
Moravie, R.M. and Corset, J.: 1976, submitted for publication to
   Spectrochimica Acta.
Mumoz, E. et al.: 1972, Molecular Mechanisms of antibiotic Action on
   Protein Biosynthesis and Membranes, Proc. Symp. in Granada Sec. III,
   Elsevier.
Ovchinnikov, Yu A. and Ivanov, V.T.: 1975, Tetrahedron 31, 2177.
Perricaudet, M.: 1973, Thèse de Doctorat de 3ème cycle - Université
   Paris VI.
Perricaudet, M. and Pullman, A.: 1973, Febs Letters 34-2, 222.
Piret, P. et al.: 1963, Bull Coc. Chim. Franc. 205.
Piret, P. et Mesureur, C.: 1965, J. Chim. Phys. 62, 287.
Piret, P. et al.: 1965, Acta Cryst. 20, 482.
Pullman, A.: 1974, Int. J. Quantum Chem.: Quantum Biology Symp. 1, 33-42.
Ramachandran, G.N.: 1967, Conformation of Biopolymers, Vols. 1 - 2.
   Academic Press, Madras, India.
Regis, A. and Corset, J.: 1973, Can J. Chem. 51, 3577.
Roche, J.P.: 1971, Thèse de Docteur Ingénieur - Université Bordeaux I -
   No 150.
Ross, S.D.: 1962, Spectrochim. Acta 18, 225.
Smith, D.W.: 1972, Structure and Bonding, Vol. 12, p. 49, Springer-Verlag
Von Hippel, P. et al.: 1969, Structure and Stability of Biological
   molecules, M. Decker.
Wilson, E.B. et al.: 1955, Molecular Vibrations, McGraw Hill, New York.

DISCUSSION

Rode: Most of your interpretation of the vibrational spectra of the
amide complexes is based merely on cation-amide interaction. We have
found, that the use of different anions also influences strongly the
frequency shifts observed in the amide spectrum sometimes even change
their direction. How can we explain this anion effect, especially in
non-hydrogen-bonding amides, and to which extent would the anion-amide
interaction influence the interpretation of the IR + RAMAM data?

   Corset: Our interpretation is based not only on the ion-amide interactions, but also on the amide-amide interactions and we take into
account the ion-ion interactions. We use ClO_4^- anion because it is
possible to determine its symmetry lowering due to interactions with
cations through observation of its internal vibrations. Furthermore,
usually this anion does not tend to associate strongly with cations.
Nevertheless we have observed direct anion-cation contact in the first
solvation sphere for the mixture $2LiClO_4$ 3 esters. With more basic
anions such ion pairing may occur and change the arrangement of solvent
molecules in the first solvation shell.

When speaking of $\nu(C=O)$ frequency decrease or increase you must take care if couplings are taken or not into account. I should be surprised if you had observed a ν_0 increase through complexation with a cation when compared to the free $\nu(C=O)$ carbonyl frequency measured in a non-polar solvent.

CONCEPT OF MOLECULAR DESIGN IN RELATION TO THE METAL-BINDING SITES OF PROTEINS AND ENZYMES

BIBUDHENDRA SARKAR
*The Research Institute, The Hospital for Sick Children,
Toronto, and The Department of Biochemistry, The University
of Toronto, Toronto, Canada*

1. INTRODUCTION

Proteins associated with metallic cations are found throughout the biological system. Consequently, studies of metal protein interactions
have developed as an important research area in biochemistry. When metal
protein interactions are viewed in the context of the varied phenomena
of metals in enzymology, a panorama of the extensive role of metals in
biological systems unfolds.

In recent years, the three dimensional structures of several enzymes, proteins and hormones have been elucidated at the atomic level.
Some of these have metal in the active site. These studies have provided
a precise three dimensional atomic description which clearly reveals
the molecular architecture, binding ligands, the associated geometry
and the microenvironment at the binding site. The amino acid residues
at the binding site may originate from a short linear sequence in the
polypeptide chain; alternatively, they may be assembled from distant
parts of the polypeptide chain. In the latter case, it is the secondary
and the tertiary structures which bring together and maintain the necessary residues in the appropriate geometry. Once the array of the
binding residues and other parameters mentioned above are known, one
should be able to design and synthesize a simple molecule which would
mimic the native metal-binding site but exclude the large extraneous
section of the protein molecule.

In this paper, I shall first outline the concept of molecular design, then discuss its application in designing and synthesizing molecules mimicking the native metal binding sites. I shall also make an
attempt to project the future implications of molecular design in relation to new approaches and therapeutic applications.

2. MOLECULAR DESIGN CONCEPT

The concept of molecular design to mimic the metal-binding site of a
protein molecule entails abstracting the minimum requirements which
must be retained in a molecule in order to maintain the parameters

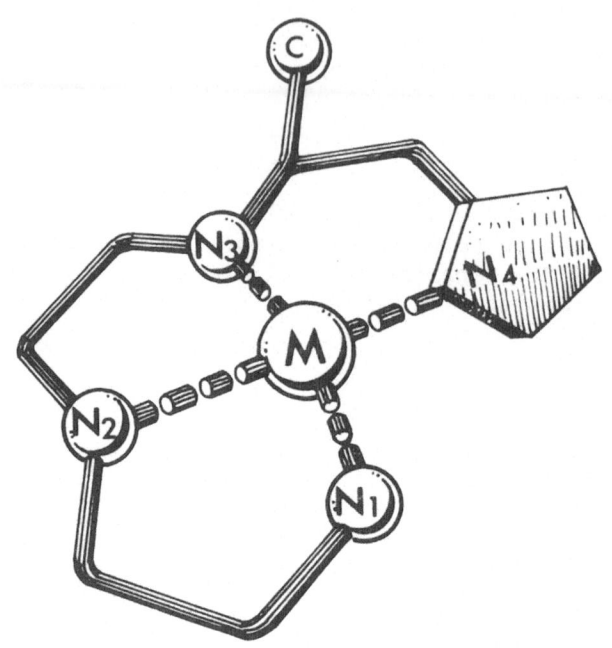

Fig. 1. Design of a peptide molecule mimicking the metal-
binding site on a short linear sequence. Metal (M) is bound
to the amino terminal nitrogen (N_1), two peptide nitrogens
(N_2, N_3) and an imidazole nitrogen (N_4) of the third position
histidine residue. The carboxyl terminal (C) is blocked.

controlling the geometry and the nature of the metal-binding ligands
and the micro-environment at the metal-binding site. The ligands are
composed of amino acid residues originating from different parts of the
polypeptide chain. The design scheme should attempt to closely retain
the geometry of the binding residues, although acceptable limits of
tolerance will vary in specific instances and require definition for
each binding site. These residues with the associated geometry can be
incorporated into a suitable backbone structure. Consequently, it may
not be nescessary to entirely maintain the native sequence in order to
preserve the geometry near the binding site. The preferred geometry of
the metal in question is important in the consideration of a suitable
molecule. However, the descriptions, 'square planar', 'regular
octahedral' and 'tetrahedral' represent idealizations, deviations from
which are more the rule than the exception. The design of the metal-
binding sites can be divided into two categories, as follows:

2.1. *Binding Site on a Short Linear Sequence*

The metal binding residues in a protein molecule may be part of a short
linear sequence. In such a case, a linear peptide can be designed, in-

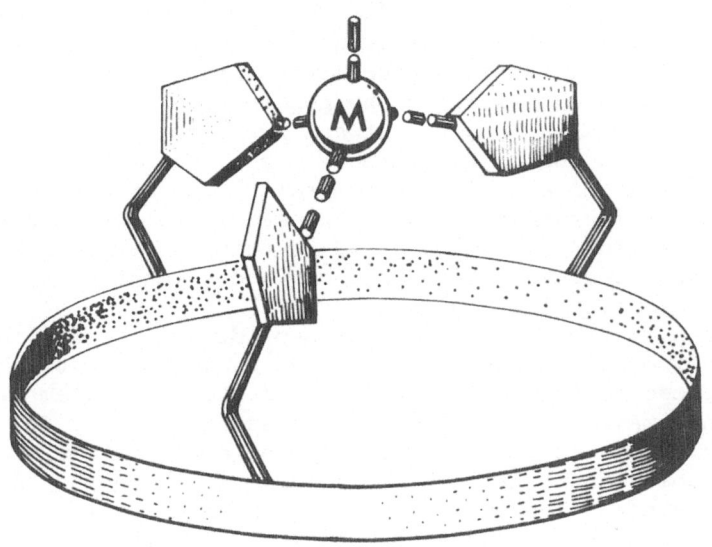

Fig. 2. Design of a cyclic peptide molecule mimicking the
native metal-binding site where ligands originate from the
distant parts of the polypeptide chain. Metal (M) is shown
bound to ligands originating from the cyclic peptide backbone
structure.

corporating the essential binding residues interspaced by amino acid
residues which do not possess liganding type side chains (Figure 1).
In this type of binding site, it is quite possible that the peptide nitrogens or carbonyl oxygens are also involved in the binding in addition
to other liganding groups. The C-terminal and the N-terminal of such
a linear peptide should be protected if not involved in the metal-binding.
The essential binding residues must be the same amino acid residues that
are binding metal in the native molecule.

2.2. *Binding Site Composed of Residues from Distant Parts*

The majority of the metal-binding sites are not located on a short
linear sequence in a polypeptide chain. The binding sites are composed
of amino acid residues originating from different parts of the polypeptide
chain. The secondary and tertiary structures of the protein molecule
bring together and maintain the necessary residues in the required
geometry.
     Experience suggests that short linear peptides have little chance of
forming a stable matrix. The large number of possible conformational
states available to a linear peptide chain is substantially reduced by
cyclization. Thus, a cyclic peptide should be designed, incorporating
liganding amino acid residues in strategic positions to provide the
metal-binding site. Such a design is illustrated in Figure 2.
     A proline residue can give a suitable bend which may help orient the

peptide backbone for cyclization or even to put the peptide backbone
to certain desired geometry. This property arises from the rigidity of
the pyrolidine ring which restricts the possible values of the angle of
rotation ϕ around the C^α-N bond to between 105° and 125°, thus diminshing
their allowed space on the conformational maps. Deber et al. (1976)
have carried out extensive studies with proline-containing cyclic peptides
having high sequential symmetry containing 2 to 12 amino acid residues.
By using proline residues, they obtained certain preferred conformers.
Their results show clearly the consequences and advantages of using
proline residues in a cyclic peptide.

One can gain further knowledge from peptide antibiotics, such as
enniatin and valinomycin, both of which have cyclic structures with
repeating sequences containing LD or DL pairs. Venkatachalam (1968)
studied the conformations of peptide chains and reported a folded type
of conformation, known as β-bend. This type of conformation has the
property of assuming a special type of folding in which the three peptide
units linked together produce a reversal in chain direction. Further
studies by Ramachandran and Chandrasekaran, (1972) with alternating
L- and D-amino acids revealed that a sequence of a L-residue followed
by a D-residue, or *vice versa,* produces a conformation which has low
energy and is further stabilized by a hydrogen bond between the NH
group of the third peptide unit and the carbonyl oxygen of the first
peptide unit. The occurrence of a reversal of chain direction is a
key feature for the closure of the ring in cyclic structures. From
these studies, it appears that in peptides (i.e. antibiotics) containing
both L- and D-residues, a mixed sequence such as LD or DL will be the
likely site for the bend leading to cyclization. The bending of the
peptide chain is thus of great importance in cyclic oligopeptides,
wherein a chain reversal could readily be made use of for closure of
the ring. By using sequence LLDDLLDD...cyclic peptides with different
diameters can be obtained.

As is well known, disulfide linkage can render rigidity to the
backbone structure. Disulfide bonds serve to stabilize the conformation
by reducing the structural fluctuations of alternative forms. In proteins, it forms a nucleation site for the purpose of folding and thus
becomes stabilized, perhaps by formation of β-turns (Anfinsen and
Scheraga, 1975). Several cyclic peptides with six or more residues
linked by a -Cys-S-S-Cys- bridge are known. In all these structures, it
can be clearly seen that the disulfide linkage provides a certain
rigidity in the molecule and a judicious use of disulfide bridges may
yield the proper geometry of the backbone structure of the designed
molecule. In a backbone structure, it is important to maintain the
bends in order that the required geometry may be produced. The sequence
used must be such as to preserve the bends.

Inclusion of α,β-unsaturated amino acid residues in the backbone structure may be of advantage. Recent crystallographic data (Pieroni et al.,
1975) provide the evidence for a degree of rigidity in the dehydro
amino acid system. It is shown that, in the doubly unsaturated peptides,
steric hindrance forces the two unsaturated groups to stay in two
distinct skewed planes. Also, the native metal binding site may have
a hydrophobic environment. Consideration could be given to provide

hydrophobic environment by introducing in appropriate geometry aromatic residues, such as tryptophan, phenylalanine or several known non-peptide molecules which can provide lipophilic properties.

Model building, as a first step, can be advantageously supplemented by conformational calculations, whenever possible. The necessary framework for predicting molecular geometry has been established as a consequence of recent advances in theoretical conformational analysis (Ramachandran and Sasisekharan, 1968; Pullman and Pullman, 1974; Anfinsen and Scheraga, 1975). The lack of an adequate representation of solvent, however, has been the major shortcoming of conformational energy calculations. For sterical systems, such as small cyclic peptides, the variables such as bond lengths, bond angles and dihedral angles are highly interdepent and they can be adjusted to produce an energy minimum. But, for a larger molecule, such an energy minimization would require phenomenal amounts of computer time. For small cyclic peptides, many of the variables can be fixed at standard values obtained mainly from X-ray diffraction studies of crystals. Thus, only rotations about single bonds remain to be considered. It should be noted that the assumptions of the standard bond lengths and angles are precisely those made in constructing molecular models. However, with a computer and a set of potential functions, one can systematically explore many more conformers than is feasible with a set of molecular models.

3. DESIGN OF A SITE ON SHORT LINEAR SEQUENCE: COPPER(II)-TRANSPORT SITE OF HUMAN ALBUMIN

3.1. *Consideration for the Design*

The Cu(II)-transport site of human albumin is one of the best delineated metal-binding sites of a protein molecule in solution. Interestingly, the site is located on a short linear sequence: H_2N-Asp-Ala-His-....., at the NH_2-terminal segment of the albumin molecule (Figure 3). The Cu(II)-binding site involves the α-amino nitrogen, two intervening peptide nitrogens and the imidazole nitrogen of the histidine residue in the third position (Peters and Blumenstock, 1967; Shearer et al., 1967; Bradshaw et al., 1968). Albumins from bovine, human and rat show a characteristic preferential binding of one Cu(II) ion. The transport form of Cu(II) in normal serum is known to be in this specific albumin bound form. However, unlike human albumin, dog albumin failed to exhibit the characteristics of a specific first binding site for Cu(II) (Appleton and Sarkar, 1971). Subsequently, the NH_2-terminal peptide fragment (1-24) of dog albumin was obtained by limited peptic hydrolysis of whole albumin. The complete sequence of this fragment was obtained by dansyl-Edman degradation on the whole peptide and by characterization of its tryptic peptide. The result showed that the important histidine residue in the third position was replaced by a tyrosine residue (Dixon and Sarkar, 1974).

The histidine in the third position does appear to be mandatory for the specificity. Several interesting features emerged from the amino acid sequences at the NH_2-terminii of bovine, rat, human and dog

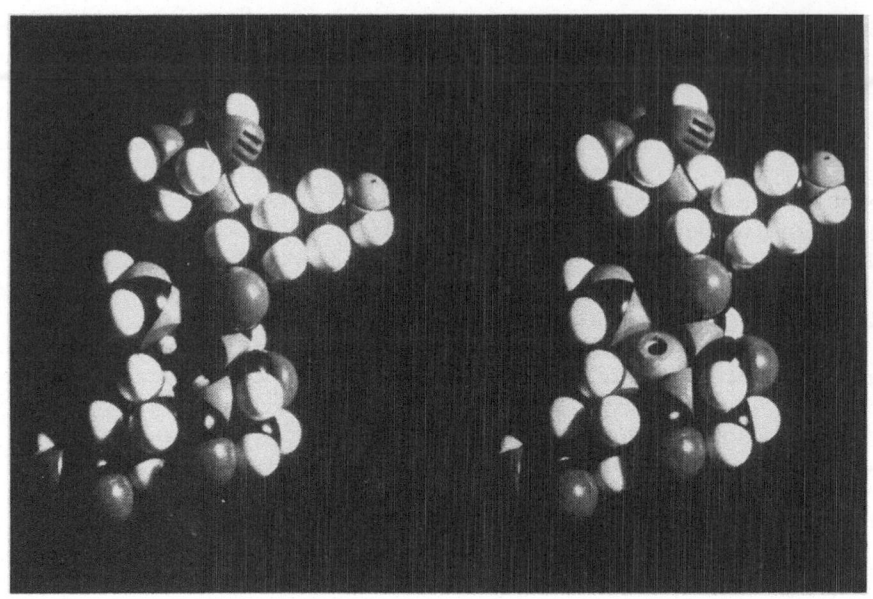

Fig. 3. Space filling models showing the Cu(II)-transport
site located on a short linear sequence at the NH$_2$-terminal
segment of human albumin. The binding site involves the α-
amino nitrogen, two intervening peptide nitrogens and the
imidazole nitrogen of the histidine residue in the third
position.

albumins (Figure 4). The albumins of the first three species do have
a functionally similar and specific Cu(II)-binding site and they all
have a histidine residue in the third position. Dog albumin has a
tyrosine residue instead and hence lacks the specificity. On the other
hand, different amino acid sequences of the first two residues do not
seem to alter the Cu(II)-binding properties of albumins of the first
three species. With this idea in mind, a simplified peptide was designed
Glycylglycyl-L-histidine-N-methyl amide, to mimic the native copper
transport site of human albumin. According to the criteria discussed

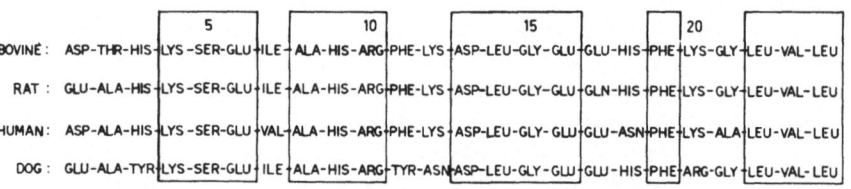

Fig. 4. Comparison of the amino acid sequence of the NH$_2$-
terminal peptides (1-24) of bovine, rat, human and dog serum
albumins. Residues *enclosed in boxes* are identical in the
four fragments.

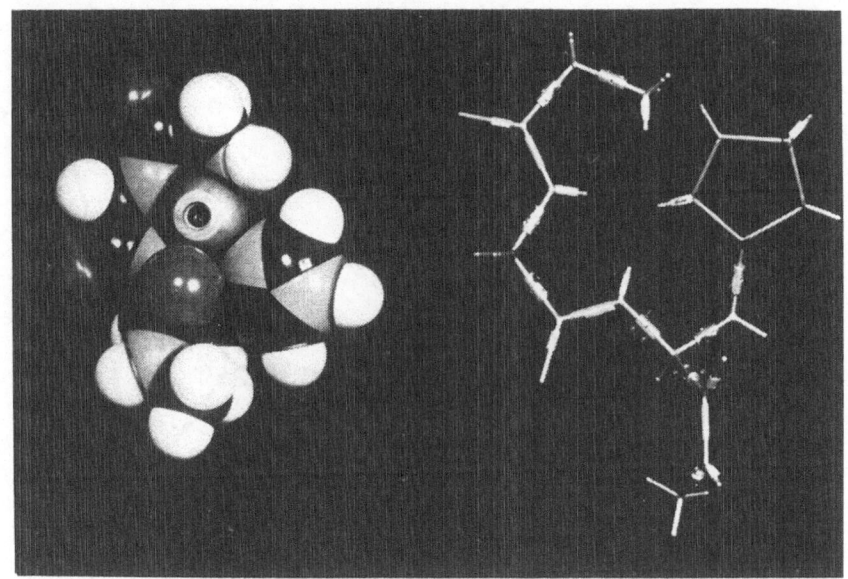

Fig. 5. Models of the designed peptide, glycylclycyl-L-histidine-N-methyl amide, showing the square planar geometry of the nitrogen ligands and the peptide backbone.

above, the carboxyl group was derivatized to N-methyl amide form to resemble more closely the protein molecule. The α-NH$_2$ was left unprotected since this group is required for the Cu(II)-binding. This peptide clearly showed a square planar geometry of the nitrogen ligands around Cu(II), as seen in both the space filling model and the Kendrew model (Figure 5).

3.2. *Theoretical Calculation of the Native and the Designed Sites*

The decision on the designed molecule is based primarily on the consideration of the coordination geometry of the metal ion at the binding site and the nature of the coordinating ligands. It is considered that a comparative study of the theoretical conformational analysis of L-Asp-L-Ala-L-His-NHMe and Gly-Gly-L-His-NHMe would be of value in the molecular design.

The total conformational energy U is expressed as a sum of the energies due to various interactions, as follows,

$$U = U_{non-bonded} + U_{electrostatic} + U_{torsion} + U_{solvent} \qquad (1)$$

$U_{non-bonded}$ was calculated using Buckingham's potential function with parameters, suggested by Ramachandran and Sasisekharan, (1968). $U_{electrostatic}$ was calculated using Coulomb's Law, assuming a dielectric

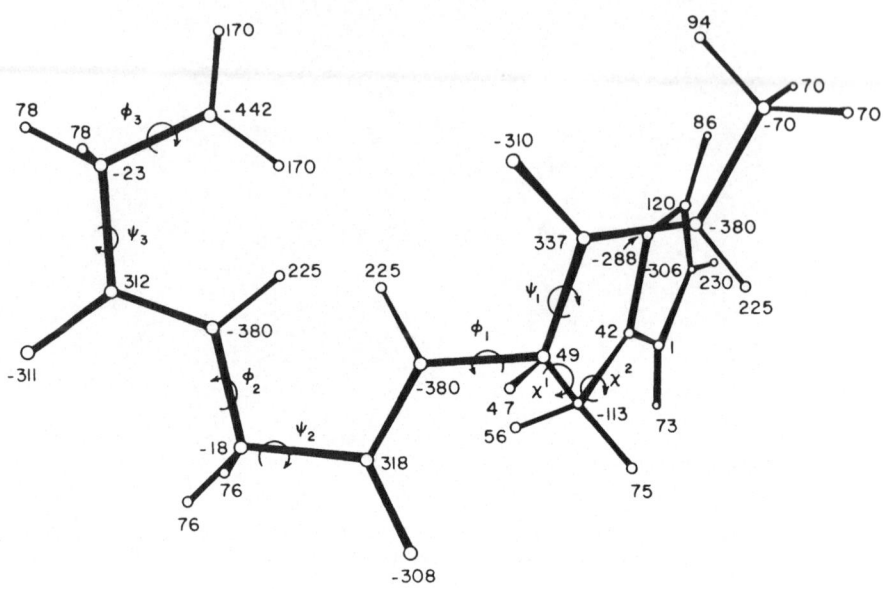

Fig. 6. A perspective of an extended conformation of glycyl-glycyl-L-histidine-N-methyl amide showing the definition of torsion angles and the *ab initio* (STO-3G) monopole charges in 10³ electron units.

constant of ten (Ponnuswamy et al., 1971). A high dielectric constant was required to suppress the large electrostatic energy arising from the interaction of the negatively charged L-aspartyl side chain of the native site with the rest of the peptide molecule. For the sake of consistency, the same value for the dielectric constant was used for the designed peptide. Ab initio minimal basis set (Hehre et al., 1969) (STO-3G) monopole charges for the native site and the designed peptide were obtained by suitable fragmentation of the molecule. The monopole charges were used in the calculation of electrostatic interactions. A threefold torsional barrier was assumed for torsions about C^α-N and C^α-C' bonds. The torsional barriers, V^0_ϕ and V^0_ψ were assumed to be 0.6 and 0.2 kcal mole^{-1} respectively (Scott and Scheraga, 1966). The torsional barrier V^0_χ for side chain single bonds was assumed to be 2.8 kcal mole^{-1} (Ponnuswamy et al., 1971). Solvent was not considered in this study.

Conformation of the backbones of the two tripeptides may be described by a set of torsion angles, ϕ_1, ψ_1, ϕ_2, ψ_2, ϕ_3, (Figures 6 and 7). ψ_1 and ψ_3 describe the orientation of N-methyl amide and terminal amino groups respectively. Since these two torsion angles do not play a direct role in anchoring the metal atom to the peptide as a first approximation, explicit consideration of these two angles was ignored in the theoretical calculations. The side chain conformations for the designed peptide are described by χ^1 and χ^2 about the C^α-C^β and C^β-C^χ bonds respectively. Similarly, the side chain conformation for the

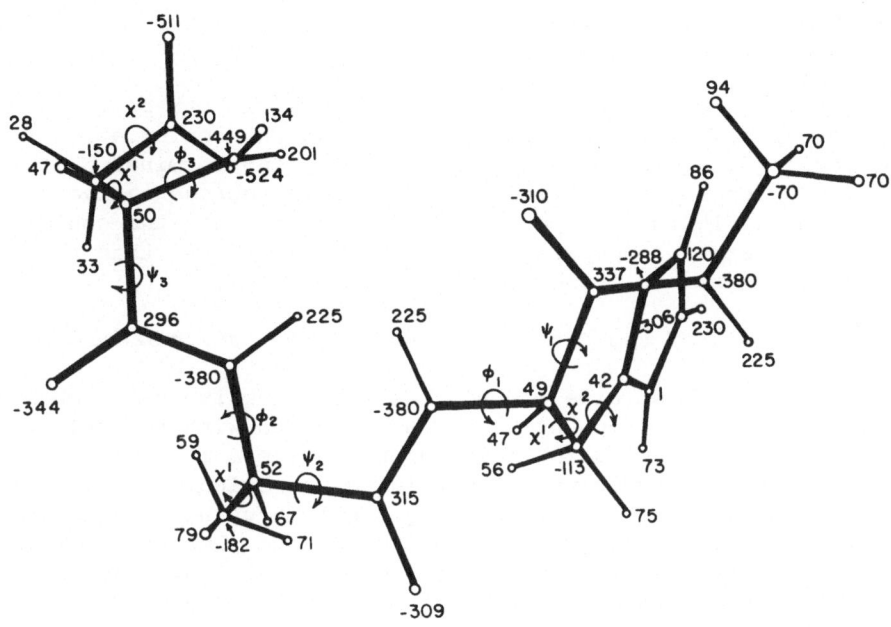

Fig. 7. A perspective of an extended conformation of L-aspartyl-L-alanyl-L-histidine-N-methyl amide showing the definition of torsion angles and the *ab initio* (STO-3G) monopole charges in 10^3 electron units.

native sequence peptide can be described by a set of torsional angles: χ^1 and χ^2 about $C^{\alpha}-C^{\beta}$ and $C^{\beta}-C^{\chi}$ bonds of L-histidyl residue, χ^1 about $C^{\alpha}-C^{\beta}$ bond of L-alanyl residue and χ^1 and χ^2 about $C^{\alpha}-C^{\beta}$ and $C^{\beta}-C^{\chi}$ bonds of L-aspartyl residues respectively. Since the side chain torsional angles, χ^n cluster around a narrow range of values corresponding to staggered and eclipsed conformations of the side chain (Ramachandran and Sasisekharan, 1968) as a first approximation, the side chains were assumed to be frozen in either staggered or eclipsed conformation. However, for the histidyl side chain, due to its direct involvement in the metal-binding χ^1 and χ^2 were varied over a closer grid of points in the conformational calculations.

Rotations about $C^{\alpha}-C'$ and $C^{\alpha}-N$ bonds were performed systematically, proceeding from the fully extended conformation of the tripeptides. The fully extended conformations, where $\phi = \psi = 0^{\circ}$, were constructed from standard dimensions of the peptide group. The peptide units were assumed to exist in a planar transconformation. Tetrahedral bond angles were assumed for sp^3 carbon atoms and C-H bond lengths were taken to be 1.09 Å. Geometry of the histidine side chain was constructed, using the bond lengths and bond angles determined by Donohue and Caron (1964).

$c^{\alpha}-c^{\beta}$, $c^{\beta}-c^{X}$ and $c^{X}-o^{X}$ bonds in L-aspartyl side chain were assumed to be 1.53 Å, 1.53 Å and 1.24 Å respectively (Momany et al., 1975). L-Alanyl side chain was constructed with the same dimensions as used by Renugopalakrishnan et al., (1975). The refinement was done by total minimization.

A large number of folded conformations of both peptides were found to occur from the calculations, possessing the conformational properties for forming a square planar complex, utilizing terminal N atom, two peptide N atoms and the imidazole N atom. A representative conformation is shown in Figure 8. However, it was noticed that the conformation of the native sequence peptide was far more restricted than the designed peptide, due to the presence of the side chains.

3.3. *Synthesis of the Designed Sites and the Native Sites Sequence*

Glycylglycyl-L-histidine was synthesized by active ester method. CBZ-glycylglycine p-nitrophenyl ester was reacted with L-histidine free base in equal molar ratio in 50% dioxane-water. The reaction mixture containing CBZ-glycylglycyl-L-histidine was dried under reduced pressure and fractionated on a silica gel column using 1-butanol-acetone-water-acetic acid (4:3:2:1,V/V) as the eluent. The CBZ-glycylglycyl-L-histidine was hydrogenated in methanol for 2 h with 10% palladium-charcoal. The filtrate after hydrogenation was evaporated and the residue was purified from methanol-water.

Glycylglycyl-L-histidine-N-methyl amide was synthesized by active ester method. L-Histidine methyl ester was reacted with CBZ-glycylglycine p-nitrophenyl ester in $CHCL_3$. The purity of the product, CBZ-glycylglycyl-L-histidine methyl ester was checked by thin-layer chromatography, m.p. 198°C. Monomethyl amine gas was passed through the ethanolic solution of this compound to obtain CBZ-glycylglycyl-L-histidine-N-methyl amide. Hydrogenation was carried out for 4 h to assure complete removal of the CBZ-group. Glycylglycyl-L-histidine-N-methyl amide was purified and recrystallized from cold 90% CH_3CN in water.

The native sequence peptide, L-Asp-L-Ala-L-His-N-methyl amide was synthesized employing the N-hydroxysuccinimide ester method. Treatment of the N-hydroxysuccinimide ester of CBZ-L-alanine with L-histidine methyl ester produced the dipeptide. CBZ-L-Ala-L-His-OMe which was converted to its N-methyl amide with monomethyl amine. The α-NH_2 protecting group on the dipeptide N-methyl amide was cleaved and the resulting compound was reacted with the N-hydroxysuccinimide ester of β-OBu^t-CBZ-L-aspartic acid to give the tripeptide, β-OBu^t-CBZ-L-Asp-L-Ala-L-His-N-methyl amide. Removal of the α-amino and side chain protecting groups on the aspartyl residue followed by purification of the product by gel filtration and countercurrent distribution yielded the pure tripeptide N-methyl amide.

All the above peptides were checked for purity by thin layer chromatography, electrophoris and amino acid analyses.

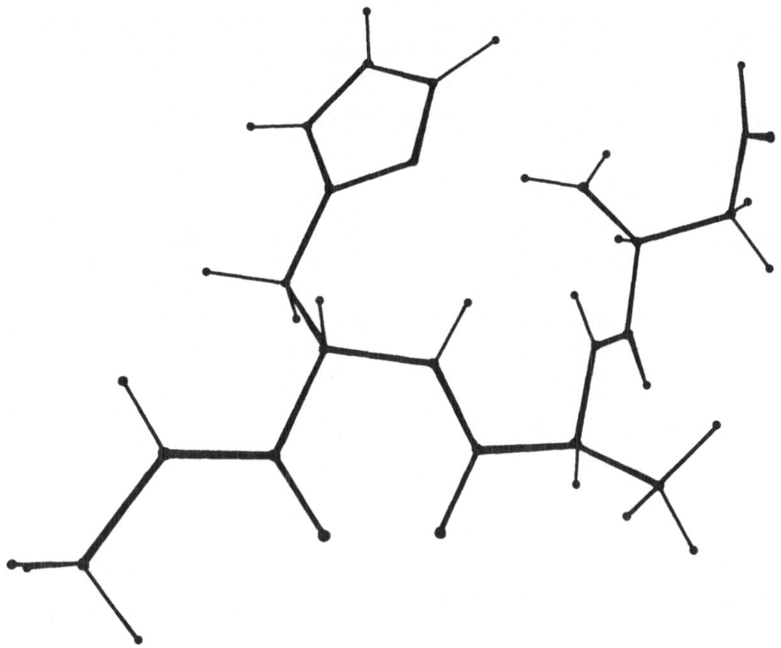

Fig. 8. One of the plausible folded conformations of the
native sequence peptide. Similar conformations are also noted
for the designed peptide to provide an array of four nitrogen
ligands in an approximately square planar geometry for the
purpode of Cu(II)-binding.

3.4. *Comparison of Albumin with the Designed Site*

3.4.1. Structural Studies

A complete analysis of the metal complex equilibria is essential to study
the structure of the species in solution. The Cu(II)-complex equilibria
in the systems containing Cu(II) and all the peptides were studies by
Analytical Potentiometry (Sarkar and Kruck, 1973; Kruck and Sarkar, 1973a,
1973b, 1973 c). The complexation reactions occurring between C_M moles
of metal ion M, C_H moles of hydrogen ion H, C_A moles of ligand anion
A, and C_B moles of ligand anion B, can be represented by the general
equilibrium reaction,

$$pM + qH + rA + sB \rightleftharpoons M_pH_qA_rB_s \qquad (2)$$

where p, q, r and s are numbers of M, H, A, and B respectively. The stabilities of the species formed are measured by the stoichiometric equilibrium constants β_{pqrs} expressed in terms of concentrations at constant
ionic strength, temperature and pressure,

$$\beta_{pqrs} = \frac{[M_p H_q A_r B_s]}{m^p h^q a^r b^s} \tag{3}$$

where m, h, a, and b are the concentrations of free metal ion, hydrogen
ion, ligand A and B respectively. The following sets of equations define
the total system,

$$C_M = m + \Sigma\ p\ \beta_{pqrs} m^p h^q a^r b^s \tag{4}$$

$$C_H = h - oh + \Sigma\ q\ \beta_{pqrs} m^p h^q a^r b^s \tag{5}$$

$$C_A = a + \Sigma\ r\ \beta_{pqrs} m^p h^q a^r b^s \tag{6}$$

$$C_B = b + \Sigma\ s\ \beta_{pqrs} m^p h^q a^r b^s \tag{7}$$

where oh represents the amount of free hydroxyl ions. The experimental
data and titration curves ($-\log h = f(base)$) were obtained from solutions
of defined concentrations of C_M, C_H, C_A and C_B. The following relationships were used to obtain the values for the unbound portions of metal
and the different ligands throughout the titration:

$$pM = pM_0 + \int_{pH_0}^{pH_i} \frac{\delta H_1^+}{\delta C_M}\ dpH \tag{8}$$

$$pA = pA_0 + \int_{pH_0}^{pH_i} \frac{\delta H_1^+}{\delta C_A}\ dpH \tag{9}$$

$$pB = pB_0 + \int_{pH_0}^{pH_i} \frac{\delta H_1^+}{\delta C_B}\ dpH \tag{10}$$

where pM = $-\log$ [free metal M], pA = $-\log$ [free ligand A], pH = $-\log$ H,
pB = $-\log$ [free ligand B], H_1^+ = moles of OH^- consumed to titrate hydrogen
ions liberated through the complexation reactions. Subscript O denotes
an initial known state of the system. The mathematical analyses of the
data were performed by the sequential use of three computer programs
(Sarkar and Kruck, 1973; Kruck and Sarkar, 1975, 1976). The species,
their stability constants and the associated pK_a values of the donor
groups are listed in Table I. The definition of Rossoti and Rossoti
(1961) was followed in identifying the species. In all systems studied,
there is one major species existing in physiological pH.

TABLE I

Comparison of log stability constants (log β_{pqr}, log β'_{pqr} and log β''_{pqr}) of the complex species $M_pH_qA_r$ (M = Cu(II), A = glygly-L-his-N-methyl amide, A' = glygly-L-his, A" = L-asp-L-ala-L-his-N-methyl amide) in 0.15 M NaCL at 25 °C

q	r	log β_{pqr} ($M_pH_qA_r$)	log β'_{pqr} ($M_pH_qA'_r$)	log β''_{pqr} ($M_pH_qA''_r$)
3	1	–	17.50	17.267
2	1	14.47	14.78	14.286
1	1	8.00	8.04	7.731
-2	1	-0.479	-1.99	-0.565[a]

a Unrefined data

TABLE II

Spectral properties of the Cu(II)-transport site of human albumin, Cu(II)-complex of the native sequence peptide and the designed peptides

Cu(II)-complex	λ_{max}	ε_{max}
human albumin-Cu(II)	525	101
glygly-L-his-Cu(II) ($MH_{-2}A$)	525	103
glygly-L-his-NHMe-Cu(II) ($MH_{-2}A$)	525	103
L-Asp-L-ala-L-his-NHMe-Cu(II) ($MH_{-2}A$)	525	103

Considering the fact that there is one species present at neutral pH, the spectral parameters of that species can be obtained rather easily. The results are listed in Table II. It is clear that the spectral parameters of Cu(II)-albumin and Cu(II)-peptides are almost identical, indicating a structural similarity.

The proton displacement patterns of the Cu(II)-complexes of the designed peptides provide additional support for the involvement of -NH_2, two peptide nitrogens and an imidazole nitrogen in the Cu(II)-binding. Furthermore, crystal of Cu(II)-glycylglycyl-L-histidine-N-methyl amide was obtained at physiological pH and the structure has recently been solved by Camerman et al. (1976) as shown in Figure 9. The peptide binds Cu(II) in exactly the same way for which it has been designed, i.e. to mimic the Cu(II)-transport site of human albumin. Each Cu(II) is tetradentately chelated by the amino terminal nitrogen, the next two peptide nitrogens and the histidyl nitrogen of a single tripeptide molecule in a slightly distorted square planar arrangement.

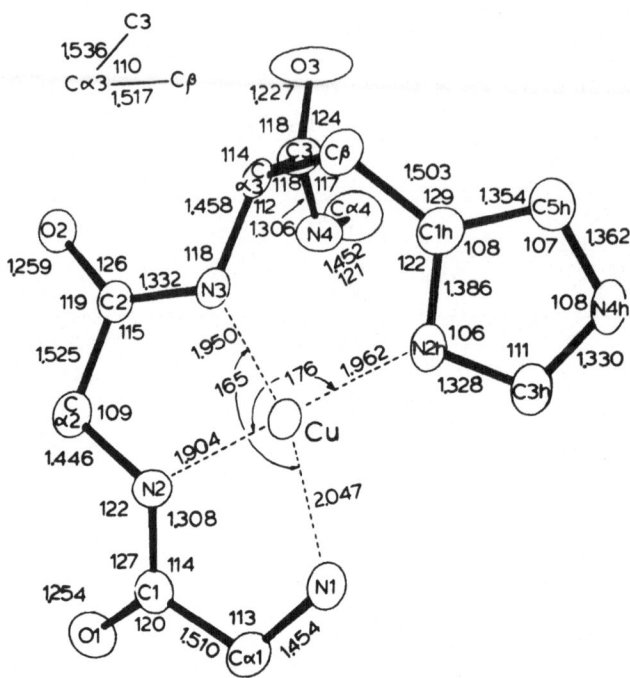

Fig. 9.   Structure of the Cu(II)-complex of the designed
peptide, glycylglycyl-L-histidine-N-methyl amide by X-ray
crystallographic analysis.

The Cu...N distance range between 1.90 - 2.05 Å, with N...Cu...N angles
between 165° and 176°.

3.4.2. Functional Studies

Copper(II) bound to albumin and amino acids are known as the transport
form of Cu(II) in human blood (Sarkar and Kruck, 1966; Neuman and
Sass-Kortsak, 1967). Interestingly, a ternary complex albumin-Cu(II)-
amino acid was also detected by Sarkar and Wigfield, (1968) which
raised the possibility of its functional role in the transportation
of Cu(II). It has been proposed that the following equilibria may be
operative in the physiological state (Lau and Sarkar, 1971),

Copper(II) + amino acid $\rightleftharpoons$ Cu(II)-amino acid   (11

Cu(II)-amino acid + albumin $\rightleftharpoons$ albumin-Cu(II)-amino acid   (12

Albumin-Cu(II)-amino acid $\rightleftharpoons$ albumin-Cu(II) + amino acid   (13

The above mechanism is thought to play an important role in the exchange
of Cu(II) between a macromolecule and a low molecular weight substance
which, in turn, can readily be transported across the biological membrane

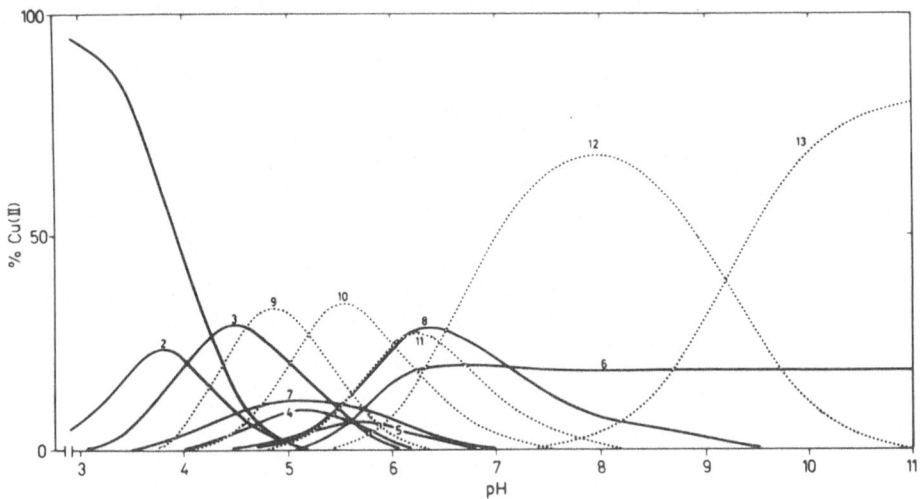

Fig. 10. Species distribution in L-histidine-Cu(II)-glycylglycyl-L-histidine system as a function of pH. Curve 1, unbound Cu(II); Curve 2, MHB; Curve 3, MB; Curve 4, MA; Curve 5, $MH_{-1}A$; Curve 6, $MH_{-2}A$; Curve 7, MHB_2; Curve 8, MB_2; Curve 9, MH_2AB; Curve 10, MHAB; Curve 11, MAB; Curve 12, $MH_{-1}AB$; Curve 13, $MH_{-2}AB$.

For the above reasons, it was important to investigate whether the designed molecule also formed the ternary complexes. At the same time, it was considered that a small molecule, which is the simplification of the natural binding site sequence, can be subjected to many important studies which are otherwise not feasible with a protein molecule. A detailed study was undertaken to delineate the species in the ternary systems: glycylglycyl-L-histidine-Cu(II)-L-histidine and glycylglycyl-L-histidine-N-methyl amide-Cu(II)-L-histidine. The species distributions are presented in Figures 10 and 11. Their stability constants are presented in Table III. Indeed, the designed molecules also formed ternary complexes as does albumin. Although it was not possible to distinguish between different protonated species of the ternary complexes of albumin, this could be clearly established in the case of a small peptide.

In order to characterize each species by their individual spectral parameters, calculated spectra were obtained by utilizing the following equation,

$$A_\lambda = \Sigma\ \varepsilon_{pqrs,\lambda}\ [M_pH_qA_rB_s] \tag{14}$$

where A_λ is the total absorbance at wavelength λ, 1 cm light path, and ε_{pqrs} is the molar extinction of the species $M_pH_qA_rB_s$ of concentration $[M_pH_qA_rB_s]$ at that wavelength. The calculated spectra are presented in Figure 12.

Fig. 11. Species distribution in the ternary system L-histidine-Cu(II)-glycylglycyl-L-histidine-N-methyl amide as a function of pH. Curve 1, unbound Cu(II); Curve 2, MHB; Curve 3, MB; Curve 4, MHB_2; Curve 5, MB_2; Curve 6, $MH_{-2}A$; Curve 7, MH_2AB; Curve 8, $MH_{-1}AB$; Curve 9, $MH_{-2}AB$.

TABLE III

Log stability constants (log β_{pqrs} and log β'_{pqrs}) of the ternary complex species $M_pH_qA_rB_s$ (M = Cu(II), A = glycyl-glycyl-L-histidine-N-methyl amide, A' = glycylglycyl-L-histidine, B = L-histidine) in 0.15 M NaCl at 25 °C

p	q	r	s	log β_{pqrs} ($M_pH_qA_rB_s$)	log β'_{pqrs} ($M_pH_qA'_rB_s$)
1	2	1	1	27.75	28.83
1	1	1	1	–	23.61
1	0	1	1	–	17.56
1	-1	1	1	11.31	11.05
1	-2	1	1	3.02	1.70

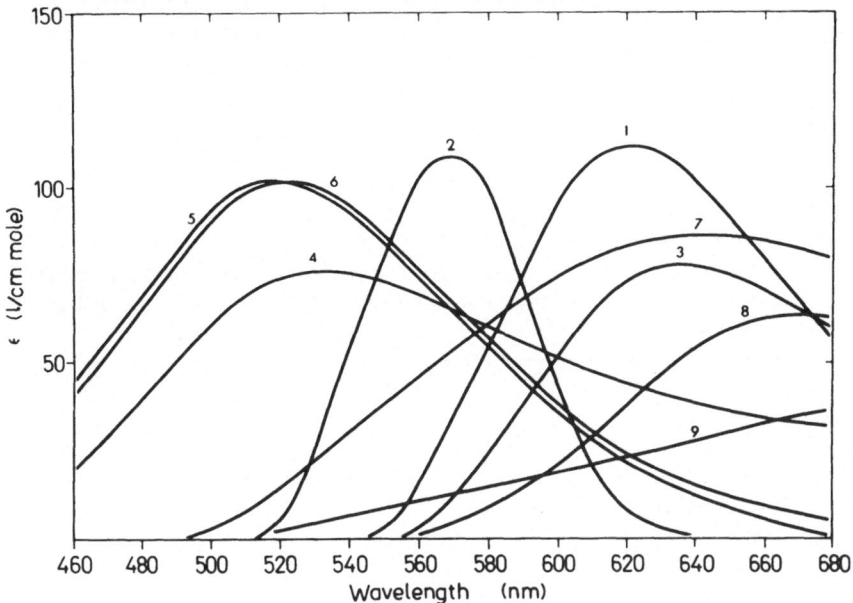

Fig. 12.   Computed spectra of the species in the L-histidine-
Cu(II)-glycylglycyl-L-histidine system. Curve 1, MH_2AB; Curve
2, MHAB; Curve 3, MAB, Curve 4, $MH_{-1}AB$; Curve 5, $MH_{-2}AB$;
Curve 6, $MH_{-2}A$; Curve 7, MB_2; Curve 8, MB; Curve 9, MHB.

Besides elucidating the species in the individual peptide system, it
is also necessary to compare the dissociation constant of the Cu(II)-
designed peptide with that of the Cu(II)-albumin obtained under equivalent conditions and identical definition of the constant. The dissociation of the Cu(II)-peptide complex and the competition of Cu(II)
between albumin and peptide can be expressed (Lau and Sarkar, 1974) as,

$$Cu(II)\text{-peptide} \xleftrightarrow{K_{D,Cu(II)\text{-peptide}}} Cu(II) + \text{peptide} \qquad (15)$$

and

$$Cu(II)\text{-albumin} + \text{peptide} \xrightleftharpoons{K_{eq}} Cu(II)\text{-peptide} + \text{albumin} \qquad (16)$$

therefore,

$$K_{D,Cu(II)\text{-peptide}} = \frac{[Cu(II)][\text{peptide}]}{[Cu(II)\text{-peptide}]} = \frac{K_{D,Cu(II)\text{-albumin}}}{K_{eq}} \qquad (17)$$

The $K_{D,Cu(II)}$-peptide values were determined from the equilibrium dialysis data utilizing $^{67}Cu(II)$ as a tracer. The results presented in Table IV show a close similarity in the values of the dissociation constants.

TABLE IV
Comparison of dissociation constants of human albumin-Cu(II), native sequence peptide-Cu(II) and designed peptide-Cu(II) complexes

Complex	K_D $(10^{17}$ M)
Human albumin-Cu(II)	6.61
Glygly-L-his-Cu(II)	11.80
Glygly-L-his-NHMe-Cu(II)	2.07
L-Asp-L-ala-L-his-NHMe-Cu(II)	1.04

TABLE V
Comparison of the Cu(II)-exchange rates for the reaction between excess of Cu(II)-L-histidine$_2$ and albumin or peptides at 20° and pH 7.53 in 0.1 M N-ethylmorpholine-HCl buffer

Reactant	k_{obs} (s^{-1})
Human albumin	0.67
Glygly-L-his	0.42
Glygly-L-his-NHMe	0.56
L-Asp-L-ala-L-his-NHMe	2.70

Another important functional property of albumin in relation to its Cu(II)-transportation is its Cu(II)-exchange rate (Lau and Sarkar, 1975). The Cu(II)-exchange reactions were studied under pseudo first-order conditions in 0.1 M N-ethylmorpholine-HCl buffer at pH 7.53 and ionic strength 0.16. The reaction observed by mixing Cu(II)-L-histidine$_2$ and albumin or peptide derivatives were followed on a Durrum-Gibson stopped-flow spectrophotometer at 20°. The changes of absorption at both wavelengths 640 nm (λ_{max} for Cu(II)-L-histidine$_2$) and 525 nm (λ_{max}

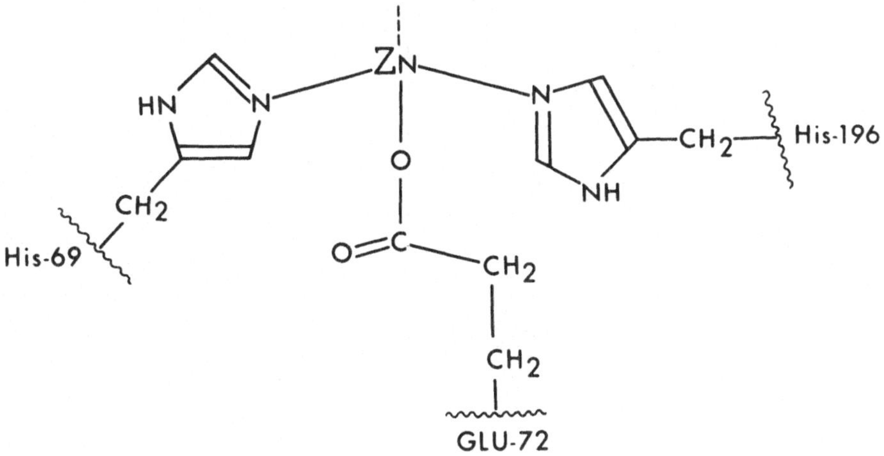

Fig. 13.   Zinc(II)-binding site of carboxypeptidase A.

for Cu(II)-albumin) were followed as the reaction proceeded. The results are presented in Table V. A very similar exchange rate is observed between albumin and the designed peptides.

4. DESIGN OF A SITE COMPOSED OF RESIDUES FROM DISTANT PARTS: ZINC(II)-BINDING SITE OF CARBOXYPEPTIDASE A

4.1. *Consideration for the Design*

Carboxypeptidase A is a Zn(II)-containing metalloenzyme which catalyzes the hydrolysis of free carboxyl C-terminal of peptides and esters (Vallee and Neurath, 1955). If Zn(II) is removed from the native enzyme, a catalytically inactive apoenzyme results. The unique role of Zn(II) and other transition metals in the activity and specificity of this enzyme has been the subject of many investigations (Vallee, et al., 1960; Coleman and Vallee, 1960). The X-ray crystallographic studies at 2 Å resolution by Lipscomb et al. (1969) have demonstrated the three dimensional structure of the enzyme and revealed the molecular architecture around the active center. The ligands involved in the Zn(II)-binding have been identified as His[69], Glu[72] and His [196] (Lipscomb et al., 1969; Bradshaw et al., 1969), (Figure 13).

In designing the Zn(II)-binding site of carboxypeptidase A, a very interesting feature has emerged. A critical survey of several metallo-enzymes, whose crystal structure has been elucidated (Lipscomb et al., 1969; Matthews et al., 1972; Liljas et al., 1972; Kannan et al., 1975; Herriot et al., 1970) reveals the natural occurrence of one to three amino acid residues in between two ligands for the metal at the active site (Table VI). Other ligands originate from a distant part of the molecule. Taking into account these features as well as the geometry of the Zn(II)-binding according to the criteria discussed above, a

TABLE VI
Metal-binding amino acid residues in metalloenzymes

Enzyme	Metal	Amino acid residues binding metal	References
Carboxypeptidase A	Zn(II)	His-69, Glu-72, His-196	Lipscomb, et al. (1969)
Thermolysin	Zn(II)	His-142, His-146, Glu-166	Matthews, et al. (1972)
Carbonic Anhydrase B	Zn(II)	His-94, His-96, His-119	Liljas, et al. (1972)
Carbonic Anhydrase C	Zn(II)	His-93, His-95, His-118	Kannan, et al. (1973)
Rubredoxin	Fe(III)	Cys-6, Cys-9, Cys-38, Cys-41	Herriot, et al. (1970)

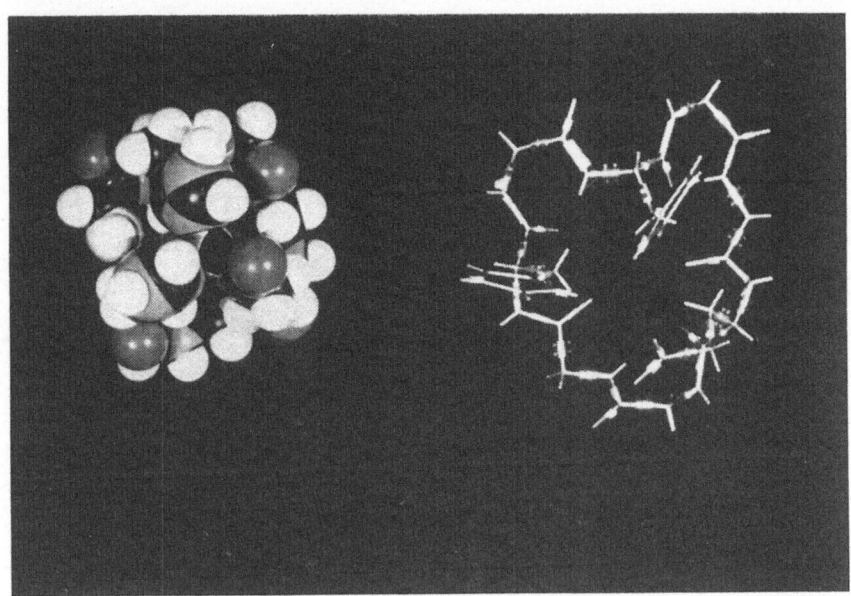

Fig. 14. Models showing the design of a cyclic octapeptide to mimic the Zn(II)-binding segment of the active site of carboxypeptidase A. Kendrew models show the native Zn(II)-binding ligands as side chains above the cyclic peptide backbone. CPK models show Zn(II)-binding to the designed octapeptide.

cyclic octapeptide of the sequence

Gly-L-Glu-Gly-Gly-L-His-Gly-L-His-Gly
    |_____|

was designed. Both CPK and Kendrew models indicated that the peptide provided the satisfactory requirements for the Zn(II)-binding geometry (Figure 14). This molecule retains the native Zn(II)-ligands interspaced by glycine residues. The γ-carboxyl moiety of the glutamate and the imidazole residues of twe histidines in the model seemed to interact well with the tetrahedral Zn(II). The spaces in between the ligand amino acid residues were filled with glycyl residues. Glycyl residues were chosen mainly due to their being structurally the simplest of residues.

4.2. *Theoretical Conformational Analysis*

The first phase of the study consisted of generating an acceptable set of atomic coordinates for the linear peptide. This was obtained by using a cartesian coordinate program from standard bond lengths, bond angles and dihedral angles. Standard tetrahedral geometry was assumed for sp^3 carbon atoms. A contact scan was made, using the procedure of Nemethy and Scheraga (1965) to select sterically allowed conformations, whose

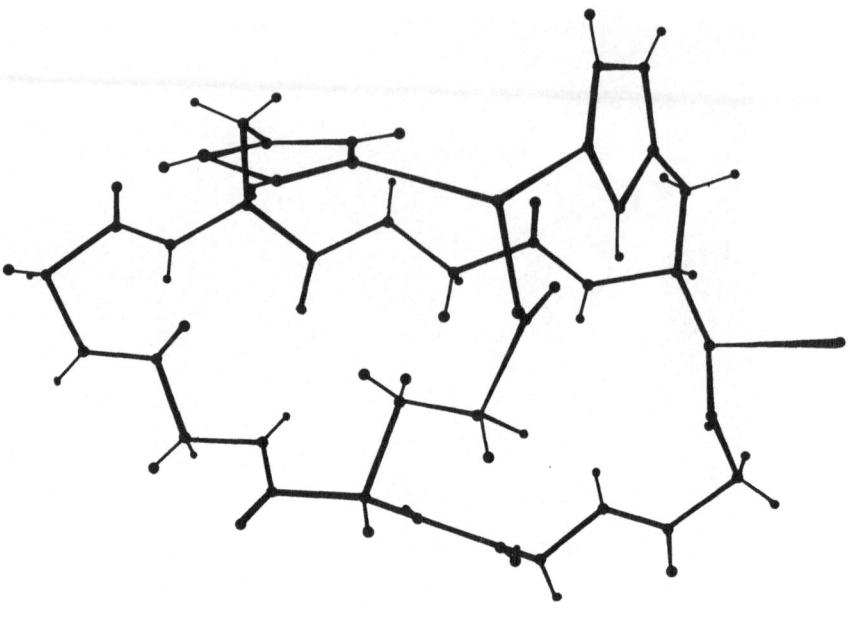

Fig. 15. One of the minimum energy conformations of the
Zn(II)-cyclic octapeptide designed to mimic the Zn(II)-
binding segment of the active site of carboxypeptidase A.

end to end distance, i.e., between the two loose ends of the octapeptide
were within 1.5 Å to 4.0 Å. Ring closure was attempted by systematically
incrementing the backbone angles, ϕ and ψ, until the two loose ends
were within reasonable bonding distance to each other, taking care not
to violate the steric criteria. Selected conformations satisfying the
ring closure constraints were minimized, using a total optimization
of molecular geometry program, based on a modified Newton-Raphson
method. The total conformational energy was assumed to consist of contributions from non-bonded, electrostatic, torsional bond angle defor-
mation and bond stretching interaction. Solvent contribution was not
considered in this calculation.

It was realized, however, that in order to obtain a reasonable
geometry for the metal complex with standard metal ligand bond distances
and angles (Freeman, 1967), a slight distortion of the conformation of
the free cyclic octapeptide was necessary. Fixing Zn(II) at the origin
of a right handed coordinate system, a systematic attempt to increment
the backbone and side chain torsional angles (i.e., ϕ_i, ψ_i and χ_i^n) in
order to achieve standard Zn(II)-ligand bond lengths and bond angles,
resulted in several conformations of the complex. By repeating the
minimization process on the satisfactory conformations, a minimum
energy conformation for metal complex was obtained. One of the minimum
energy conformations obtained is schematically shown in Figure 15. The
conformation is stabilized by intramolecular hydrogen bonding. The Zn(II

TABLE VII

Comparison of the bond angles and bond lengths of Zn(II)-binding in the
native carboxypeptidase A; those predicted from the crystal structures
of Zn(II)-complexes and those calculated for the Zn(II)-complex of the
designed cyclic octapeptide

Parameters	Native carboxy-peptidase (Ludwig et al., 1973)[a]	Predicted from crystal struc-tures of Zn(II)-complexes (Freeman, 1973)[b]	Calculated values for the Zn(II)-complex of the designed cyclic octapeptide
Bond angles			
$N_{69} - Zn - N_{196}$	$86°$	$104°$	$102°$
$N_{69} - Zn - O_{72}$	$99°$	$80°$	$110°$
$N_{196} - Zn - O_{72}$	$143°$	$160°$	$150°$
Bond lengths			
$Zn - O_{72}$		2.1 Å	2.05 Å
$Zn - N_{69}$		2.0 Å	2.01 Å
$Zn - N_{196}$		2.0 Å	2.07 Å

[a] According to these authors, even allowing for errors of $\pm 10°$ in the
bond angles, the large distortion from tetrahedral geometry is significant.

[b] These values are predicted by Freeman on the basis of the structural
parameters of Zn(II)-complexes of amino acids and peptides. There is
the possibility that the carboxyl group may coordinate the Zn(II)
through both of its O(Carboxyl) atoms. If this is the case, then the
second bond is about 2.6 Å long and the angle between the two Zn—O
(carboxyl) bonds is about $60°$.

has a distorted tetrahedral coordination geometry. In Table VII, the
bond lengths and bond angles obtained by calculation are compared with
those obtained by Lipscomb et al., (1973) in carboxypeptidase A and
those predicted by Freeman, (1973) from crystal structure of Zn(II)-
complexes.

4.3. *Synthesis of the Designed Molecule*

The designed cyclic octapeptide (I) was obtained from the corresponding linear octapeptide BOC-Gly-γ-OBu^t-L-Glu-Gly-Gly-L-His-Gly-L-His-
Gly-OBzlNO$_2$ (II) which was synthesized according to the scheme shown
in Figure 16. Mixed anhydride procedure (4-methylmorpholine and iso-
butylchloroformate) was used exclusively to build the fragment BOC-

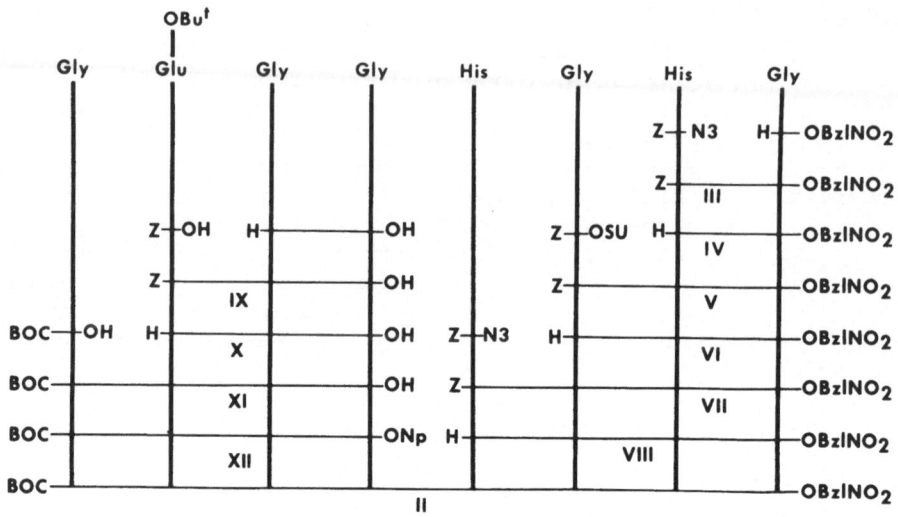

Fig. 16. Scheme of synthesis of the linear octapeptide II.

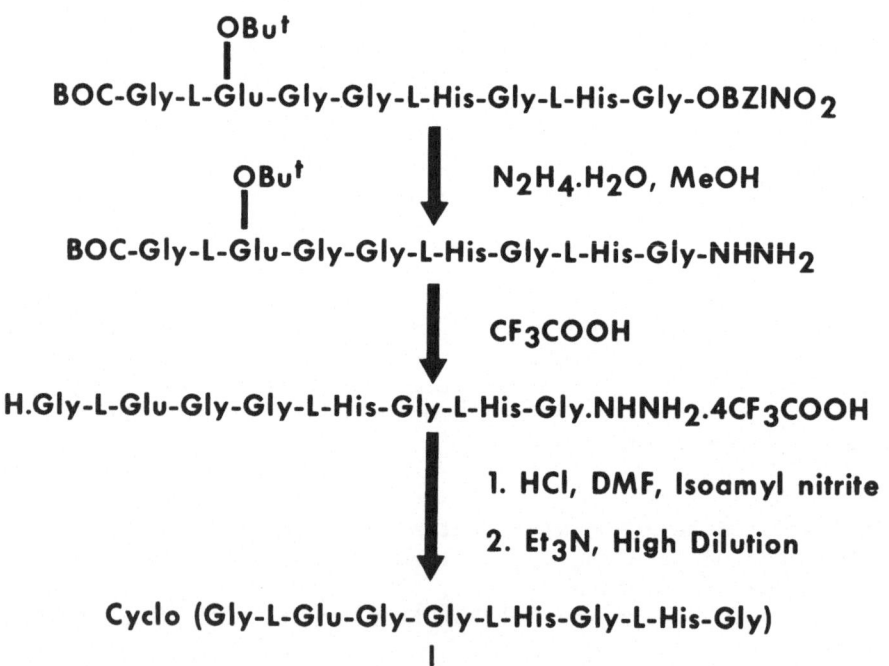

Fig. 17. Synthesis of the cyclic octapeptide I.

Gly-γ-OBu^t-L-Glu-Gly-Gly-ONp (XII). Condensation of glycylglycine with the mixed anhydride of N-CBZ-γ-OBu^t-L-glutamic acid followed by catalytic hydrogenation of the product afforded the tripeptide X. Repetition of the coupling using X and the mixed anhydride of N-t-BOC-glycine

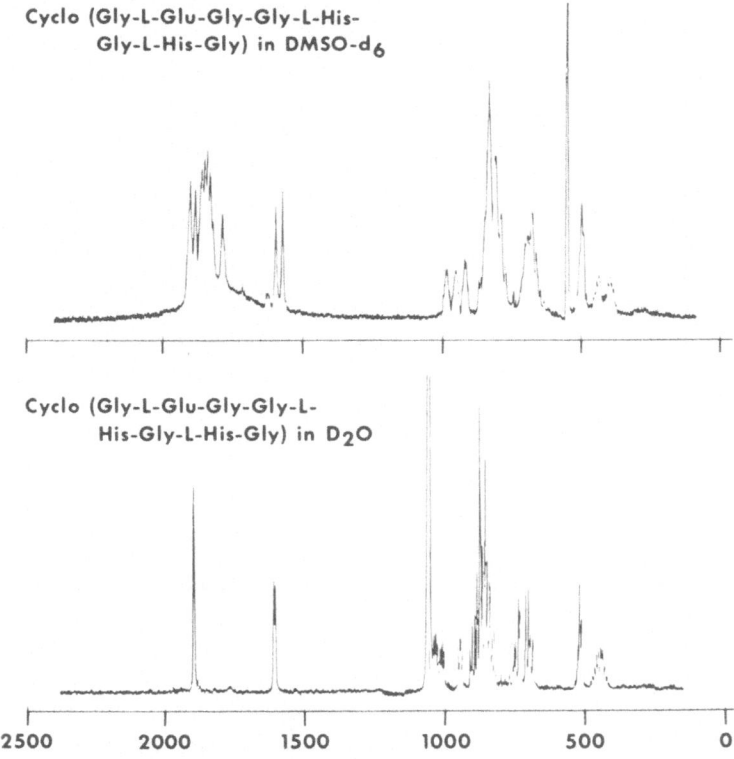

Cyclo (Gly-L-Glu-Gly-Gly-L-His-
Gly-L-His-Gly) in DMSO-d$_6$

Cyclo (Gly-L-Glu-Gly-Gly-L-
His-Gly-L-His-Gly) in D$_2$O

2500   2000   1500   1000   500   0

Fig. 18.   220 MHz PMR spectra of the cyclic octapeptide I in
DMSO-d$_6$ and D$_2$O.

urnished the tetrapeptide XI which was converted to the active ester
II by treatment with p-nitrophenol and DCC. Both the histidine residues
n the fragment L-His-Gly-L-His-Gly-OBzlNO$_2$ (VIII) were introduced
hrough the intermediates of N-CBZ-L-histidyl azide and the glycine via
ts N-hydroxysuccinimide ester. Hydrogen bromide in acetic acid was
sed to cleave the CBZ group used to protect the α-amino function of
he various intermediates in the synthesis of this fragment. Coupling
f the tetrapeptide fragments XII and VIII proceeded smoothly to produce
he linear octapeptide II in good yield.

The linear octapeptide II was converted to the corresponding hy-
razide which was cyclized via the azide method following the procedure
f Kopple et al. (1973). Figure 17 illustrates the various steps involved
n the cyclization of the linear peptide. A homogeneous material was
solated from the cyclization reaction by repeated gel filtration fol-
owed by purification by countercurrent distribution. The product ob-
ained was found to be ninhydrin negative and Pauly's positive. Further
onfirmation of the structure of this material was obtained from the
20 MHz proton magnetic resonance spectrum (Figure 18) of this material
hich, upon integration, indicated the proper ratios of the various
inds of protons to be expected of the cyclic octapeptide I.

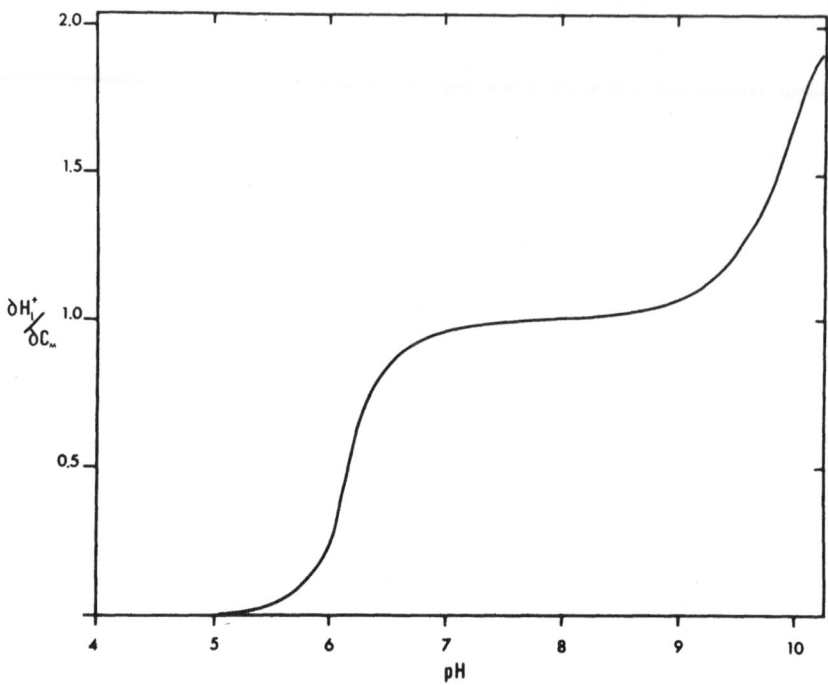

Fig. 19. Proton displacement $\delta H_1^+/\delta C_M$ as a function of pH for
the Zn(II)-cyclic octapeptide system.

4.4. *Zinc(II)-Binding Studies*

The Zn(II)-binding studies with the cyclic octapeptide has just begun
and only preliminary results are available. The unrefined pK_a values
of the side chain residues are: pK_{a_1} (Glu-COOH) $= 4.20$, pK_{a_2} (His-ImH$^+$)
$= 6.16$ and pK_{a_3} (His-ImH$^+$) $= 7.18$. The proton displacement in the Zn(II)
cyclic octapeptide system due to metal as a function of pH, is presented
in Figure 19. It is clear that the complexation starts around pH 5.2.
In the absence of the complete species distribution, it is not possible
to draw any conclusion about the possible mode of Zn(II)-binding to the
peptide. However, from the nature of the $\delta H_1^+/\delta C_M$ vs pH curve, it is
certain that Zn(II) facilitates proton displacement from protonated
ligands having pK_a values of 6 or higher.
      In an independent investigation by thin-layer chromatography, Zn(II
cyclic octapeptide complex at pH ~ 7 showed Pauly's negative reaction.
Since Pauly's reaction is a sensitive test for the histidine imidazole
group, it appears that the Zn(II) is bound to both histidine residues
in the cyclic octapeptide. The involvement of the carboxyal side chain
of the glutamic acid residue remains to be determined. Detailed structural studies are currently in progress. Attempts are also underway to

TABLE VIII
Metal-dependent diseases

Metal	Deficiency disease	Diseases associated with metal accumulation and intoxication
Iron	Anemia	Hemochromatosis
Copper	Anemia Kinky-hair syndrome	Wilson's disease
Zinc	Dwarfism Gonadal failure	Metal-fume fever
Cobalt	Anemia	Heart failure Polycythemia
Manganese	Gonadal dysfunction Skeletal absormalities	Ataxia
Chromium	Abnormal glucose metabolism	
Selenium	Liver necrosis White muscle disease	Blind staggers (cattle)
Lead		Anemia Encephalitis Peripheral neuritis
Nickel		Cancer
Arsenic		Cancer
Cadmium		Nephritis
Mercury		Encephalitis Peripheral neuritis Minamata disease

crystallize the Zn(II)-cyclic octapeptide complex for crystallographic
investigation.

5. APPLICATION AND FUTURE POSSIBILITIES

5.1. *Metal-Related Diseases*

There are many metabolic reactions that are related to metal ions which
maintain normal body functions. These are essential metals. However,
excess of the metals cause severe toxicity in humans. Table VIII shows
the metal related diseases caused by both deficiency states and toxic
accumulations. The chelating agents which are presently used to treat
metal toxicity have the serious drawback of being non-specific. As a
consequence, some other essential metals are also removed from the
body while removing one particular metal causing the toxic effects.
Also, these agents often cause many undesirable side effects. They
often have delayed action due to the lack of specificity. A certain
degree of selectivity and specificity of binding must be achieved to
overcome this problem. The foregoing sections have shown that, once the
nature of a biological metal-transport mechanism is known and the metal-

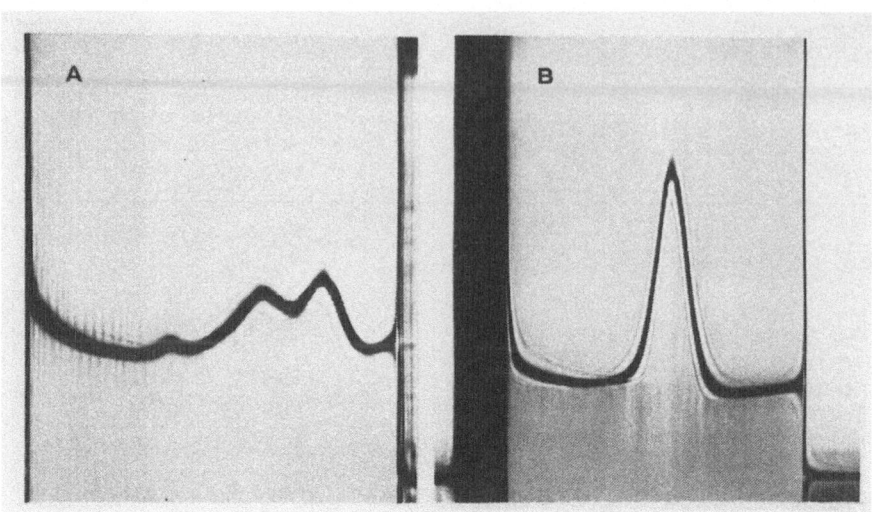

Fig. 20. Albumin polymerization by Cu(II) and depolymerization
by glycylglycyl-L-histidine as studied by velocity ultracentrifugation. (A) the presence of monomer, dimer and tetramer
in a solution containing 1:14 molar ratio of albumin and Cu(II)
in 0.5 M NaClO$_4$ at 8°; (B) the same solution in the presence
of 4 molar equivalents of glycylglycyl-L-histidine showing
depolymerization to monomer only.

binding site, usually on a carrier protein, is identified, one can design and synthesize a small molecule which mimics the specific macromolecular binding site. This approach of designing and using a small
molecule to mimic and thus assume certain functions of a macromolecule
could have important biomedical applications in situations where over-
loading by metals is a health problem.

The design and synthesis of the Cu(II)-transport site of human
albumin opens up the possibility of mobilizing Cu(II) in patients
suffering from Cu(II)-loaded situations, such as in Wilson's disease.
This is a genetic disease, leading to an excessive accumulation of
Cu(II) in the liver, kidney and brain, which causes liver failure,
malfunction of kidney and to various neurological abnormalities, and
death if unrecognized or untreated. At present, much of the suffering
can be avoided if the patient receives early treatment with D-penicillamine. However, there are many problems associated with its use.

The designed peptide glycylglycyl-L-histidine and its methyl amide
derivative have been shown to mobilize Cu(II) in vitro. For example,
excess Cu(II) induces polymer formation of human altumin (Österberg,
et al. 1975). Addition of the designed peptide completely depolymerized
the Cu(II)-albumin polymers, indicating that the designed peptide can
compete effectively with the specific Cu(II)-binding site of human
albumin. The depolymerization of Cu(II)-albumin by the peptide is shown
in Figure 20. Results from an in vitro experiment also show that glycyl-

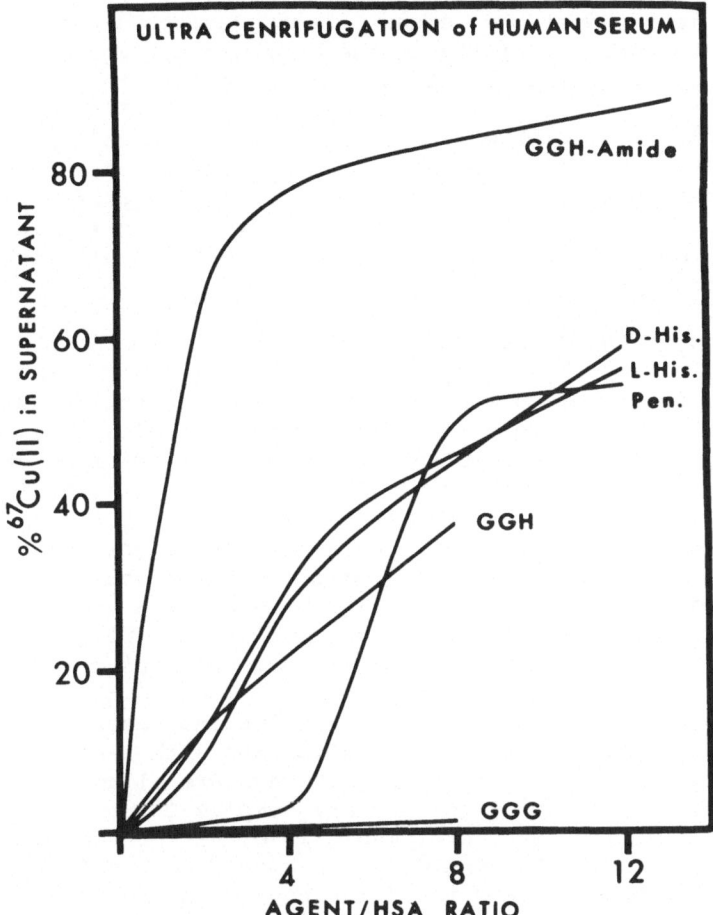

Fig. 21. The ability of D-penicillamine, D- and L-histidines, triglycine, glycylglycyl-L-histidine and glycylglycyl-L-histidine-N-methyl amide to bind Cu(II) in the presence of whole human serum. Copper(II) in the low molecular weight fraction of serum was obtained by ultracentrifugation and expressed as % ^{67}Cu(II) in the supernatant using different agents and ^{67}Cu(II) in the serum. Varying ratios of agent to albumin were used to study the effect of these agents as a function of concentration in a fixed concentration of albumin.

glycyl-L-histidine-N-methyl amide removes about 80% of the Cu(II) from the serum components at a concentration range where D-penicillamine has scarcely removed any (Figure 21). Since the peptide molecule may be degraded in the body, several possibilities such as chemical modification of the peptide, use of D-amino acid residues instead of L-amino acids, or even designing non-peptide molecules incorporating four nitrogen ligands in a slightly distorted square planar geometry, can be explored. This may increase their stability in the body without changing their Cu(II)-binding characteristics. It must be emphasized that, apart from a high degree of selectivity and affinity, the designed molecules must have stability in vivo, with minimal or no toxicity problems associated and rapid excretion of the complex in order to qualify for any future therapeutic applications. Apart from the removal of a particular metal in metal overloaded situations, the possibility also exists in the use of the designed molecules in making essential metals available in a deficiency state as well.

5.2. *Elucidation of the Mechanism*

The design of a small molecule mimicking the native metal-binding site of a protein makes it possible to implement many studies that are not possible with a large protein molecule. Through the detailed study of the designed peptides which mimic the Cu(II)-transport site, it has been possible to detect the species and study their structural features and Cu(II)-exchange characteristics. The physiological significance of the ternary complex L-histidine-Cu(II)-albumin in the biological transport of Cu(II) has been mentioned earlier. The studies with the designed peptides indeed demonstrate the rate controlling role played by the ternary complex in the exchange of Cu(II) between albumin and L-histidine in order to maintain the equilibrium concentrations. A postulated scheme of the exchange pathways can be derived, taking into account known structural features of several ternary complexes with L-histidine, Cu(II) and the designed peptide. The exchange pathways are illustrated in Figure 22. In the formation of the ternary complex C from Cu(II)-L-histidine$_2$ (A), the first step presumably involves the fast dissociation of one histidine molecule to form Cu(II)-L-histidine$^+$, and allow the NH$_2$-terminal amino nitrogen and the first peptide nitrogen of the protein to enter as an anchoring point to form the intermediate B. This is followed with the displacement of the amino nitrogen of the remaining histidine by the second carbonyl oxygen of the protein to form the ternary complex C which is then converted to another ternary complex D through structural rearrangement. Since both ternary complexes C and D are thermodynamically favourable, this latter process would be expected to be rather sluggish. Once the ternary complex D is formed, three coordination sites of Cu(II) are occupied by the groups from the albumin, the displacement of the imidazole nitrogen of histidine by the imidazole nitrogen of albumin from the histidine in the third position (E and F) would seem to be much more facilitated due to both the chelation and steric effects involved. In the reverse reaction, the same rational is also implied.

Even though the three dimensional structures of several metalloenzymes are known, the role of metals in many of these enzymes is not clearly understood. It is envisaged that designing small molecules to mimic the native metal-binding site is an important step in this direction. Work along these lines is underway with carboxypeptidase A discussed here, as well as with carbonic anhydrase A and C (Appleton and Sarkar, 1974, 1975a, 1975b).

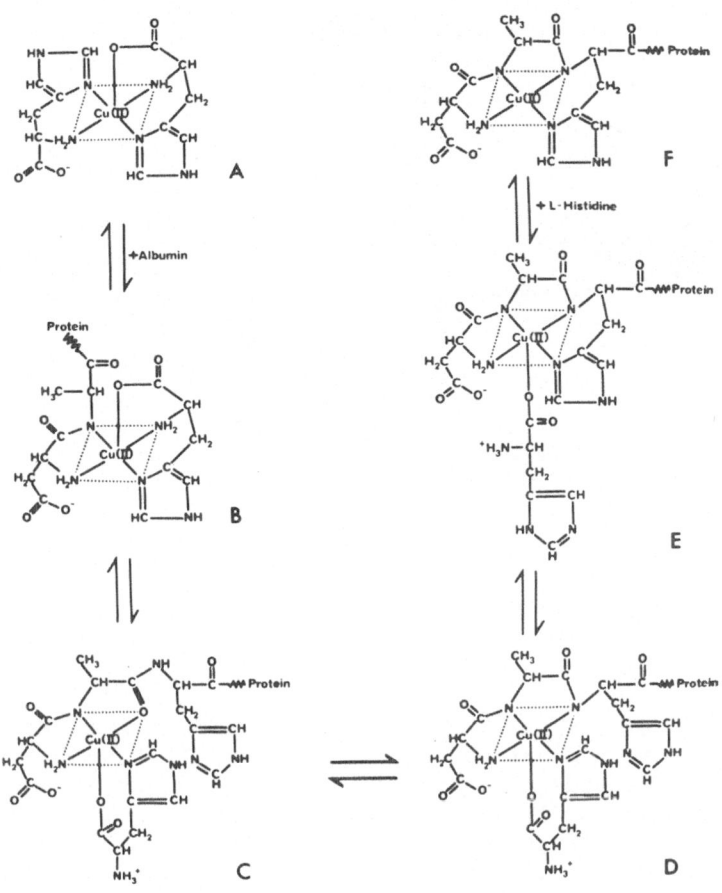

Fig. 22. The proposed reaction paths for the Cu(II)-exchange from L-histidine to albumin.

5.3. *Design of the Total Active Site of a Metalloenzyme*

Further knowledge and investigation will be necessary before we attain the design of the total active site of a metalloenzyme. However, when the substrate binding site, the catalytic residues, the associated geometry and the microenvironment at the active site of a metalloenzyme are firmly established, there is a possobility of designing the total active site of the enzyme. Metals play various roles in a metalloenzyme, i.e. as a structural component, as an anchorpoint for substrate binding or as an active catalytic entity. Furthermore, metals may be involved in any or all of these functions at the same time. Incorporation of a metal ion in the designed molecule will have to be judged on the basis of its intended functional role at the active site. The position of the metal ion would have to be decided on the basis of the relative geometry of the other functional residues of the intended active site in question.

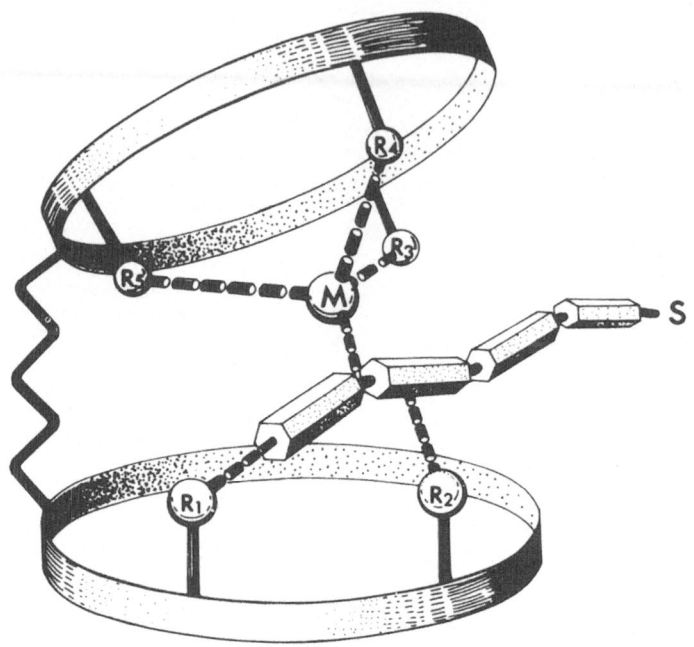

Fig. 23. Incorporating metal in the active site. Metal (M)
is bound to the native ligands (R_3, R_4, R_5) originating from
a cyclic peptide. The cyclic peptide is joined as a branch
with another cyclic peptide backbone structure. Substrate (S)
is attached simultaneously to the metal and functional residues (R_1, R_2) originating from the cyclic peptide backbone.

The design scheme should attempt to closely retain the native geometry
of the functional residues, although the acceptable limits of tolerance
will vary in specific instances and require definition for each active
site. Figure 23 shows a schematic presentation of a design involving a
metal ion in the active site. As active sites of many enzymes possess
the hydrophobic environment, Figure 24 depicts a design of the active
site which includes the hydrophobic environment.

6. CONCLUSION

The design concepts I have discussed are aimed at simulating the native
geometry and microenvironment of the active residues found at the active
site of proteins and enzymes. Application of this concept will contribute
significantly to our understanding of the role of metals in metalloenzymes
and metal transport proteins. This approach also opens up the possibility
of new approaches to therapeutic applications dealing with the functional
sites of biological macromolecules. The increasing body of knowledge
from sequence determination, X-ray crystallography, conformational probes

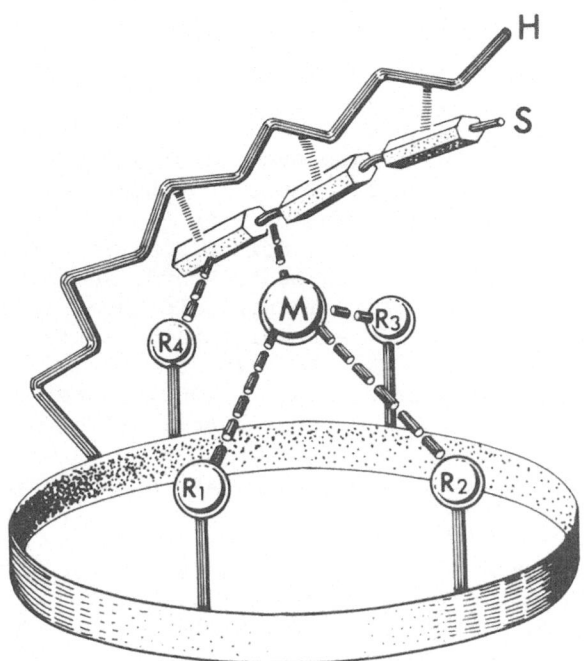

Fig. 24.   Design of an active site having both metal and a
hydrophobic environment. A wide area of substrate (S) is
interacting with the hydrophobic environment (H) provided
by a lipophilic long chain which is attached to the backbone
structure. The substrate is interacting with the metal (M)
which is bound to ligands (R_1, R_2, R_3) originating from the
cyclic peptide backbone.

nd theoretical conformational analyses greatly facilitates the present
ask. A molecular design chemist lacks the time span of nature to produce
ophisticated and interesting molecules, but surely he has the great
dvantage of learning from nature what his designed molecule should be
o perform a certain function.

CKNOWLEDGEMENT

he author acknowledges the excellent collaboration, enthusiastic support
nd fruitful discussions with Dr. S. Lau, Mr. T.P.A. Kruck, Mrs. J.W.
ixon Parkes, Dr. D.W. Appleton, Dr. K.S.N. Iyer and Dr. V. Renugopalak-
ishnan. Financial assistance was provided by the Medical Research
ouncil of Canada.

REFERENCES

Anfinsen, C.B. and Scheraga, H.A.: 1975, 'Experimental and Theoretical
  Aspects of Protein Folding', Ad. Protein Chem. 29, 205-300.
Appleton, D.W. and Sarkar, B.: 1971, 'The Absence of Specific Copper(II)-
  Binding Site in Dog Albumin', J. Biol. Chem. 246, 5040-5046.
Appleton, D.W. and Sarkar, B.: 1974, 'The Activity Related Ionization
  in Carbonic Anhydrase', Proc. Nat. Acad. Sci. U.S.A. 71, 1686-1690.
Appleton, D.W. and Sarkar, B.: 1975a, 'Carbonic Anhydrase: Studies of
  Uncharged Metal Directed Inhibitors', Bioinorganic Chem. 4, 309-321.
Appleton, D.W. and Sarkar, B.: 1975b, 'A Model for the Active Site of
  Co(II)-Carbonic Anhydrase', Proc. Can. Fed. 18, 5.
Bradshaw, R.A., Shearer, W.T., and Gurd, F.R.N.: 1968, 'Sites of Binding
  of Copper(II) Ion by Peptide (1-24) of Bovine Serum Albumin', J. Biol.
  Chem. 243, 3817-3825.
Bradshaw, R.A. and Peters, T. Jr.: 1969, 'The Amino Acid Sequence of
  Peptide (1-24) of Rat and Human Serum Albumins', J. Biol. Chem. 244,
  5582-5589.
Camerman, N., Camerman, A., and Sarkar, B.: 1976, 'Molecular Design to
  Mimic the Copper(II)-Transport Site of Human Albumin. The Crystal and
  Molecular Structure of Copper(II)-Glycylglycyl-L-Histidine-N-Methyl
  Amide Mono Aquo Complex', Can. J. Chem. 54, 1309-1316.
Coleman, J.E. and Vallee, B.L.: 1960, 'Metallocarboxypeptidases', J.
  Biol. Chem. 235, 390-395.
Deber, C.M., Madison, V., and Blout, E.R.: 1976, 'Why Cyclic Peptides:
  Complementary Approaches to Conformation', Accounts Chem. Res.
  (in press).
Dixon, J.W. and Sarkar, B.: 1974, 'Isolation, Amino Acid Sequence and
  Copper(II)-Binding Properties of Peptide (1-24) of Dog Serum Albumin',
  J. Biol. Chem. 240, 5972-5977.
Donohue, J. and Caron, A.: 1964, 'Refinement of the Crystal Structure
  of Histidine Hydrochloride Monohydrate', Acta Crystallographica 17,
  1178.
Freeman, H.C.: 1967, 'Crystal Structures of Metal-Peptide Complexes',
  Adv. Protein Chem. 22, 258-424.
Freeman, H.C.: 1973, 'Metal Complexes of Amino Acids and Peptides', in
  G.L. Eichhorn (ed.), Inorganic Biochemistry, vol. 1, Elsevier,
  Amsterdam, London and New York, 1973, pp. 121-166.
Hehre, W.G., Ditchfield, R., Stewart, R.F., and Pople, J.A.: 1970,
  'Self-Consistent Molecular Orbital Methods. IV. Use of GAUSSIAN
  Expansions of Slater-Type Orbitals. Extension to Second-Row Molecules',
  J. Chem. Phys. 52, 2769-2773.
Herriot, J.R., Sieker, L.C., Jensen, L.H., and Lovenberg, W.: 1970,
  'Structure of Rubredoxin: An X-Ray Study to 2.5 Å Resolution', J.
  Molecular Biol. 50, 391-406.
Kannan, K.K., Notstrand, B., Fridborg, K., Lövgren, S., Ohlsson, A.,
  and Petef, M.: 1975, 'Crystal Structure of Human Erythrocyte Carbonic
  Anhydrase B. Three Dimensional Structure, a Nominal 2.2 Å Resolution',
  Proc. Nat. Acad. Sci. U.S.A. 72, 51-55
Kopple, K.D., Go, A., Schamper, T.J., and Wilcox, C.S.: 1973, 'Confor-
  mation of Cyclic Peptides. VII. Cyclic Hexapeptides Containing the
  D-Phe-L-Pro Sequence', J. Amer. Chem. Soc. 95, 6090-6096.

Kruck, T.P.A. and Sarkar, B.: 1973a, 'Equilibria of the Simultaneously Existing Multiple Species in the Copper(II)-L-Histidine System', Can. J. Chem. 51, 3549-3554.

Kruck, T.P.A. and Sarkar, B.: 1973b, 'Ternary Coordination Complexes Between Copper(II) and Amino Acids and an Appraisal of the Enhancement of Ternary Complex Stability', Can. J. Chem. 51, 3555-3562.

Kruck, T.P.A. and Sarkar, B.: 1973c, 'Structure of the Species in the Copper(II)-L-Histidine System', Can. J. Chem. 51, 3563-3571.

Kruck, T.P.A. and Sarkar, B.: 1975, 'Equilibria and Structures of the Species in the Ternary System of L-Histidine, Copper(II) and Diglycyl-L-Histidine, a Peptide Mimicking the Copper(II)-Transport Site of Human Serum Albumin', Inorganic Chem. 14, 2383-2388.

Kruck, T.P.A., Lau, S., and Sarkar, B.: 1976, 'Molecular Design to Mimic the Copper(II) Transport Site of Human Albumin: Studies of Equilibria Between Copper(II) and Glycylglycyl-L-Histidine-N-Methyl Amide and Comparison with Human Albumin', Can. J. Chem. 54, 1300-1308.

Lau, S. and Sarkar, B.: 1971, 'Ternary Coordination Complex Between Human Serum Albumin, Copper(II), and L-Histidine', J. Biol. Chem. 246, 5938-5943.

Lau, S., Kruck, T.P.A., and Sarkar, B.: 1974, 'A Peptide Molecule Mimicking the Copper(II) Transport Site of Human Serum Albumin', J. Biol. Chem. 249, 5878-5884.

Lau, S. and Sarkar, B.: 1975, 'Kinetic Studies of Copper(II)-Exchange from L-Histidine to Human Serum Albumin and Diglycyl-L-Histidine, a Peptide Mimicking the Copper(II)-Transport Site of Albumin', Can. J. Chem. 53, 710-715.

Liljas, A., Kannan, K.K., Bergsten, P.C., Waara, I., Fridborg, K., Strandberg, B., Carlbom, U., Järup, L., Lövgren, S., and Petef, M.: 1972, 'Crystal Structure of Human Carbonic Anhydrase C', Nature, New Biology 235, 131-137.

Lipscomb, W.N., Hartsuck, J.A., Quiocho, F.A. and Reeke, G.N. Jr.: 1969, 'The Structure of Carboxypeptidase A. IX. The X-Ray Diffraction Results in the Light of the Chemical Sequence', Proc. Nat. Acad. Sci. U.S.A. 64, 28-35.

Ludwig, M.L. and Lipscomb, W.N.: 1973, 'Carboxypeptidase A and Other Peptidases' in G.L. Eichhorn (ed.), Inorganic Biochemistry, vol. 1, Elsevier, Amsterdam, London and New York, 1973, pp. 438-487.

Matthews, B.W., Jansonius, J.N., Colman, P.M., Schoenborn, B.P., and Dupourque, D.: 1972, 'Three Dimensional Structure of Thermolysin', Nature, New Biology 238, 37-41.

Momany, F.A., McGuire, R.F., Burgess, A.W., and Scheraga, H.A.: 1975, 'Energy Parameters in Polypeptides. VII. Geometric Parameters, Partial Atomic Charges, Non-Bonded Interactions, Hydrogen Bond Interactions, and Intrinsic Torsional Potentials for the Naturally Occurring Amino Acids', J. Phys. Chem. 79, 2361-2381.

Nemethy, G. and Scheraga, H.A.: 1965, 'Theoretical Determination of Sterically Allowed Conformations of a Polypeptide Chain by a Computer Method', Biopolymer 3, 155-184.

Neumann, P.Z. and Sass-Kortsak, A.: 1967, 'The State of Copper in Human Serum: Evidence for an Amino Acid-Bound Fraction', J. Clinical Investig. 46, 646-658.

Österberg, R., Branegård, B., Ligaarden, R., and Sarkar, B.: 1975, 'Copper(II) Induced Polymerization of Human Albumin and its Depolymerization by Diglycyl-L-Histidine. A pH Static and Ultracentrifugation Study', Bioinorganic Chem. 5, 149-165.

Peters, T. Jr. and Blumenstock, F.A.: 1967, 'Copper-Binding Properties of Bovine Serum Albumin and its Amino-terminal Peptide Fragment', J. Biol. Chem. 242, 1574-1578.

Pieroni, O., Montagnoti, G., Fissi, A., Merlino, S., and Ciardelli, F.: 1975, 'Structure and Optical Activity of Unsaturated Peptides', J. Amer. Chem. Soc. 97, 6820-6826.

Ponnuswamy, P.K. and Sasisekharan, V.: 1971 'Studies on the Conformation of Amino Acids', Intl. J. Peptide Protein Res. 3, 1-18.

Pullman, B. and Pullman, A.: 1974 'Molecular Orbital Calculations on the Conformation of Amino Acid Residues of Proteins' Adv. Protein Chem. 28, 348-526.

Ramachandran, G.N. and Sasisekharan, V.: 1968 'Conformation of Polypeptides and Proteins', Adv. Protein Chem. 23, 283-437.

Ramachandran, G.N. and Chandrasakaran, R.: 1972 'Studies on Dipeptide Conformation and on Peptides with Sequences of Alternating L and D Residues with Special Reference to Antibiotic and Ion Trnasport Peptides', in S. Lande (ed.), Progress in Peptide Research, Gordon and Beach Science Publishers Inc., New York, 1972 vol. 2, 195-215.

Renugopalakrishnan, V., Nir, S., and Rein, R.: 1975, 'Theoretical Studies on the Conformation of Peptides in Solution', in B. Pullman (ed.), Environmental Effects on Molecular Structure and Properties, D. Reidel Publ. Co. Dordrecht, Boston, 1975, 109-133.

Rossotti, F.J.C. and Rossetti, H.: 1961, The Determination of Stability Constants, McGraw Hill Inc., New York, 1961.

Sarkar, B. and Kruck, T.P.A.: 1966, 'Copper-Amino Acid Complexes in Human Serum' in J. Peisach, P. Aisen, and W. Blumberg (eds.), Biochemistry of Copper, Academic Press, New York, 1966, 183-196.

Sarkar, B. and Wigfield, Y.: 1968 'Evidence for Albumin-Cu(II)-Amino Acid Ternary Complex', Can. J. Biochem. 46, 601-607.

Sarkar, B. and Kruck, T.P.A.: 1973 'Theoretical Considerations and Equilibrium Conditions in Analytical Potentionmetry. Computer Facilitated Mathematical Analysis of Equilibria in a Multicomponent System', Can. J. Chem. 51, 3541-3548.

Scott, R.A. and Scheraga, H.A.: 1966 'Conformational Analysis of Macromolecules. III. Helical Structures of Polyglycine and Poly-L-Alanine', J. Chem. Phys. 45, 2091-2101.

Shearer, W.T., Bradshaw, R.A., Gurd, F.R.N. and Peters, T. Jr.: 1967, 'The Amino Acid Sequence and Copper(II)-Binding Properties of Peptide (1-24) of Bovine Serum Albumin', J. Biol. Chem. 242, 5451-5459.

Vallee, B.L. and Neurath, H.: 1955 'Carboxypeptidase, A Zinc Metalloenzyme' J. Biol. Chem. 217, 253-261.

Vallee, B.L., Rupley, J.A., Coombs, T.L. and Neurath, H.: 1960, 'The Role of Zinc in Carboxypeptidase' J. Biol. Chem. 235, 64-69.

Venkatachalam, C.M.: 1968, 'Stereochemical Criteria for Polypeptides and Proteins. V. Conformation of a System of Three Linked Peptide Units' Biopolymers 6, 1425-1436.

METAL COMPLEXES IN PROTEINS

LYLE H. JENSEN
Dept. of Biochemistry and Dept. of Biological Structure
University of Washington, Seattle, Wash. 9895, U.S.A.

The metalloproteins are widely distributed in living systems, and they
function in key cellular processes. The metal complexes which characterize these proteins have been studied by a variety of physical methods,
and the complete three-dimensional structures of a number of metalloproteins have been determined by single crystal, X-ray analysis. The
electron density maps on which the X-ray models are based seldom reach
atomic resolution, however, and important structural features of the
complexes may not be clear in the maps. The purpose of this account is
to indicate the detail that can be visualized at several resolutions,
to summarize the structural information on complexes in metalloproteins
that have been studied by single crystal, X-ray methods, and to discuss
and compare the complexes in several iron proteins.

1. EVALUATION OF ELECTRON DENSITY MAPS

The deductions of protein crystallographers have been based mainly on
electron density maps calculated by the Fourier series

$$\rho(x,y,z) = 1/V \Sigma\Sigma\Sigma F_{hkl} e^{i\alpha_{hkl}} e^{-2\pi i(hx + ky + lz)}$$

where $\rho(x,y,z)$ is the electron density at the point x, y, z in the unit
cell, V is the volume of the unit cell, h, k, l are the indices of the
reflections, F_{hkl} are the amplitudes of the reflections and α_{hkl} are
their phases.
   The resolution of an electron density map will depend primarily
on the minimum interplanar spacing d_{min}, of the reflections used in
the Fourier series, but it also depends on the precision of both the
reflection amplitudes and the phases. Theoretically, the limit is
0.72 d_{min} (James, 1948), but in practice protein crystallographers take
the limit as d_{min}. They speak, for example, of a 2 Å resolution data
set as one containing reflections with d_{min} = 2 Å and the map based on
such a data set as a 2 Å resolution map or simply a 2 Å map.

Data sets for crystals of small molecules can usually be readily observed to d_{min} values of 0.77 Å, the limit of CuK_α radiation. For protein crystals, on the other hand, intensities of the reflections decrease rapidly at the higher reflection angles, and in only a few cases have protein data sets been sufficiently extensive to resolve covalently bonded C, N, and O atoms. In fact, most protein sets are truncated at resolutions in the range 3 — 2 Å, often before the practical limit of the data has been reached, and in adverse cases the data may fade out at even higher d_{min} values.

Whatever the particular case, the limited data sets for protein crystals exacts its toll in the information available in the set. Thus, a 2 Å data set contains no more than $(0.77/2)^3 \approx 1/18$ of the number of reflections that would be available if all reflections to the limit of CuK_α radiation could be observed, and if the limit of a set is 3 Å, it has no more than $(0.77/3)^3 \approx 1/59$.

The iron-sulfur protein rubredoxin with an FeS_4 complex in the molecule illustrates the kinds of detail to be expected at several different resolution limits for moderately heavy atoms in a metalloprotein. In the 3 Å map, the greatest electron density corresponds to the position of the Fe atom, but because of the limited resolution, the peak is not as high as might have been expected, the electron density being no more than half again as great as in many regions of the main chain (Herriott et al., 1970). Nevertheless, the coordination was clearly seen to be four-fold, the four cysteine Sγ atoms being assumed to coordinate to the Fe atom, and the complex was evidently tetrahedral although it appeared to be considerable distorted.

At 2.5 Å resolution three of the four cysteine Sγ atoms were resolved even though the expected Fe—S bond length of 2.3 Å was less than d_{min}. An approximate position could be assigned to the unresolved cysteine S atom, Sγ42, and bond lengths and angles in the FeS_4 complex were estimated by independent observers. The Fe—S bonds ranged in length from 2.21 Å to 2.42 Å and the S—Fe—S bond angles ranged from 96° to 120°. Mean values were 2.30 Å and 109°, respectively. The range of the individual estimates were rather large, however, 0.9 Å in bond lengths and 43° in bond angles.

In the 2 Å resolution electron density map, Sγ42 was finally resolved from the Fe atom (Watenpaugh et al., 1971). Figures 1 and 2 illustrate the electron density in the 2.5 Å and 2 Å maps at the level of the metal complex. Both figures show the same 2.7 Å slice of the molecule containing the Fe atom and three of the four sulfur atoms in the FeS_4 complex along with other parts of the molecule. Sγ9 and Sγ39 in the figure are at almost the same level as the Fe atom, while Sγ6 is at a different level, overlapping the Fe atom. Sγ42 is not included in the slice. Inspection of the figures shows that both Sγ6 and Sγ39 are resolved from the Fe atom in both maps, i.e. the contours clearly show a waist between the Fe and S atoms. These sections illustrate the appearance of moderately heavy atoms at these resolutions.

Although C, N, and O atoms are not resolved in the 2 Å map, approximate positions for most of them can be assigned and these were used to initiate refinement of the model in the usual crystallographic sense, i.e., coordinates were adjusted to minimize the differences between the observed and calculated F's. The initial refinement with the 1.5 Å res-

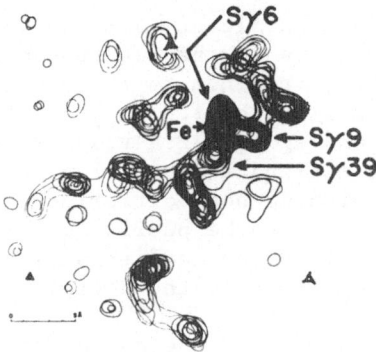

Fig. 1.   A Composite of sections from the 2.5 Å resolution electro
density map of rubredoxin. These sections represent a 2.7 Å
slice through the molecule at the level of the FeS₄ complex.
Sγ9 and Sγ39 stand out from the overlapped Fe and Sγ6 atoms.
The fourth S atom, Sγ42 lies outside the slice of electron
density shown.

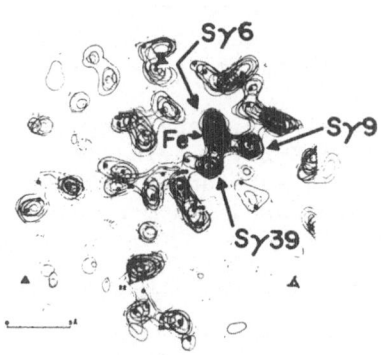

Fig. 2.   The same slice of electron density shown in Figure 1
but from the 2 Å resolution map of rubredoxin. Comparison of
the contours in the region between Sγ9 and Fe with the corresponding region in Figure 1 illustrates the improved resolution
at 2 Å. (Modified from Watenpaugh et al., 1971).

solution data set greatly improved the precision of the model, giving
standard deviations of ∼0.03 Å in the Fe-S bond lengths and ∼1-2° in
S-Fe-S bond angles (Watenpaugh et al., 1973).
    A surprising result of the refinement with the 1.5 Å data set was
the short Fe-Sγ42 bond length, less by 0.25 Å than the 2.30 Å average
of the other three Fe-S bonds. This difference exceeds 8σ, and on a

purely statistical basis would be highly significant (Jensen, 1974). The observation that Sγ42 remained unresolved in the electron density maps until the resolution reached 2 Å lent credence to the short Fe-Sγ42 bond length that emerged from refinement with the 1.5 Å data set.

A much more extensive data set extending to 1.2 Å resolution has been collected and the rubredoxin model refined further. The new results have a mean Fe-S bond length of 2.28 Å with no bond differing from the mean by more than 0.06 Å (Watenpaugh et al., to be published). The new S-Fe-S bond angles do not differ greatly from the earlier ones and their precision establishes more firmly the magnitude of the deviations from the tetrahedral value.

It is clear from the rubredoxin example that the coordination of the Fe atom was evident at 3 Å resolution, but it was not until 2 Å resolution was reached that the bond lengths and angles of the Fe-S$_4$ complex could be established with any degree of precision. At resolutions beyond 2 Å, the positions of the Fe and S atoms were determined with improved precision, but the discrepancy between the lengths of the Fe-Sγ42 bond from the 1.5 Å and 1.2 Å data sets is disquieting. The source of the difficulty is not yet known, but either one or both of the data sets must suffer systematic error(s), impairing the validity of the estimated standard deviations.

In extrapolating from the rubredoxin example, one should be mindful that light ligands such as N and O atoms in other complexes will be less precisely determined than the heavier S atoms in the rubredoxin. Furthermore, M-L distances for the lighter ligand atoms will be less than for the heavier ones and higher resolution data is required to resolve them.

In the field of protein structure analysis, considerable efforts are now being directed toward refining the X-ray models and assessing the precision of the results. We can expect substantially improved models as improved refinement methods are developed and more widely applied.

2. METALLOPROTEIN COMPLEXES

The three-dimensional structures of more than twenty metalloproteins have been determined. Most of these have reached resolutions sufficient for following the polypeptide chains and constructing models. Although it is not exhaustive, the following tabulation indicates the diversity of the metalloprotein structures that have thusfar been solved.

2.1. *Heme Proteins*

2.1.1. Myoglobin

This protein was the first one to be solved by the methods of X-ray crystal structure analysis, first at 6 Å resolution and subsequently at higher resolution (Kendrew, 1963; Watson, 1969). It is the oxygen carrying protein of muscle and is similar to the subunits in hemoglobin. The single Fe atom in the molecule of weight ~17 000 daltons is octa-

hedrally coordinated, four ligands being the four N atoms of the heme.
The fifth ligand is one of the N atoms of a histidine side chain, and
the sixth one in the ferrimyoglobin analyzed is a water O atom (Kendrew,
1972).

2.1.2. Hemoglobin

The molecule is composed of four subunits, two with α chains and two
with β chains, the subunits being similar to but somewhat smaller than
the myoglobin molecule. Each subunit contains one heme group, the Fe
in general being octahedrally coordinated as in myoglobin (Perutz, 1969).
      Many mutants of hemoglobin have been studied by crystallographic
methods, and in one of these the α subunits have ferric iron in five
coordination (Pulsinelli et al., 1973).

2.1.3. Erythrocruorin (insect hemoglobin)

The coordination of Fe in meterythrocruorin is octahedral, similar to
that of Fe in myoglobin (Huber et al., 1969). In deoxyerythrocruorin,
however, the water ligan is absent and the Fe is penta-coordinate (Huber
et al., 1970).

2.1.4. Cytochrome

The coordination of Fe in the cytochromes which have been solved is
octahedral (Dickerson et al., 1971; Mathews et al., 1971; Ashida et al.,
1973; Salemme et al., 1973). Of particular interest is cyotchrome b_5
since the inertness of the iron in this structure distinguishes it from
cytochrome c and myoglobin. In fact, the structure shows the Fe atom
to be coordinated on opposite sides of the porphyrin ring by two firmly
placed histidine residues, explaining the inertness of the iron.

2.2. *Nonheme Iron Proteins*

2.2.1. Myohemerythrin

Just as myoglobin is similar to the subunits in hemoglobin, so myohemerythrin is similar to the subunit in hemerythrin, the oxygen carrying
protein found in the red cells of some marine worms, among other species.
In the case of hemerythrin, however, the molecule is composed of eight
subunits, and despite its name, contains no heme.
      The structure of myohemerythrin was solved at 5.5 Å resolution, and
the exceptional quality of the electron density map enabled Hendrickson
and his colleagues to describe the structure in considerable detail
(Hendrickson et al., 1975).

2.2.2. Hemerythrin

With the knowledge of the structure of a myohemerythrin, Ward et al.,
(1975) solved the structure of hemerythrin from Phascolopsis (syn.
Golfingia) gouldii at 5.5 Å resolution, and by means of a computer they

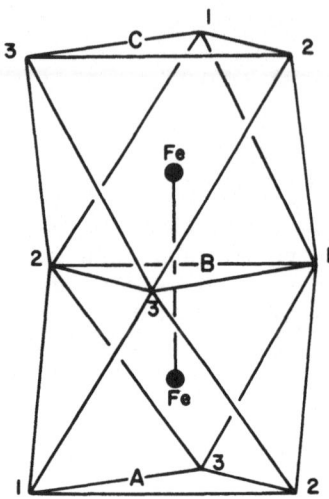

Fig. 3.   Idealized representation of the iron complex in
T. dyscritum hemerythrin (from Stenkamp et al., 1976b).

fit slightly different models to both structures (Hendrickson and Ward,
1975).
    In an independent study, Stenkamp et al. (1976a) solved the struc-
ture of hemerythrin from Themiste dyscritum at 5 Å resolution and extended it to a nominal resolution of 2.8 Å, maps at both resolutions being
based on a single isomorphous derivative (Stenkamp et al., 1976b). At
the higher resolution, the coordination of the two Fe atoms can be described in terms of two trigonal antiprisms about the iron atoms, the
two antiprisms appearing to share a common face, Figure 3.

2.3. *Iron-Sulfur Proteins*

Until recently the iron-sulfur proteins have been classed as nonheme
iron proteins, which, in fact, they are. However, in view of their
widespread distribution and the number of them that have been characterized in recent years, it is desirable to class them separately as
iron-sulfur proteins.

2.3.1. Rubredoxin

Except for the unusual rubredoxin from P. elsdenii, these proteins are
of low molecular weight, ~6 000 daltons, and have one Fe atom in the
molecule with four tetrahedrally coordinated cysteine S atoms. In the
two rubredoxins solved to date, coordination is tetrahedral (Herriott
et al., 1970; Adman et al., to be published).

2.3.2. High Potential Iron Protein (HiPIP)

As the name implies, this protein has a relatively high redox potential $E_0' \approx +350$ mV. There are four Fe atoms and four labile S atoms at opposite corners of a cube-like cluster within the molecule and four cysteine S atoms coordinated to the Fe atoms, Figure 4 (Carter et al., 1974a).

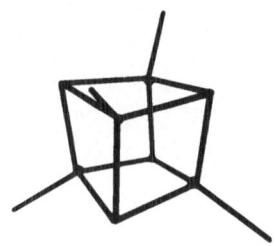

Fig. 4.   The cube-like iron-sulfur complex found in Chromatium high potential iron protein and in P. aerogenes ferredoxin.

2.3.3. Ferredoxin

The structure of ferredoxin from Peptococcus aerogenes has been solved. There are two HiPIP-like, iron-sulfur clusters per molecule of weight $\sim$6 000 daltons (Adman et al., 1973).

2.4. *Zn Proteins*

2.4.1. Insulin

A rhombohedral, two Zn form of insulin has been described in detail (Blundel et al., 1971). The two Zn ions in the asymmetric unit each have a coordinated histidine, the octahedron about each Zn ion appearing to be completed by solvent molecules.

2.4.2. Carboxypeptidase A

The coordination of the single Zn ion is tetrahedral, the ion binding two histidine residues, a glutamic acid residue, and a water molecule or hydroxyl ion (Lipscomb, 1971).

2.4.3. Liver Alcohol Dehydrogenase

The two Zn atoms in the molecule are both tetrahedrally coordinated. The one at the catalytic site is coordinated to two cysteine S atoms, one histidine side chain, and one water molecule or hydroxyl ion. The other Zn atom is coordinated by four cysteine S atoms, with striking similarities to the coordination of S atoms to Fe in both rubredoxin and ferredoxin (Eklund et al., 1976).

2.4.4. Carbonic anhydrase

The single Zn ion is coordinated by three protein side. chians and a
fourth ligand, presumably a water molecule or hydroxyl ion, in a distorted tetrahedral configuration (Liljas et al., 1972).

2.5. *Ca Proteins*

2.5.1. Extracellular Nuclease

The Ca ion is coordinated by four carboxylate side chains and two water
molecules in an approximate octahedral arrangement (Arnone et al., 1971;
Kretsinger, personal communication).

2.5.2. Ca Binding Protein

Both Ca ions in the molecule can be described as octahedrally coordinated, but the octahedra are considerably distorted. Ligands are primarily O atoms of protein side chains (Kretsinger and Nockolds, 1973).

2.5.3. Trypsin

The single Ca is octahedrally coordinated, the ligands being two carbonyl O atoms, two glutamic acid side chains, and two water molecules
(Bode and Schwager, 1975).

2.6. *A Mg Protein*

2.6.1. Bacteriochlorophyll Protein

Seven bacteriochlorophyll molecules, each with its associated Mg, lie
within a protein envelope. Each Mg is coordinated to four nitrogen atoms
of the porphine ring system and a fifth is indicated by electron density
protruding from the plane of ring, essentially a square pyramidal arrangement (Fenna and Matthews, 1975).

2.7. *Mixed Metal Proteins*

2.7.1. Thermolysin

Each molecule contains one Zn ion tetrahedrally coordinated to two
histidine residues, one glutamic acid residue, and either a water molecule or hydroxyl ion (Coleman et al., 1972). The protein molecule binds
four Ca ions which do not appear to be necessary for the activity of
this enzyme, but at least three of which are essential for its thermal
stability. None of the Ca ions are closer than 13 Å to zinc, but two of
them are only 2.8 Å apart. One of the latter binds to four aspartic or
glutamic acid side chains and the other to three of these four. The remaining two Ca ions each bind to a single aspartic acid side chain.
Presumably the coordination of the Ca ions is octahedral with water or
hydroxyl ions filling out the octahedra.

2.7.2. Concanavalin A

The molecule contains one manganous and one Ca ion separated by a distance of 5.3 Å, each ion apparently being octahedrally coordinated to four main chain or side chain oxygen atoms and two water molecules or hydroxyl ions. The two octahedra share a common face (Edelman et al., 1972).

2.7.3. Superoxide Dismutase

The subunit of the dimeric molecule contains one cupric and one Zn ion separated by a distance of ~6 Å. The cupric ion coordinates four histidine in what appears to be a slighty distorted square array and the Zn ion coordinates three histidine and an aspartic acid in an approximately tetrahedral arrangement, one of the histidine side chains being coordinate to both ions (Richardson et al., 1975).

3. DISCUSSION

I shall limit this discussion to metalloproteins that have been studied by X-ray methods in our laboratory at the University of Washington and some related ones with similar structural features studied in other laboratories.

3.1.1. HiPIP and Ferredoxin

Differences in properties between proteins with structurally similar complexes raise some intriguing questions. Thus, when the structure of the bacterial ferredoxin from P. aerogenes was solved and the similarity of its iron-sulfur complexes to the one in HiPIP was recognized, a number of questions were immediately raised (Sieker et al., 1972): for example, why are the redox potentials of the proteins so different? To what extent does the rest of the protein influence the properties of the iron-sulfur clusters? How similar are the complexes in the two proteins and are they distorted in any systematic way?

The difference in the redox potential between HiPIP and ferredoxin was formally explained by a three-state hypothesis (Carter et al., 1972) based on magnetic and spectroscopic data indicating the identity of the complex in reduced HiPIP with those in oxidized ferredoxin. This state of the complex was designated C, the other two being C^+ and C^-. Since the redox potential for HiPIP refers to the reaction $C^+ + e^- \leftrightarrows C$ and the one for ferredoxin to the reaction $C + e^- \leftrightarrows C^-$, the difference between the two potentials is explained. But this serves to emphasize the question of the influence of the protein on the properties of the complexes. Indeed, Cammack (1973) has shown that reduced HiPIP can be further reduced under appropriate conditions to the C^- state with a potential considerably below that for the reduction of oxidized ferredoxin, suggesting a substantial effect of the protein (Carter, to be published).

The complexes in both proteins lie in hydrophobic pockets, and in HiPIP five of the aromatic side chains are close to the complex. In

ferredoxin, on the other hand, there are only two tyrosine residues,
each lying next to one of the complexes. Another difference in the two
proteins is found in the number and type of N—H...S bonds between main
chain amide groups and the S atoms of the complex, ferredoxin having
almost twice as many such bonds as HiPIP. Furthermore, four of the nine
per complex in ferredoxin involve labile S atoms (Carter et al., 1974b;
Adman et al., 1975).

Refinement of models for both oxidized and reduced HiPIP (Freer
et al., 1975) and oxidized ferredoxin (Adman et al., 1976) has established the essential dimensional identity of the complexes, but the
precision is insufficient for the models to reveal a significant difference between the complexes in the oxidized forms of HiPIP and ferredoxin.
However, systematic differences of approximately 0.1 Å in the Fe—S_{labile}
bond lengths in the complex can be observed in the difference electron
density map between the oxidized and reduced forms of HiPIP (Carter et
al., 1974b).

Holm and his associates have synthesized a number of complexes
that are genuine analogues of the complexes in the iron-sulfur proteins.
Magnetic and spectroscopic properties clearly show one of these to be
a close analogue of the complex in reduced HiPIP and oxidized ferredoxin
(Herskovitz et al., 1972). Not only have the analogue structures contributed greatly to understanding the nature of the complexes in the proteins
but they also lend themselves to experiments which would be difficult
or impossible with the proteins.

3.1.2. Hemerythrin

On the basis of a 2.8 Å resolution electron density map, Stenkamp et al.
(1976b) have proposed the model shown in Figure 3 for the binuclear iron
complex in hemerythrin from T. dyscritum. The coordination of Fe appears
to be octahedral with protein side chains attached at A1-3, B1 and B2,
C1-3. The identity and sequence number of the coordinating residues cannot be established with certainty from the present electron density map
and the sequence of this hemerythrin has not yet been completed. However,
comparison with the P. gouldii sequence which is known (Klippenstein
et al., 1968) suggests the following correspondences: A-1, His 25; A-2,
His 54; A-3, Tyr 109; B-1, Gln 58; B-2, Asp 106; C-1, His 73; C-2, His 77;
C-3, His 101.

In contrast to the Fe complex described by Stenkamp et al.,
Hendrickson and Ward have proposed a model for the complex in myohemerythrin and in hemerythrin from P. gouldii in which three protein side
chains are coordinated to each Fe atom. The proposed ligand side chains
in hemerythrin are His 25, His 54 and Tyr 109 for one Fe atom and Tyr 67,
His 73 and His 101 for the other. Each Fe atom may be considered to fill
a fourth ligand position of the other Fe atom in each pair. Thus, although five of the ligands in the two proposed hemerythrin complexes
are the same, the models are basically different.

The model for the P. gouldii complex was derived from a 5.5 Å resolution electron density map and uncertainty must remain unless the model
can be substantiated at higher resolution. Much more detail is evident
in the 2.8 Å resolution T. dyscritum electron density map, but even here

better phases and increased resolution would be desirable to confirm the present model.

One additional point involving the complex shown in Figure 3 deserves comment. In their study of myohemerythrin, Hendrickson et al. (1975) reported the Fe-Fe distance as 3.44 ± 0.05 Å based on anomalous scattering data for the native protein to 2.8 Å resolution. Use of this distance in the model of Figure 3 with an assumed Fe-ligand distance of ~2 Å leads, however, to impossibly close contacts for ligands at B-1 and B-2.

In the 2.8 Å resolution T. dyscritum electron density map, Fe atoms in three of the four crystallographically independent subunits were resolved. The average Fe-Fe distance in the resolved pairs was $\sim d_{min}$, but peak positions in barely resolved electron density maps are unreliable. However, comparison with the rubredoxin example cited earlier also suggests an Fe-Fe distance $\sim d_{min}$. In that example three of the four S atoms coordinated to the Fe atom at a distance of ~2.3 Å were resolved in the 2.5 Å resolution map. Thus the fact that Fe atoms in three of the four hemerythrin pairs were resolved in the 2.8 Å map implies an Fe-Fe distance $\sim d_{min}$, a distance that gives an acceptable separation of ligands at B-1 and B-2, Figure 3. Nevertheless, the Fe-Fe distance suggested by the results for T. dyscritum hemerythrin to date cannot be regarded as definitive, and further work will be necessary to establish the details of the Fe complex

ACKNOWLEDGEMENTS

I wish to acknowledge the many contributions of my colleagues, particularly L.C. Sieker, K.D. Watenpaugh, E.T. Adman, and R.E. Stenkamp, and the support of our work on protein structure under grants GM-13366 and AM-3288 from the National Institutes of Healt.

REFERENCES

Adman, E.T., Sieker, L.C., and Jensen, L.H.: 1073, The structure of a Bacterial Ferredoxin, J. Biol. Chem., 248, 3987 — 3996.

Adman, E.T., Watenpaugh, K.D., and Jensen, L.H.: 1975, NH...S Hydrogen Bonds in Peptococcus aerogenes Ferredoxin, Clostridium pasteurianum Rubredoxin, and Chromatium High Potential Iron Protein, Proc. Nat. Acad. Sci. U.S.A. 72, 4854-4858

Adman, E.T., Sieker, L.C., and Jensen, L.H.: 1976, Structure of P. aerogenes Ferredoxin: Refinement at 2 Å Resolution, J. Biol. Chem. 251, 3801 — 3806

Arnone, A., Bier, C.J., Cotton, F.A., Day, V.W., Hazen, Jr., E.E., Richardson, D.C., Richardson, J.S., and in part, Yonath, A.: 1971, A High Resolution Study of an Inhibitor Complex of the Extracellular Nuclease of Staphylococus aureus, J. Biol. Chem., 246, 2302 — 2316.

Ashida, T., Tanaka, N., Yamane, T., Tsukihara, T., and Kakudo, M.: 1973, The Crystal Structure of Bonito (Katsuo) Ferrocytochrome c at 2.3 Å Resolution, J. Biol. Chem. 73, 463 — 465.

Blundel, T.L., Cutfield, J.F., Dodson, E.J., Dodson, G.G., Hodgkin, D.C., and Mercola, D.A.: 1971, The Crystal Structure of Rhombohedral 2 Zinc Insulin, Cold Spring Harbor Symposia on Quantitative Biology, vol. XXXVI, pp. 233 — 241.

Bode, W. and Schwager, P.: 1975, The Refined Crystal Structure of Bovine
    β-Trypsin at 1.8 Å Resolution, J. Molecular Biol. 98, 693 — 717.
Cammack, R.: 1973, ' Super-Reduction of Chromatium High Potential Iron-
    Sulfur Protein in the Presence of Dimethyl Sulfoxide', Biochem. Biophys.
    Res. Commun. 54, 548 — 554.
Carter, Jr., C.W., Kraut, J., Freer, S.T., Alden, R.A., Sieker, L.C.,
    Adman, E., and Jensen, L.H.: 1972, 'A Comparison of Fe_4S_4* Clusters
    in High-Potential Iron Protein and in Ferredoxin', Proc. Nat. Acad.
    Sci. U.S.A. 69, 3526 — 3529.
Carter, Jr., C.W., Kraut, J., Freer, S.T., Xuong, N., Alden, R.A., and
    Bartsch, R.G.: 1974a, 'Two-Ångstrom Crystal Structure of Oxidized
    Chromatium High Potential Iron Protein', J. Biol. Chem. 249, 4212 — 4225
Carter, Jr., C.W., Kraut, J., Freer, S.T., and Alden, R.A.: 1974b:
    'Comparison of Oxidation-Reduction Site Geometries in Oxidized and
    Reduced Chromatium High Potential Iron Protein and Oxidized Peptococcus
    aerogenes Ferredoxin' 249, 6339 — 6346.
Coleman, P.M., Jansonius, J.N., and Matthews, B.W.: 1972 'The Structure
    of Thermolysin: An Electron Density Map at 2.3 Å Resolution', J. Mo-
    lecular Biol. 70, 701 — 724.
Dickerson, R.E., Tanako, T., Eisenberg, D., Kalli, O.B., Samson, L.,
    Cooper, A., and Margoliash, E.: 1971 'Ferricytochrome c I', J. Biol.
    Chem. 246, 1511 — 1535.
Edelman, G.M., Cunningham, B.A., Reeke, Jr., G.N., Becker, J.W., Waxdal,
    M.J., and Wang, J.L.: 1972, 'The Covalent and Three-Dimensional Struc-
    ture of Concanavalin A', Proc. Nat. Acad. Sci., U.S.A. 69, 2580 — 2584.
Eklund, H., Nordström, B., Zeppezauer, E., Söderland, G., Ohlsson, I.,
    Boiwe, T., Söderberg, B.-O., Tapia, O., and Brändén, C.-I.: 1976, 'The
    Three-Dimensional Structure of Horse Liver Alcohol Dehydrogenase at
    2.4 Å Resolution', J. Molecular Biol., in press.
Fenna, R.E. and Matthews, B.W.: 1975, 'Chlorophyll Arrangement in a
    Bacteriochlorophyll Protein from Chlorobium limicola', Nature 258,
    573 — 577.
Freer, S.T., Alden, R.A., Carter, Jr., C.W., and Kraut, J.: 1975,
    'Crystallographic Structure Refinement of Chromatium High Potential
    Iron Protein at Two Ångstrom Resolution', J. Biol. Chem. 250, 46 — 54.
Hendrickson, W.A., Klippenstein, G.L., and Ward, K.B.: 1975, 'Tertiary
    Structure of Myohemerythrin at Low Resolution', Proc. Nat. Acad. Sci.,
    U.S.A. 72, 2160 — 2164.
Hendrickson, W.A. and Ward, K.B.: 1975, 'Atomic Models for the Polypeptide
    Backbones of Myohemerythrin and Hemerythrin', Biochem. Biophys. Res.
    Commun. 66, 1349 — 1356.
Herriot, J.R., Sieker, L.C., Jensen, L.H., and Lovenberg, W.: 1970,
    'Structure of Rubredoxin: An X-ray Study to 2.5 Å Resolution', J.
    Molecular Biol. 50, 391 — 406.
Herskovitz, T., Averill, B.A., Holm, R.H., Ibers, J.A., Phillips, W.D.,
    and Weiher, J.F.: 1972, 'Structure and Properties of a Synthetic Anal-
    ogue of Bacterial Iron-Sulfur Proteins', Proc. Nat. Acad. Sci., U.S.A.
    69, 2437 — 2441.
Huber, R., Epp, O., and Formanek, H.: 1969, 'The Environment of the Haem
    Group in Erythrocruorin (Chironomus thummi)', J. Molecular Biol. 42
    591 — 594.
Huber, R., Epp, O., and Formanek, H.: 1970, 'Structures of Deoxy- and
    Carbonmonoxy Erythrocruorin', J. Molecular Biol. 52, 349 — 354.

James, R.W.: 1948, 'False Detail in Three-Dimensional Fourier Represen-
   tations of Crystal Structures', Acta Crystallographica 1, 132 — 134.
Jensen, L.H.: 1974, 'Protein Model Refinement Based on X-ray Data',
   Ann. Rev. Biophys. Bioengin. 3, 81 — 93.
Kendrew, J.C.: 1962, 'Side-Chain Interactions in Myoglobin', Brookhaven
   Symposia in Biology, No. 15, 216 — 228.
Kendrew, J.C.: 1963, 'Myoglobin and the Structure of Proteins', Science
   139, 1259 — 1266.
Klippenstein, G.L., Holleman, J.W., and Klotz, I.M.: 1968, 'The Primary
   Structure of Golfingia gouldii Hemerythrin. Order of Peptides in Frag-
   ments Produced by Tryptic Digestion of Succinylated Hemerythrin.
   Complete Amino Acid Sequence', Biochem. 7, 3868 — 3878.
Kretsinger, R.H. and Nuckolds, C.E.: 1973, 'Carp Muscle Calcium-Binding
   Protein' J. Biol. Chem. 248, 3313 — 3326.
Liljas, A., Kannan, K.K., Bergsten, P-C., Waara, I., Friborg, K.,
   Strandberg, B., Carlstrom, U., Jäpur, L., Lövgren, S., and Petef, M.:
   1972, 'The Crystal Structure of Human Carbonic Anhydrase C.', Nature,
   New Biology 235, 131 — 137.
Lipscomb, W.N.: 1971, 'Structures and Mechanisms of Enzymes', The Robert
   A. Welch Foundation Conferences on Chemical Research, vol. XV, pp.
   131 — 182.
Mathews, F.S., Argos, P., and Levine, M.: 1971 'The Structure of Cyto-
   chrome b_5 at 2.0 Å Resolution', Cold Spring Harbor Symposium on Quan-
   titative Biology, vol. XXXVI, pp. 387 — 395.
Perutz, M.F.: 1969, 'The Haemoglobin Molecule', Proc. Roy. Soc., B 173,
   113 — 140.
Pulsinelli, P.D., Perutz, M.F., and Nagel, R.L.: 1973, 'Structure of
   Hemoglobin M Boston, a Variant with a Five-Coordinated Ferric Heme',
   Proc. Nat. Acad. Sci., U.S.A. 70, 3870 — 3874.
Richardson, J.S., Thomas, K.A., Rubin, B.H., and Richardson, D.C.: 1975,
   'Crystal Structure of Bovine Cu, Zn Superoxide Dismutase at 3 Å Reso-
   lution: Chain Tracing and Metal Ligands', Proc. Nat. Acad. Sci., U.S.A.
   72, 1349 — 1353.
Salemme, F.R., Freer, S.T., Xuong, Ng.H., Alden, R.A., and Kraut, J.:
   1973, 'The Structure of the Oxidized Cytochrome c_2 of Rhodospirillum
   rubrum', J. Biol. Chem. 248, 3910 — 3921.
Sieker, L.C., Adman, E., and Jensen, L.H.: 1972, 'Structure of the Fe-S
   Complex in a Bacterial Ferredoxin', Nature 235, 40 — 42.
Stenkamp, R.E., Sieker, L.C., Jensen, L.H., and Loehr, J.S.: 1976a,
   'Structure of Methemerythrin at 5 Å resolution', J. Molecular Biol.
   100, 23 — 34.
Stenkamp, R.E., Sieker, L.C., and Jensen, L.H.: 1976b, 'Structure of
   the Iron Complex in Methemerythrin', Proc. Nat. Acad. Sci., U.S.A. 73,
   349 — 351.
Ward, K.B., Hendrickson, W.A., and Klippenstein, G.L.: 1975, 'Quaternary
   and Tertiary Structure of Hemerythrin', Nature 257, 818 — 821.
Watenpaugh, K.D., Sieker, L.C., Herriott, J.R., and Jensen, L.H.: 1971,
   'The Structure of a Non-Heme Iron Protein: Rubredoxin at 1.5 Å Resolu-
   tion', Cold Spring Harbor Symposium on Quantitative Biology, vol. XXXVI,
   pp. 359 — 367.
Watenpaugh, K.D., Sieker, L.C., Herriott, J.R., and Jensen, L.H.: 1973,
   'Refinement of the Model of a Protein: Rubredoxin at 1.5 Å Resolution',
   Acta Crystallographica B29, 943 — 956.

Watson, H.C.: 1969, 'The Stereochemistry of the Protein Myoglobin',
  Progr. Stereochem. 4, 299 — 333. London, Butterworth.

DISCUSSION

Sue Hanlon: In your structure of hemerythrin, where does the O_2 bind,
and what are the other ligands about the iron atoms?
     Jensen: We do not known where O_2 binds. It is tempting to speculate,
however, that it is at B-3, Figure 3. Although the metal complex is
burried within the subunit, B-3 appears to be accessible by way of a
channel to the outside of the molecule.
     Werber: You have discussed the structure of 4 Fe and 8 Fe sulfur-
proteins as forming 1 or 2 clusters. Could you extrapolate your data to
the case of 2Fe sulfur proteins, which should form only half a cluster?
Also would you comment on the fact that the 8Fe ferrodoxins have molecular weights of about 6000 daltons, whereas the 2Fe ferredoxins have
higher molecular weights of about 12 000 daltons?
     Jensen: The three-dimensional structure of no two-iron ferredoxin
has not yet been done, but the proposed model with two labile sulfur
atoms bridging the two iron atoms and two cysteine sulfur atoms coordinated to each iron atom in a tetrahedral arrangement satisfactorily
accounts for the properties of the complex. With respect to the molecular
weights, I find it interesting that the sizeable complexes in the eight-
iron ferredoxin are accommodated in a molecule of no more than $\simeq$ 6000
daltons. In fact the chain is barely sufficient to cover the two complex.
The two-iron ferredoxins, on the other hand, have a substantially smaller
complex, and then only one in a molecule of $\simeq$ 12 000 daltons.
     Sarkar: In designing polypeptide molecules to mimic the native metal
binding sites of metalloenzymes and metalloproteins, a very interesting
feature has emerged. A survey of several metalloenzymes whose crystal
structures has been elucidated, reveals the natural occurance of one
to three amino acid residues in between two ligands for the metal. While
the other ligands originate from a distant part of the polypeptide chain
For example, Carboxypeptidase A: His(69), Glu(72), His(196); Thermolysin
His(142), His(146), Glu(166); Carbonic anhydrase B: His(94), His(96),
His(119); Carbonic anhydrase C: His(93), His(95), His(118); Rubredoxin:
Cys(6), Cys(9), Cys(38), Cys(41), Now, this is an observation. Can you
or anyone in the audience illuminate on this subject?
     Jensen: Your list can be expanded. The eight-iron ferredoxins have
two runs of the sequence with Cys XX Cys XX Cys XXX Cys where the X's
are residues other than Cys. In each complex three of the four ligands
are the first three Cys in each run and the fourth is far removed in
the sequence.
     In liver alcohol dehydrogenase, Eklund et al. have noted the simila-
rity to the ferredoxins in the case of the second Zn atom, even to the
configuration of the attached ligands. Also, high potential iron protein
has two residues between the first two Cys ligands, but the other two
are considerably removed.
I do not know the significance of your observation, but it would appear
reasonable that the limited flexibility of closely spaced residues may
be necessary to inotiate the formation of the proper complex.

GIUSEPPE ROTILIO, LAURA MORPURGO, LILIA CALABRESE, ALESSANDRO
FINAZZI AGRÒ
*Institute of Biological Chemistry and CNR Center for Molecular
Biology, University of Rome, Rome, Italy*

and

BRUNO MONDOVÌ
*Institute of Applied Biochemistry, University of Rome, Rome,
Italy*

The object of this paper is to discuss some aspects of metal-ligand
interactions in copper proteins on the basis of recent experimental
data which concern mononuclear Cu(II) centers accessible to the solvent.
The restriction of the experimental approach to this class of copper
centers is justified by the fact that changing any property included
in the above definition implies also changing chemical reactivity, spectroscopic features and biological functions of the protein-bound copper.
In fact all species of protein-bound copper have the same general role,
that is electron transfer to oxygen, but this function is fulfilled
through different specialized reactions. Even the oxygen transport by
the hemocyanins can be envisaged as a particular type of electron transfer from a pair of cuprous ions to O_2 (Morpurgo and Williams, 1968).
In the majority of cases, however, copper acts, as Cu(II), as a sink
of electrons, which are then forwarded—with or without an intermediate
formal change of the metal valence — to an ultimate electron acceptor.
This is, as far as present knowledge is concerned, oxygen in any case;
in other words no copper enzyme has been reported to be involved in
systems other than oxidases or oxygenases. A special case is superoxide
dismutase, where the electron acceptor is O_2^- (for general information
on copper enzymes see Malkin and Malmström, 1970 and Malmström et al.,
1975).
      Electron donation to copper occurs either by direct contact with
the substrate or indirectly through the protein which provides other
groups — amino acid side chains or prosthetic groups — as mediators
between the substrate and copper. This difference results in great structural and spectroscopic differences between the copper centers. In particular the class requiring mediators is buried deep inside the protein
and is intensely blue (Type 1 copper; Malkin and Malmström, 1970), while
the former class, which directly interacts with substrates, is open to
solvent and is much less intensely chromophoric (Type 2 copper). In
multicopper oxidases there is a third class of copper centers (Type 3)

which does not give electron paramagnetic resonance (EPR) signals and
which seems to consist of couples of cupric ions (Malkin and Malmström,
1970). The spectroscopic properties and the function of this class are
not yet well characterized.

Electron transfer from copper to oxygen can involve binding between
oxygen and copper — which is probably the case for the only oxidase
containing just a single copper as the prostetic group, i.e. galactose
oxidase (Malmström et al., 1975), and is also likely for amine oxidases
(Mondovi et al., 1971) — or may require as mediators others copper or
non-copper centers inside the same protein, as in the multi-copper oxidases laccase, ascorbate oxidase and ceruloplasmin, and cytochrome oxidase, or even other enzymes, as in the case of the single copper blue
proteins plastocyanin and azurin. The experimental evidence so far available (Malström et al., 1975) indicates that Type 1 copper does not participate in direct electron donation to O_2, which could involve Type 2
or Type 3.

The most relevant data we shall discuss concern bovine superoxide
dismutase and Cu(II) bovine carbonic anhydrase. The former (32 000 M.W.)
contains one Cu(II) and one Zn(II) on each of two identical subunits
and catalyzes the dismutation of O_2^- into O_2 and H_2O_2 with the intermediate reduction of the copper (Fridovich, 1974). The latter is a monomeric
Zn(II) enzyme (29 000 M.W.) which catalyzes the reversible hydration of
CO_2 to H_2CO_3 and gives an inactive derivative on substitution of Cu(II)
for the zinc, though the two metals are bound at the same site by the
same ligands (Lindskog et al., 1971). For both enzymes X-ray data are
available (Richardson et al., 1975; Lindskog et al., 1971) from which
the two copper sites can be outlined as shown in Figure 1.

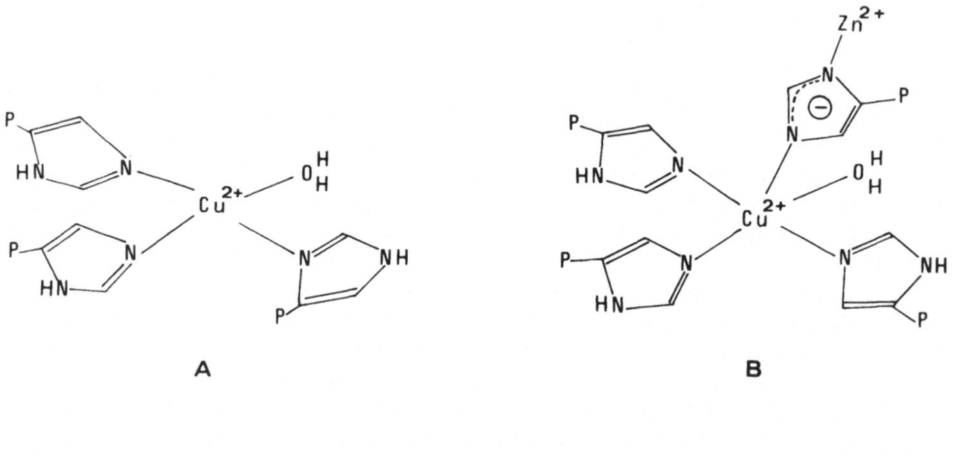

Fig. 1. Schematic representation of the copper sites of bovine carbonic anhydrase (A) and bovine superoxide dismutase
(B).

The two sites are similar in that both have histidine imidazoles and
water as copper ligands. Moreover in both enzymes the water molecule

plays a major role in the interaction with substrates and inhibitors. For carbonic anhydrase there is evidence that the metal-bound water participates in the catalytic reaction (Lindskog and Coleman, 1972; Woolley, 1975). In the case of superoxide dismutase exchange of substrate and inhibitors with the copper-bound water molecule appears to be a step in the interaction between the metal and these reagent. This is strongly suggested by EPR (Rotilio et al., 1972a) nuclear magnetic resonance (NMR) (Fee and Gaber, 1972; Terenzi et al., 1974) and activity data (Rotilio et al., 1972b). In particular kinetic measurements (Rigo et al., 1975a) have shown that inhibition of the enzyme by OH^- and CN^-, which coordinate to the copper in the place of the water (Rotilio et al., 1971; Rotilio et al., 1972a; Fee and Gaber, 1972; Terenzi et al., 1974) is of the competitive type.

The effect of ionic strength on the activity (Rigo et al., 1975b) also suggests the formation of a complex between O_2^- and copper during catalysis.

Beside the many analogies, a profound difference between the Cu(II) sites of superoxide dismutase and carbonic anhydrase is that the coordination number is 5 for the former enzyme and 4 for the latter. This feature, together with the presence of a copper-zinc imidazolate bridge in superoxide dismutase (see Figure 1), brings about a very different reactivity in the two cases.

First of all the superoxide dismutase copper undergoes a very fast redox cycle during catalysis (Klug-Roth et al., 1973; Fielden et al., 1974) and exchanges electron quite efficiently with reductants such as sulfide, ferrocyanide and H_2O_2 (Rotilio et al., 1973a), while the carbonic anhydrase copper is not only inactive in the specific enzyme reaction but is also not easily reduced by the same reductants (Morpurgo et al., 1976a, 1976b). In this respect it resembles the aqueous Cu(II) complex of diethylenetriamine, with which it shares the axial line shape of the EPR spectrum (Morpurgo et al., 1973, 1975). On the other hand the EPR spectrum of superoxide dismutase is rhombic (Rotilio et al., 1972a) like that of crystalline Cu(II) diethylenetriamine formate, which was proposed as a pentacoordinate model for the superoxide dismutase copper (Morpurgo et al., 1973) before the pentacoordinate nature of the enzyme site was apparent from X-ray data. Cyanide changes the EPR line shape of the superoxide dismutase copper to axial (Rotilio et al., 1972a) and at the same time stabilizes the oxidized form of the enzyme in the reaction with ferrocyanide and H_2O_2 (Rotilio et al., 1973a).

Beside the rhombic pentacoordinate structure, the presence of the copper-zinc imidazolate bridge is relevant to the molecular mechanism underlying the redox properties of copper in superoxide dismutase. In this respect the use of Co(II) — substituted superoxide dismutase, where Co(II) replaces zinc (Calabrese et al., 1972) has proved particularly useful. It suggested (Rotilio et al., 1974) the structure of the bimetal center before the X-ray structure became available. In the Co(II)-Cu(II) protein the cobalt chromophore is characterized by a fairly intense visible absorption in the 500 — 630 nm region with a maximum at 600 nm (Figure 2) and is EPR silent because of the magnetic coupling of the two metal spin systems (Calabrese et al., 1972; Rotilio et al., 1974). If copper is either reduced or removed the EPR signal of Co(II) is ob-

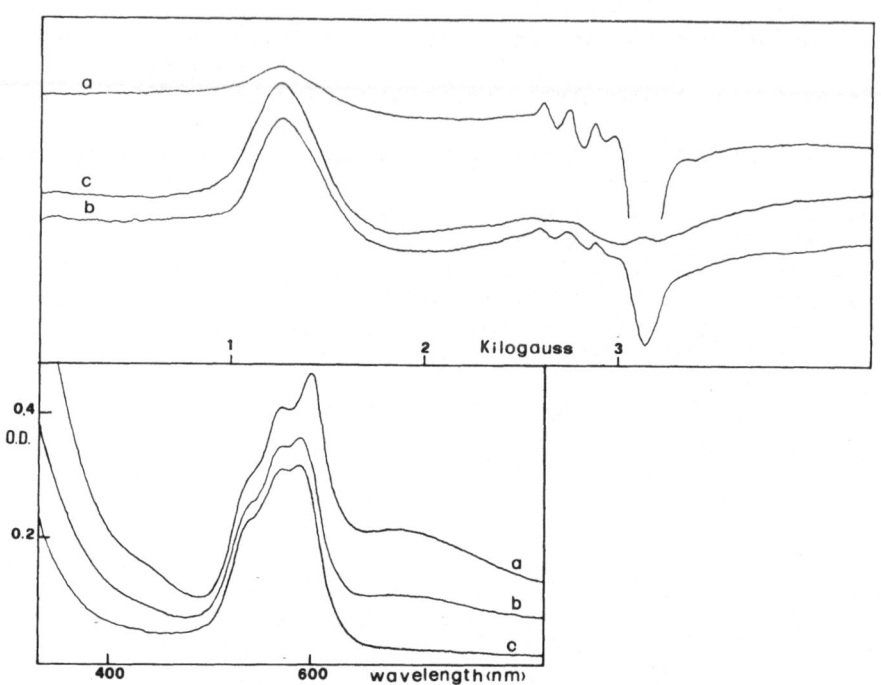

Fig. 2. Optical and EPR spectra of Co(II)-bovine superoxide
dismutase. Cobalt was substituted for zinc by an exchange
method (Calabrese et al., 1972) which results in 50 to 60%
exchange of Co(II) for Zn (II). A. Optical spectra. Curve a:
0.8 mM Co(II) – Cu(II) protein, 0.8 mM Co(II); curve b, the
same, immediately after addition of a tenfold excess of H_2O_2
which reduces approximately 50% of the copper; curve c:
0.8 mM copper-free protein, 0.9 mM Co(II). B. EPR spectra
of the samples of Figure 2a. The spectra were obtained with
a Varian V-4502 spectrometer in the following conditions:
temperature, 10 K; microwave frequency, 9.24 GHz; microwave
power, 6 mW; modulation amplitude, 10 G. The copper signal
is strongly saturated in these conditions.

served and the cobalt optical spectrum is typically modified by a shift
of the maximum to shorter wavelengths (Figure 2). The identity of optical and EPR spectra of cobalt in the absence of copper and in the
presence of Cu(I) is an evidence that copper is released from the bridging imidazolate on interaction with the reductant. The pH dependence
of the superoxide dismutase-ferrocyanide redox equilibrium (Fee and
Di Corleto, 1973) already suggested that a ligand with pK>>9 is protonated and released from the copper coordination sphere while reduction
of copper occurs.
    It has been shown (Calabrese et al., 1975) that low pH causes re-

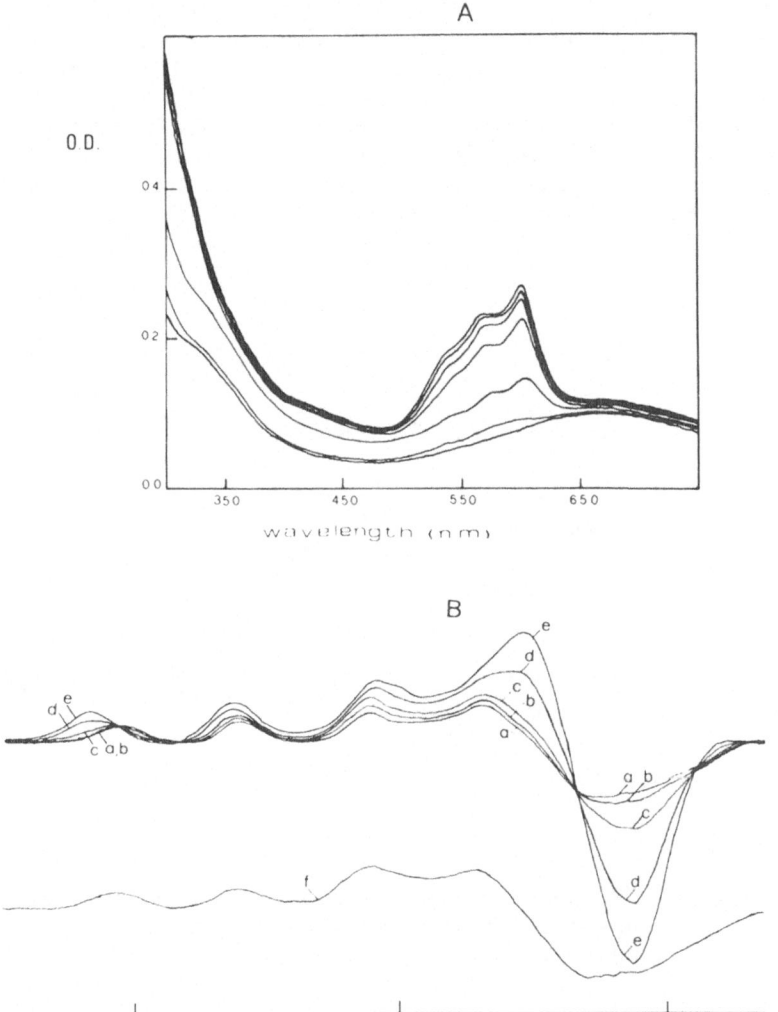

Fig. 3. Optical and EPR spectra of Co(II)-Cu(II) and Zn(II)-
Cu(II) superoxide dismutase at acid pH values. A: Optical
spectra. The protein was 0.35 mM, with a cobalt content of
1.2 Co(II)/protein. Curves from top to bottom are the Co(II)-
Cu(II) protein at pH 5.56, 4.60, 4.01, 3.62, 3.15, 2.99, and
the Zn(II)-Cu(II) protein at pH 2.96, respectively. B: EPR
spectra of Co(II)-Cu(II) superoxide dismutase at acid pH values.
Spectra (a)-(e) correspond to the solutions at pH 5.56, 4.60
3.63, 3.15, and 2.99 of Figure 3a. The bottom curve (f) is the
sample at pH 2.99 brought to pH 6 by addition of NaOH and incubated 10 min at room temperature before freezing for EPR.
Microwave frequency: 9.15 GHz; microwave power: 10 mW; modulation amplitude: 10 G; temperature: 100 K.

versible uncoupling of the Cu(II) and Co(II) spin systems. In these
conditions the cobalt optical spectrum disappears while the broad band
with maximum at 680 nm due to the copper is practically unchanged (Figure
3a). These facts can reflect breaking of the intermetal bridge at the
cobalt-imidazolate bond by protonation of the cobalt-facing nitrogen
of the imidazole (cobalt or zinc imidazolates are less stable than the
copper analog). The chemistry of the two situations, i.e. copper reduction and low pH, is shown in Figure 4.

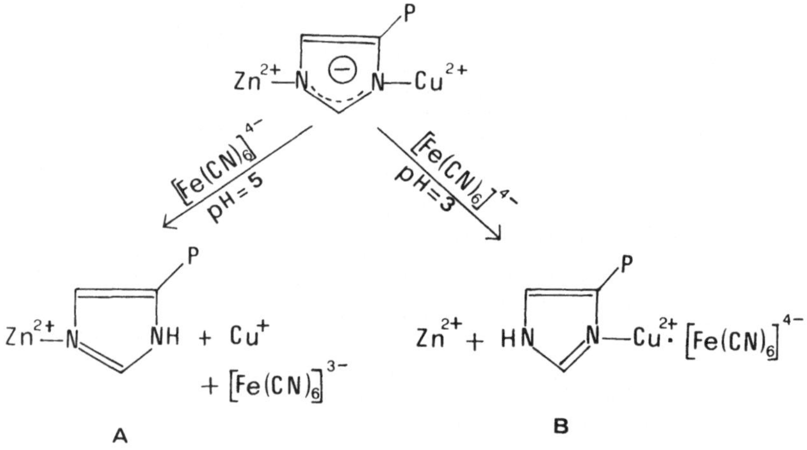

Fig. 4. Proposed structure for the zinc-copper bridge of
bovine superoxide dismutase when copper is reduced (A) and
at acid pH (B).

An important point in the low pH experiment is that in these conditions
the copper shows an axial EPR spectrum (Figure 3b) and is not reduced
by ferrocyanide, as the axial complexes of diethylenetriamine and carbonic anhydrase (Morpurgo et al., 1976a). It therefore appears that
the presence of the imidazolate bridge and of a rhombic pentacoordinate
geometry in the active site of superoxide dismutase are related to each
other and in turn determine the reducibility of the copper. External
reagents — substrate, protons, reductants, inhibitors — seem to display
their effects through selective perturbations of the whole copper-zinc
system. Anion inhibitors such as OH$^-$ and CN$^-$ do not cause breaking of
the intermetal bridge (Calabrese et al., 1976) but change the geometry
of the site in a way that makes the reduction of the copper more difficult. Protonation of the zinc-facing nitrogen of the bridging imidazole
at low pH brings to the same change of geometry and lack of reduction.
     On the other hand fast reduction of the copper takes place when
Cu(I) can find a suitable steric accomodation and this probably occurs

because a tetrahedral arrangement of ligands results from the release of one coordinating group from the pentacoordinate structure. It can be assumed that in the untreated protein at pH values between 5 and 10 — that is the protein displaying a rhombic EPR spectrum (Rotilio et al., 1971; Rotilio et al., 1972a) and full enzyme (Rotilio et al., 1972b) and redox (Fee and DiCorleto, 1973; Morpurgo et al., 1976a) activities — the bridging imidazolate is the best candidate for such a role and can therefore be protonated and released on reduction. In fact the pK of the second nitrogen of a metal bound imidazole is certainly >>9 (Sundberg and Martin, 1974) as required by the pH dependence of the redox potential of superoxide dismutase (Fee and DiCorleto, 1973). During the catalytic cycle Cu(I) is reoxidized by a second molecule of O_2 with the same rate as the reduction, and this fast reoxidation could analogously occur because Cu(II) can change from the unfavourable tetrahedral geometry of the reduced state to pentacoordinate without any slow change of position of the surrounding nuclei.

The reaction product in the reoxidation step is probably a peroxide anion, which is likely to take up the proton displaced from the imidazolate binding to the reoxidized copper. Recent unpublished pulse radiolysis experiments with the Co(II)—protein seem to indicate that the mechanism of copper release is operating also in the catalytic redox cycle.

At variance with the site of superoxide dismutase the four coordinated axial copper site of carbonic anhydrase seems not to be suited for redox reactions but appears to be rather unique in stabilizing binding of external ligands in a distorted environment (Morpurgo et al., 1975). Contrary to superoxide dismutase, binding of anion inhibitors changes the EPR spectrum from axial to rhombic, and the stability constant of the resulting complexes is much higher than that of the analogous complexes of superoxide dismutase. This extra-stability is achieved by hydrogen bonding or ion pair formation and in both cases is related to the behaviour of the enzyme which is unique in its affinity toward HCO_3^-, CO_2 and sulfonamide compounds (Morpurgo et al., 1975). For those ligands which are stabilized by ion pair formation — such as the halides — pentacoordinated structures have been suggested (Morpurgo et al., 1975). The optical spectra, characterized by intense charge transfer transitions, make these complexes possible models for the blue copper (Morpurgo et al., 1976b). As mentioned above blue or Type 1 copper belongs to the class of electron acceptor copper sites which accept electrons by outer sphere processes, being buried inside the protein. Type 1 copper is usually involved in fast valence changes (Malmström et al., 1975) and therefore a pentacoordinate structure could be the most suited for fast interconversion with the tetrahedral geometry required by the reduced state, as in the case of superoxide dismutase.

The model proposed for superoxide dismutase may also be valid for the Type 2 sites of blue oxidases, which share most of the ligand field properties of the dismutase, such as a bound water molecule and high affinity for anion inhibitors (Malkin and Malmström, 1971; Deinum and Vänngård, 1975). The affinity for anions is in some cases higher than that of superoxide dismutase, as for instance in the partially reduced proteins (Morpurgo et al., 1974). Mechanisms of stabilization analogous to those of Cu(II) carbonic anhydrase could be operating. Kinetic evidence suggests that Type 2 copper of blue oxidases is involved in the first step of catalysis, that is binding of substrate, followed by intra-molecular electron transfer to other copper sites in the protein (Holwerd and Gray, 1974, 1975). In the oxidases however, there are additional effects due: (a) to interaction between the different copper centers when they are in different oxidation states (Murpurgo et al, 1974) and (b) to the possibility of alternative pathways for electrons to reach the center with the highest redox potential inside the protein, which is generally the Type 1 copper (Malmström et al., 1975).

   Type 1 and Type 2 copper centers involved in fast valence change would thus have the same pentacoordinate structure. The great spectroscopic differences between the two copper types would rather be due to the presence of a water ligand in the Type 2 copper, which thus becomes specialized for inner sphere redox reactions, and of a sulfur ligand in a hydrophobic environment in the Type 1 copper (Finazzi et al., 1971), which renders the copper site more suited for outer sphere electron transfer.

REFERENCES

Calabrese, L., Cocco, D., Morpurgo, L., Mondovì, B., and Rotilio, G.:
   1975, 'Reversible Uncoupling of the Copper and Cobalt Spin Systems
   in Cobalt Bovine Superoxide Dismutase at Low pH', FEBS Letters 59,
   29-31.
Calabrese, L., Cocco, D., Morpurgo, L., Mondovì, B., and Rotilio, G.:
   1976, 'Cobalt Superoxide Dismutase. Reactivity of the Cobalt Cromophore
   in the Copper Containing and the Copper Free Protein', Eur. J. Biochem. 64, 465-470.
Calabrese, L., Rotilio, G., and Mondovì, B.: 1972 'Cobalt Erythrocuprein:
   Preparation and Properties', Biochim. Biophys. Acta 263, 827-829.
Deinum, J.S.E. and Vänngård: 1975, '^{17}O Hyperfine Interaction in the
   EPR Spectrum of Fungal Laccase A', FEBS Letters 58, 62-65.
Fee, J.A. and DiCorleto, P.E.: 1973, 'Observations on the Oxidation-
   Reduction Properties of Bovine Erythrocyte Superoxide Dismutase',
   Biochemistry 12, 4893-4899.
Fee, J.A. and Gaber, B.P.: 1972 'Anion Binding to Bovine Erythrocyte
   Superoxide Dismutase', J. Biol. Chem. 247, 60-65.
Fielden, E.M., Roberts, P.B., Bray, R.C., Lowe, D.J., Mautner, G.N.,
   Rotilio, G., and Calabrese, L.: 1974 'The Mechanism of Action of
   Superoxide Dismutase from Pulse Radiolysis and Electron Paramagnetic
   Resonance', Biochem. J. 139, 49-60.
Finazzi, Agrò A., Rotilio, G., Avigliano, L., Guerrieri, P., Boffi, V.,
   and Mondovì, B.: 1970, 'Environment of Copper in Pseudomonas Floures-
   cens Azurin: Flourometric Approach', Biochemistry 9, 2009-2014.
Fridovich, I.: 1974, 'Superoxide Dismutases', Adv. Enzymol. 41, 35-97.
Holwerda, R.A. and Gray, H.B.: 1974, 'Mechanistic Studies of the
   Reduction of Rhus vernicifera Laccase by Hydroquinone', J. Amer. Chem.
   Soc. 96, 6008-6022.
Holwerda, R.A. and Gray, H.B.: 1975, 'Kinetics of the Reduction of Rhus
   vernicifera Laccase by Ferrocyanide Ion', J. Amer. Chem. Soc. 97, 6036
   -6041.

Klug-Roth, D.E., Fridovich, I., and Rabani, J.: 1973, 'Pulse Radiolytic Investigations of Superoxide Catalyzed Disproportionation, Mechanism for Bovine Superoxide Dismutase', J. Amer. Chem. Soc. 95, 2786-2790.

Lindskog, S. and Coleman, J.E.: 1973, 'The Catalytic Mechanism of Carbonic Anhydrase', Proc. Nat. Acad. Sci. U.S.A. 70, 2505-2508.

Lindskog, S., Henderson, L.E., Kannan, K.K., Liljas, A., Nyman, P.O., and Strandberg, B.: 1971, 'Carbonic Anhydrase', in P. Boyer (ed.), The Enzymes, Academic Press, New York and London, 1971, vol. V, pp. 587-665.

Malkin, R. and Malmström, B.G.: 1970, 'The State and Function of Copper in Biological Systems', Adv. Enzymol. 33, 177-244.

Malmström, B.G., Reinhammar, B., and Andreasson, L.E.: 1975, 'Copper Containing Oxidases and Superoxide Dismutase', in P. Boyer (ed.), The Enzymes, Academic Press, New York and London, 1975, vol. XII, pp. 507-579.

Mondovi, B., Rotilio, G., Finazzi Agrò, A., and Antonini, E.: 1971, 'Amine Oxidases: a New Class of Copper Oxidases', in C. Franconi, (ed.), Magnetic Resonances in Biological Research, Gordon and Breach Science Publishers, New York, Londen and Paris, 1971, pp. 233-246.

Morpurgo, G. and Williams, R.J.P.: 1968, 'The State of Copper in Biologica Systems' in F. Ghiretti (ed.), Physiology and Biochemistry of Haemocyanins, Academic Press, London and New York, 1968, pp. 113-130.

Morpurgo, L., Finazzi Agrò, A., Rotilio, G., and Mondovi, A.: 1976b, 'Anion Complexes of Cu(II) Bovine Carbonic Anhydrase as Models for the Copper Site of Blue Copper Proteins', Eur. J. Biochem. 64, 453-457.

Morpurgo, L., Giovagnoli, C., and Rotilio, G.: 1973, 'Studies of the Metal Site of Copper Proteins. V.A Model Compound for the Copper Site of Superoxide Dismutase', Biochim. Biophys. Acta 322, 204-219.

Morpurgo, L., Mavelli, I., Calabrese, L., Finazzi Agrò, A., and Rotilio, G.: 1976a, 'A Ferrocyanide Charge-Transfer Complex of Bovine Superoxide Dismutase. Relevance of the Zinc Imidazolate Bond to the Redox Properties of the Enzyme', Biochem. Biophys. Res. Commun. 70, 607-614.

Morpurgo, L., Rotilio, G., Finazzi Agrò, A., and Mondovi, B.: 1974, "Anion Binding to Rhus vernicifera Laccase', Biochem. Biophys. Acta 33, 324-328.

Morpurgo, L., Rotilio, G., Finazzi Agrò, A., and Mondovi, B.: 1975, 'Anion Complexes of Cu(II) Bovine Carbonic Anhydrase', Arch. Biochem. Biophys. 170, 360-367.

Richardson, J.S., Thomas, K.A., Rubin, B.H., and Richardson, D.C.: 1975, 'Cristal Structure of Bovine Cu, Zu Superoxide Dismutase at 3 Å Resolution: Chain Tracing and Metal Ligands', Proc. Natl. Acad. Sci. U.S.A. 72, 1349-1353.

Rigo, A., Viglino, P., and Rotilio, G.: 1975a, 'Kinetic Study of O_2^- Dismutation by Bovine Superoxide dismutase. Evidence for Saturation of the Catalytic Sites by O_2^-', Biochem. Biophys. Res. Commun. 63, 1013-1018.

Rigo, A., Viglino, P., Rotilio, G., and Tomat, R.: 1975b, 'Effect of Ionic Strength on the Activity of Bovine Superoxide Dismutase', FEBS Letters 50, 86-88.

Rotilio, G., Bray, R.C., and Fielden, E.M.: 1972b, 'A Pulse Radiolysis Study of Superoxide Dismutase', Biochim. Biophys. Acta 268, 605-609.

Rotilio, G., Calabrese, L., Bossa, F., Barra, D., Finazzi Agrò, A., and Mondovì, B.: 1972c, 'Properties of the Apoprotein and Role of Copper and Zinc in Protein Conformation and Enzyme Activity of Bovine Superoxide Dismutase', Biochemistry 11, 2182–2187.

Rotilio, G., Calabrese, L., and Coleman, J.E.: 1973b, 'Magnetic Circular Dichroism of Cobalt — Copper and Zinc — Copper Bovine Superoxide Dismutase', J. Biol. Chem. 248, 3853–3859.

Rotilio, G., Calabrese, L., Mondovì, B., and Blumberg, W.E.: 1974, 'Electron Paramagnetic Resonance Studies of Cobalt — Copper Bovine Superoxide Dismutase', J. Biol. Chem. 249, 3157–3160.

Rotilio, G., Finazzi Agrò, A., Calabrese, L., Bossa, F., Guerrieri, P., and Mondovì, B.: 1971, 'Studies of the Metal Sites of Copper Proteins. Ligands of Copper in Hemocuprein', Biochemistry 10, 616–620.

Rotilio, G., Morpurgo, L., Calabrese, L., and Mondovì, B.: 1973a, 'On the Mechanism of Superoxide Dismutase. Reaction of the Bovine Enzyme with Hydrogen Peroxide and Ferrocyanide', Biochim. Biophys. Acta 302, 229–235.

Rotilio, G., Morpurgo, L., Giovagnoli, G., Calabrese, L., and Mondovì, B.: 1972a, 'Studies of the Metal Sites of Copper Proteins. Symmetry of Copper in Bovine Superoxide Dismutase and its Functional Significance', Biochemistry 11, 2187–2192.

Sundberg, R.T. and Martin, R.B.: 1974, 'Interactions of Histidine and Other Imidazole Derivatives with Transition Metal Ions in Chemical and Biological Systems', Chem. Rev. 74, 471–517.

Terenzi, M., Rigo, A., Franconi, C., Mondovì, B., Calabrese, L., and Rotilio, G.: 1974 'pH Dependence of the Nuclear Magnetic Relaxation Rate of Solvent Water Protons in Solution of Bovine Superoxide Dismutase', Biochim. Biophys. Acta 351, 230–236.

Woolley, P.: 1975, 'Models for Metal Ion Function in Carbonic Anhydrase', Nature 258, 677–682.

DISCUSSION

Sarkar: There are two types of nitrogen in an imidazole residue: one is a pyridine type, another is a pyrrole type of nitrogen. In our studies of metal binding to imidazole containing model compounds, we found that the pyrrole pK_a was considerably lowered when a metal was bound to the imidazole nitrogen (D.W. Appleton and B. Sarkar, Proc. Natl. Acad. Sci. U.S.A., 1974). I think this deprotonation of a pyrrole pK_a at a lower pH than its normal pK_a could have important significance in biological systems. In reference to superoxide dismutase an imidazole residue in a histidyl side chain behaves as a bidentate ligand. One nitrogen is bound to Zn(II) while the other is bound to Cu(II). Could the above observation of lowering the pK_a of a pyrrole nitrogen have any relevance in the action of this enzyme?

Rotilio: Yes, it could have relevance to the reversible decrease of activity between pH 10 and 12, which could reflect either deprotonation of the metal bound water or deprotonation of the pyrrole nitrogen of one or more of the imidazole ligands at a pK_a lower than its normal pK_a.

Dobson: What evidence have you that there is a water molecule coordinated to the copper ion in carbonic anhydrase? Does the copper derivative possess any activity?

Rotilio: Answer to Dr Dobson: The presence of a water molecule coordinated to the zinc or cobalt ion in carbonic anhydrase at low pH is a much debated question. Any argument either in favour or against its presence may apply to the copper derivate. The latter does not possess any activity.

Lanir: An answer to Dr Dobson in connection to this question about the existance of fast exchange water from the first coordination position of Cu-carbonic anhydrase. Fast exchange water molecule was found in the first coordination position of Mn(II)-carbonic anhydrase by us (Lanir et al., Biochemistry 14, 242, 1975) and in Co(II) and Cu(II)-carbonic anhydrase by S. Koenig at IBM Lab, at high pH values. At acid pH value no relaxation effect was observed in all metallo-carbonic anydrases. This was interpreted as the lack of water near the metal ion at low pH values.

Lanir: Is there any change of the EPR spectra of Cu(II) and Co(II) carbonic anhydrases with pH? Can you interpret the results in terms of the coordination number at high and low pH values, or the geometry around the metal ion.

Rotilio: (1) I do not have any data on the low and high pH forms in the absence of CO_2. (2) There is still the possibility that water is bound at low pH, but exchanges slowly.

Jagur-Grodzinsky: You claim that different sites must be assigned to Cu^{2+} and Co^{2+} complexation. However, your results could be simply interpreted in terms of higher affinity towards Cu^{2+}. Do you have an independant prove for your claim? What is the difference in the chemical nature of the respective sites?

Rotilio: We start from the protein containing both copper and zinc and dialyse against a large excess of cobalt. In these conditions only the zinc site is accessible to the cobalt, probably for steric reasons, being this tetrahedral symmetry of the zinc site (Richardson et al., 1975) more favourable to Co(II) than that of the copper site. In fact X-ray data do not show great differences in the chemical nature of the two sites, apart from the symmetry and the coordination

Werber: Do you think that the water molecule bound to Cu^{2+}-carbonic anhydrase is fast or slow-exchanging? If it were slow-exchanging it could explain why the Cu^{2+}-complex with halide ions becomes pentacoordinate, instead of having a water-halide ion exchange.

Answer to Dr Werber: This is exactly what we suggest, that the water molecule at the metal site of carbonic anhydrase may be slowly-exchanging at low pH, as it does not exchange with halides, which rather give rise to pentacoordinate metal species.

NUCLEAR MAGNETIC RESONANCE STUDIES OF THE INTERACTION OF LANTHANIDE CATIONS WITH LYSOZYME

C.M. DOBSON and R.J.P. WILLIAMS
Inorganic Chemistry Laboratory, South Parks Road, Oxford OX1 3QR, England

1. INTRODUCTION

Hen egg-white lysozyme is a small protein which consists of a single polypeptide chain of 129 amino acid residues, and has a molecular weight of ca. 14 400. Recent reviews (Imoto et al., 1972; Osserman et al., 1974) summarise the properties of the protein. It promotes the dissolution of bacterial cell walls, and is able to catalyse the hydrolyses of β-1,4-glycosidic linkages between residues in the polysaccharide components of the cell walls. The three-dimensional structure of the protein in the solid state has been determined, and a model of the structure constructed (Blake et al., 1967). This model shows that the molecule is roughly ellipsoidal in shape, with dimension about 45 × 30 × 30 Å. The molecule has a deep cleft on one side, divided into six sites A-F in which polysaccharide inhibitors and substrates bind. Two catalytically important ionisable groups are the carboxylic acid groups of glu 35 and asp 52 situated between sites D and E. The X-ray structure of the active site cleft is shown in Figure 1.

The outline conformation of the molecule in solution has been suggested to be similar to the X-ray structure, because of the correlation between known chemical properties of the molecule and those expected from the X-ray structure (see Imoto et al., 1972). On a quantitative basis the degree of similarity remains unknown. Indeed, the exact meaning of a protein conformation is difficult to define when the situation in solution is considered, because the constraints which could result in the existence of a single immobile conformation are less than these operating in the crystalline state. For example, even if on a time-average the solution conformation closely resembles the X-ray structure, the groups of the protein may have considerable freedom of movement, which is independent of the overall molecular tumbling. This independent freedom of movement could be of considerable importance when considering the mechanism of enzymic action, which is conventionally considered using a static model of the protein.

Diffraction methods cannot be applied to the determination of protein conformation in solution. Nuclear resonance (nmr) spectroscopy is the only technique at present which can provide in principle the detailed

B. Pullman and N. Goldblum (eds.), Metal-Ligand Interactions in Organic Chemistry and Biochemistry, part 1, 255-282. *All Rights Reserved.*
Copyright © *1977 by D. Reidel Publishing Company, Dordrecht-Holland.*

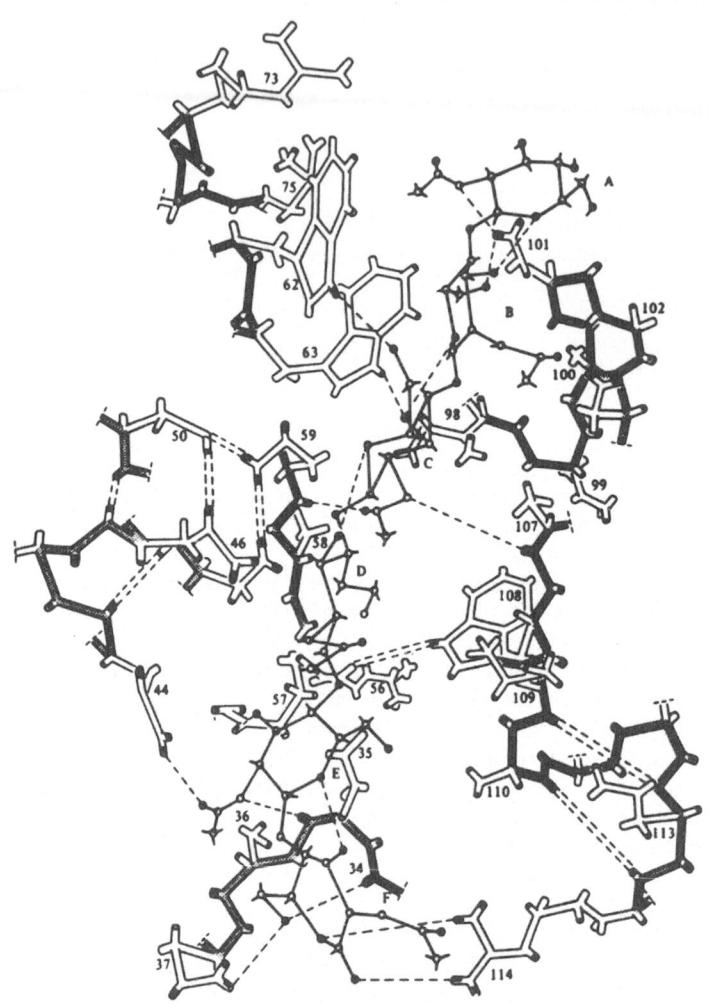

Fig. 1.   The active site cleft of the lysozyme X-ray structure
showing the proposed binding of $(GlcNAc)_6$ in sites A to F.
From Phillips (1967).

information of the type required. Through the development of techniques
for the improvement of resolution and interpretation of protein nmr
spectra (Campbell et al., 1973a, 1975a, 1975b) this has become possible
in practice.

In order to obtain conformational information from nmr spectroscopy,
the technique of paramagnetic lanthanide ion probes has been employed.
These probes have been applied extensively to determine the conformations
of small molecules in solution, and the methods have been discussed elsewhere (Barry et al., 1971, 1974; Levine and Williams, 1975; Dobson and
Levine, 1976). The principles of the application of these probes are

straightforward. When a paramagnetic lanthanide ion is bound to a ligand, the nmr spectrum of the ligand is perturbed. The perturbation may either be to the linewidths (relaxation times) of resonances, or to the frequencies (chemical shifts) at which absorbtion of energy occurs. Both types of perturbations depend, although in different manners, upon the geometrical relationship between the bound lanthanide ion and the nuclei whose resonances are perturbed. By measuring the induced perturbations, information about the geometry of the molecule may be obtained.

There are clearly a very large number of possible conformations of a polypeptide chain of 129 residues. The approach adopted by us (Campbell et al., 1973b; 1975b) is to use the conformational information from the lanthanide binding experiments to compare the solution and X-ray structures, rather than to attempt an independent structure determination. The conformation of lysozyme is of greatest interest in the region of the active site cleft, and particularly close to glu 35 and asp 52. The lanthanide cations were found (Blake and Rabstein, 1970) to bind near to these groups in the crystal, and as discussed below this occurs also in solution.

The assignment of resonances in the protein nmr spectrum will not be discussed in this paper, but has been described elsewhere (Campbell et al., 1973b, 1975a, b). The assignment procedures depend to a certain extent upon the interpretation of the data from the lanthanide perturbation experiments, and are therefore not fully separable from conformational considerations. Fortunately, it has been possible to use several independent series of observations to show that there is sufficient similarity between the X-ray structure and the solution conformation to allow these assignment procedures to be used (Campbell et al., 1975b).

In this paper, the binding of the lanthanides to lysozyme will be examined. This involves the definition of the number of binding sites, the strength of binding, the nature of the binding groups and the position of the metal ion in the protein structure.

2. LANTHANIDE INDUCED PERTURBATIONS

The lanthanide cations when bound to a ligand can induce both relaxation of resonances, and shifts of resonant frequencies. Two of the lanthanides La^{3+} and Lu^{3+} are diamagnetic. One lanthanide, Gd^{3+}, induces considerable relaxation but generally no paramagnetic shift. The remaining lanthanides all induce paramagnetic shifts, and their effects on relaxation times are relatively small. This is described fully elsewhere (Barry et al., 1971, 1974; Levine and Williams, 1975; Dobson and Levine, 1976).

2.1. *Induced Shifts*

The shifts induced by the binding of a paramagnetic lanthanide ion can arise from three main sources. First, a shift may arise from a conformational change or a pK value change resulting from the binding. This diamagnetic effect may be estimated by observing the effects of binding La^{3+} or Lu^{3+}, and the observed shifts corrected to leave only the paramagnetic contribution. There are two possible paramagnetic contributions to the shift, arising from through-bound (contact or scalar) or from through-space (pseudocontact or dipolar) mechanisms. The contact shift becomes negligible for nuclei more than two or three bonds away from the coordinated ion, and is not important in the present work. The pseudocontact shift is given (Bleaney, 1972) by

$$\Delta_p^{\;i} = \frac{D(\cos^2 \theta_i - 1)}{r_i^{\;3}} - \frac{D'(\sin^2 \theta_i \cos 2\phi_i}{r_i^{\;3}}$$

where r_i, θ_i and ϕ_i are the spherical coordinates of the nucleus (i) with the lanthanide ion as origin. The axes are determined by the magnetic symmetry properties of the system, the z axis being the direction of the principal magnetic axis of symmetry. D and D' are parameters determined by the ligand field symmetry of the bound ion.

In nearly every case studied it appears (see for example Dobson and Levine, 1976) that this equation is simplified because D' = 0. This is the situation of axial (or effective axial) symmetry, and is thought to arise from exchange processes which average rhombic symmetries to an effective axial symmetry. For this case, the ratio (R_{ij}) of the induced shift of the resonance of nucleus i to that of the resonance of nucleus j is (Barry et al., 1971)

$$R_{ij} = \Delta_p^{\;i}/\Delta_p^{\;j} = \left(\frac{3 \cos^2 \theta_i - 1}{r_i^{\;3}}\right)\bigg/\left(\frac{3 \cos^2 \theta_j - 1}{r_j^{\;3}}\right)$$

This ratio therefore contains only geometric parameters, the constants have been eliminated. A simple relationship such as this does not arise when D' $\neq$ 0.

Consider now the values of R_{ij} for a ligand, induced by binding of different lanthanides. If these values are the same for the different lanthanides, three major conclusions may be drawn. First, the complexes must be homologous, in other words the values of r, θ and ϕ must be the same for each lanthanide complex. Secondly, the shifts must be solely pseudocontact in origin. Thirdly, the symmetry must be effectively axial so that D' = 0.

2.2. *Induced Relaxation*

The slow electron relaxation time of Gd^{3+} can result in this ion causing large perturbations to the relaxation times of nearby nuclei. As with induced shifts, diamagnetic, scalar and dipolar mechanisms may operate, but only dipolar relaxation is important here. The induced relaxation in this case ($1/T_{1M}$ or $1/T_{2M}$) is given (see Solomon, 1955; Bloembergen, 1957) by

$$1/T_{1M} = f(\tau_c)/r^6 \quad \text{and} \quad 1/T_{2M} = f'(\tau_c)/r^6$$

where τ_c is the correlation time for the dipolar relaxation and r is the distance between the nucleus and the Gd^{3+} ion. The ratio of the relaxation induced in nucleus i relative to nucleus j is then

$$\frac{\left(^1/T_{2M}\right)_i}{\left(^1/T_{2M}\right)_j} = \frac{\left(^1/T_{1M}\right)_i}{\left(^1/T_{1M}\right)_j} = \left(\frac{r_j}{r_i}\right)^6$$

2.3. *Exchange Effects*

The equations given above relate to a ligand (L) fully bound to a lanthanide ion (M). In practice, an equilibrium of the type

$$M + L \rightleftharpoons ML$$

exists, where the equilibrium constant $K = [ML]/[M][L]$.

Provided that ligand exchange is rapid, the observed induced perturbations (denoted δ, $1/T_{1p}$, $1/T_{2p}$) are related to those of the fully bound system by $\delta = f\Delta$

$$1/T_{1p} = f\,(1/T_{1M}), \quad 1/T_{2p} = f\,(1/T_{2M})$$

where $f = [ML]/([L] + [ML])$ and is the fraction of ligand bound to the metal ion.

This fast exchange condition is the case for most systems, including lysozyme, where the binding constants are not very large. The measured induced shift or relaxation of a given nucleus depends upon the relative concentrations of metal and ligand. By varying these relative concentrations, information about binding can be obtained as described below.

2.4. *Requirements for Conformational Analysis*

Any approach to conformational analysis must attempt to match the nmr shift and relaxation data to (i) a molecular conformation, (ii) a metal ion position, and (iii) a symmetry axis direction. For a given set of (i) - (iii), values of $(3\cos^2\theta - 1)/r^3$ and of $1/r^6$ can readily be calculated for different nuclei, and ratios taken of each. These ratios can then be compared to the experimental shift and relaxation ratios. (i) — (iii) can then be varied in turn, to discover whether measured and calculated ratios are in agreement for any set. The computer treatment of such an approach has been fully described (Barry et al., 1971, 1974).

For lysozyme, the most straightforward approach is to compare observed shift and relaxation ratios with values of $(3\cos^2\theta - 1)/r^3$ and of $1/r^6$ calculated from the X-ray structure and the X-ray determined metal binding site. The only variable is then the direction of the symmetry axis which must be defined for the shift ratio comparison.

The procedure required in this study of lysozyme can therefore be summarized as follows. First, the number of lanthanide binding sites

on the protein must be defined. Secondly, the groups binding the metal
ions must be identified and characterised. This does not require the
complete definition of binding sites, for the exact orientation of the
lanthanide with respect to the binding groups is not needed. Thirdly,
it is necessary to measure perturbations caused by a metal ion bound
at a single site, in order to be able to define the protein atoms with
respect to this fixed point. Fourthly, the measured perturbations must
be interpreted. Each of these steps will be treated in turn. Note that
defining the metal-ligand interaction in detail results in determination
of the conformation of the protein relative to the lanthanide ion. Provided that the binding of the lanthanide ion does not perturb the native
protein conformation, this native conformation is then determined.

3. EXPERIMENTAL METHODS

The experiments described in this paper were all carried out using proton nuclear magnetic resonance. All spectra were run at 270 MHz using
the Oxford Enzyme Group Bruker Spectrometer. Full experimental details
have been given elsewhere (Campbell et al., 1973, 1975a–d). Unless
otherwise stated, all the data presented here are for 5 mM protein concentration, an ionic strength of 0.4, and a temperature of 54 $^{\circ}$C. The
spectral assignments have been given in Campbell et al. (1975b). Computation was carried out on the Oxford University ICL 1906A Computer.

4. THE NUMBER OF BINDING SITES

The binding of lanthanide cations to lysozyme in solution has been investigated by proton relaxation enhancement methods (Dwek et al., 1971;
Jones et al., 1974) and by UV methods (Secenski and Lienhard, 1974).
Both these types of study concluded that a single binding site exists
but neither would have been able to detect sites of weak binding in the
presence of this site. In the solid state, binding of lanthanide ions
to lysozyme has been studied at low resolution (Blake and Rabstein,
1970) and binding at a single site (close to asp 52 and glu 35) was reported, although more recent studies (Kurachi et al., 1975; Perkins,
1975) have indicated that several sites in this region of the protein
could exist.
     The addition of lanthanide ions to lysozyme solutions causes pertur-
bations to the nmr spectrum (Campbell et al., 1973b, 1975a–d) as Figure
2 shows. The binding in solution was investigated in the present work
by making use of these perturbations. Because the broadening induced
by Gd^{3+} is hard to measure quantitatively, the shifts induced by the
other paramagnetic lanthanides were used for this purpose.

4.1. *Binding Curves at Fixed pH*

The effect on the resonances in the nmr spectrum of lysozyme of varying
concentrations of lanthanide ions have been measured. In these titrations
the ionic strength was maintained constant by using KCl, and each exper-

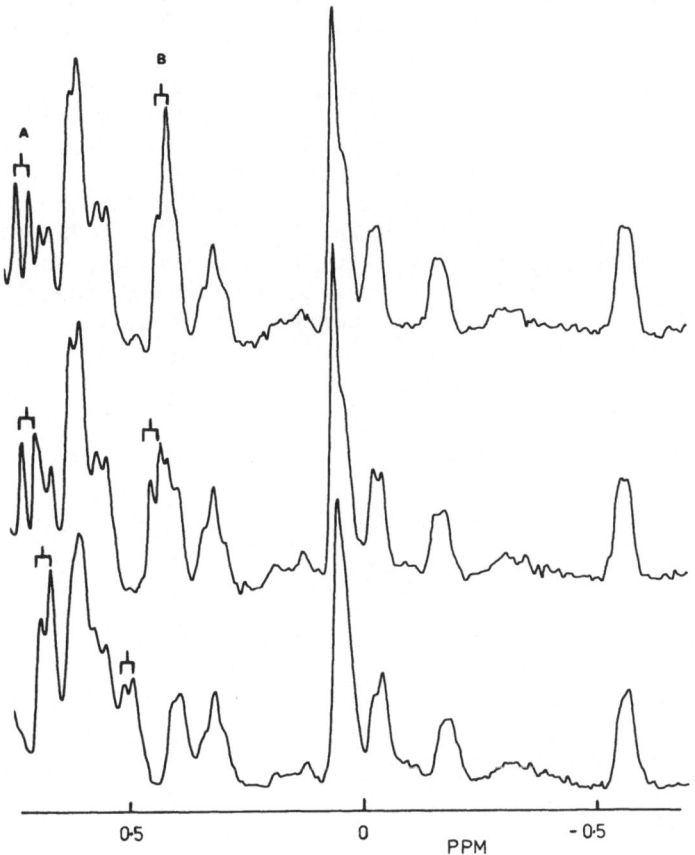

Fig. 2. The effect of Pr^{3+} at pH5.3 on the high field (CH$_3$) resonances of lysozyme. Top, no Pr^{3+}; Middle, 0.6 mM Pr^{3+}; bottom, 3.0 mM Pr^{3+}. The large shifts of ala 107 (A) and thr 51 (B) can be seen.

iment was done at a constant pH value and a constant concentration of lysozyme.

The shifts induced by La^{3+} and Lu^{3+} were small (Campbell et al., 1975b) and were subtracted at the relevant concentrations from the observed shifts induced by the paramagnetic ions. The resultant paramagnetic induced shifts were plotted against the concentration of lanthanide ions in the solution. This was carried out using the different lanthanides, and the type of behaviour of resonances with the different lanthanides was similar although the magnitude and direction of the shifts varied as expected for each ion (Campbell et al., 1975b).

Examination of the resultant curves showed that not all resonances experienced shifts which could be attributed to binding at a single site. Certain resonances suffered only small shifts which were linear in the

concentration of lanthanide ions over a wide range (0—0.2 M). These shifts
were independent of the concentration of lysozyme (1—5 mM) and of pH
(3.5—6.0). This indicates that these shifts are not connected with the
binding of lanthanide ions to ionisable groups in the protein. In fact,
all the other resonances contain these linear shifts of similar magnitude superimposed upon the true binding curves. It is likely that the
shifts arise from the weak interaction between the internal standards
(dioxan and acetone) and lanthanides in solution. In any case, the shifts
are small and readily subtracted from the binding curves described below.
       Many resonances experience large shifts which give rise to binding
curves typical of those arising from relatively strong binding at a
single site (provided that the linear correction is made), see Figure 3.

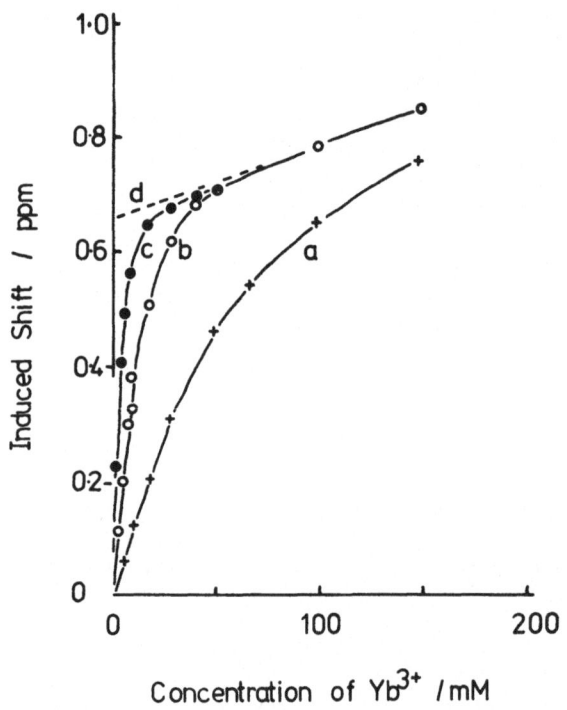

Fig. 3.   Shifts induced in the chemical shift value of thr 51
(CH_3) as a function of the concentration of Yb^{3+}. (a) pH 4.3;
(b) pH 5.3; (c) pH 5.3; (d) extrapolation of linear contribution. A positive shift is upfield.

These resonances are all assigned to groups which are close to asp 52
and glu 35. The binding curves are pH dependent (Figure 3), and a group
of pK value close to 6 is implicated in the binding.
       A number of resonances show titration curves which indicate that

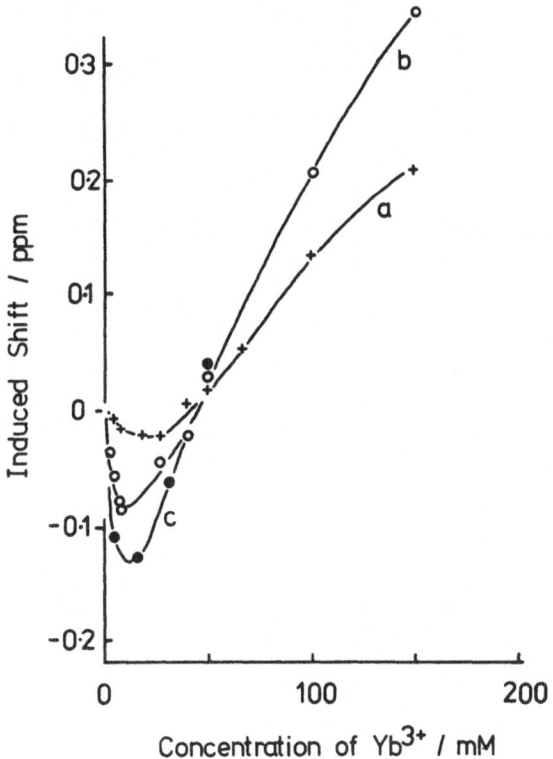

Fig. 4. Shifts induced in the chemical shift value of ile 98 (γ-CH$_3$) as a function of the concentration of Yb^{3+}. (a) pH 4.3; (b) pH 5.3; (c) pH 5.8.

they arise from lanthanide binding at weaker sites. The pH dependence implicates binding groups of pK value closer to 4 than to 6. In some cases (Figure 4) the resonances are clearly shifted by the effects of lanthanide binding at both the strong site and a weaker site.

The titration curves therefore indicate that a strong binding site exists, but also that weaker sites exist. In order to confirm this conclusion, the pH dependence of the binding curves was investigated in more detail.

4.2. *pH Titration Curves*

pH titrations were carried out by varying the pH of solutions of 5 mM lysozyme containing various concentrations of a lanthanide ion up to 50 mM. The ionic strength of all solutions was maintained at 0.4 by using relevant concentrations of KCl. We shall return to the effects of pH on the lysozyme spectrum itself later, but here only the paramagnetic shifts (the difference between the shift in the presence of a paramagnetic ion and in the presence of La^{3+} or Lu^{3+}) are considered. These were plotted against pH. Again, a small pH independent shift was observed

as a contribution for all resonances, and correction was made for this.
The resonances which are affected by binding at the main site only have
titration curves of the type shown in Figure 5, which again indicates
binding to a group with pK value of ca. 6.0. As the pH value is decreased,

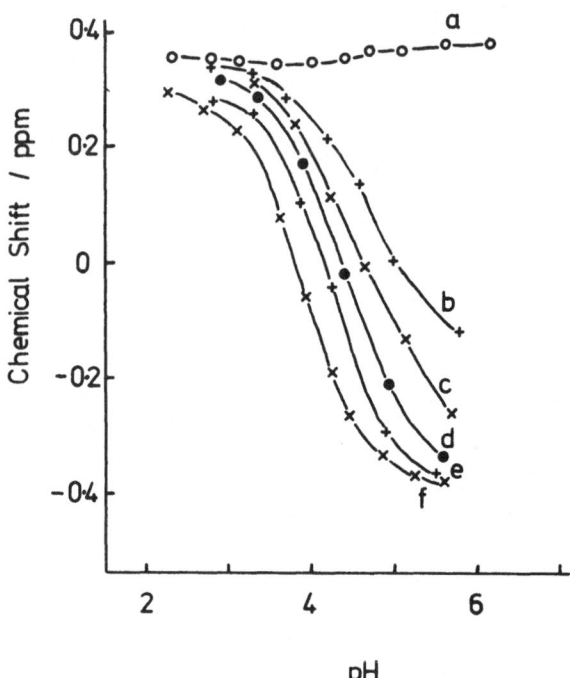

Fig. 5.   pH dependence of the chemical shift value of thr 51
(CH$_3$) in the presence of (a) 0.05M La^{3+}; (b) 0.005M Yb^{3+};
(c) 0.009M Yb^{3+}; (d) 0.015M Yb^{3+}; (e) 0.03M Yb^{3+}; (f) 0.05M
Yb^{3+}.

the fraction (f) of lysozyme bound to the lanthanide ion decreases as H$^+$
competes for the binding group. The observed induced shift ($\delta = f\Delta_i$)
therefore decreases.

The resonances which were observed to have titration curves which
at fixed pH values indicated binding to weaker sites give pH titration
curves which implicate binding groups of lower pK values. Those resonances which are affected by binding at more than one site, as expected
have complicated pH dependences as shown in Figure 6.

At this stage we may conclude that a strong binding site for lanthanide ions exists. This perturbs many of the resonances, which examination shows are those assigned to groups close to asp 52 and glu 35.
Weaker binding sites also exist however. Thus it is necessary to examine
in detail the nature of the binding sites, and the groups involved, in
order that pertubations from a single site may be measured for the conformational study.

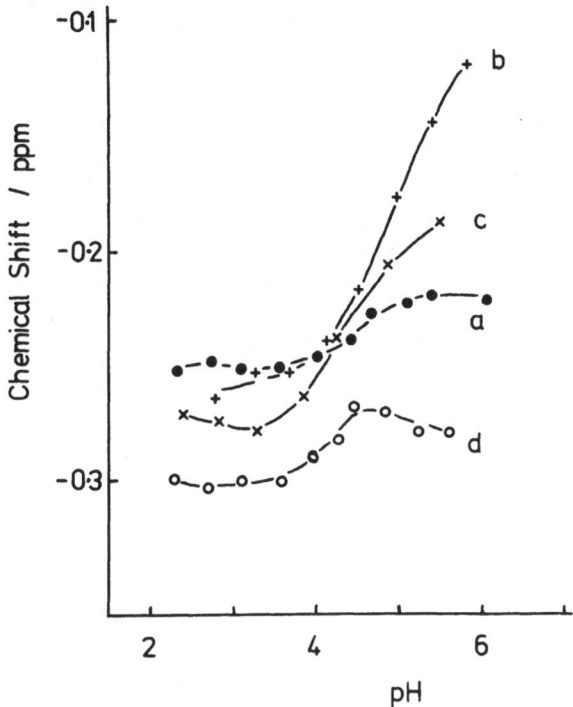

Fig. 6. pH dependence of the chemical shift value of ile 98 (γ-CH$_3$) in the presence of (a) 0.05M La^{3+}; (b) 0.005M Yb^{3+}; (c) 0.03M Yb^{3+}; (d) 0.05M Yb^{3+}.

5. CHARACTERISATION OF THE BINDING GROUPS

The data accumulated in the experiments described in the previous section may now be used to define the characterictics of the binding sites. These characteristics include the values of binding constants, the number of ionisable groups involved in binding, the pK values of these groups, and their position in the protein sequence. First, we shall consider only those groups whose resonances reflect binding at the major site alone.

5.1. *The Major Site*

The data from the binding curves recorded at fixed pH values alow the apparent binding constant to be measured at each pH value. Consider the binding to a single site, M + L ⇌ ML

$$1/\delta_i = 1/\Delta_i + 1/K \, \Delta_i \, [M]$$

using the nomenclature described above. A plot of $1/\delta_i$ vs. $1/[M]$ allows Δ_i and K to be calculated, provided that $[M] \simeq [M] + [ML]$ which is the case as here, for a large excess of M over L. Figure 7 shows plots of this type. These show that Δ_i, the fully bound shift, is independent of pH for the resonances affected by the major site, and that the binding constant increases with pH

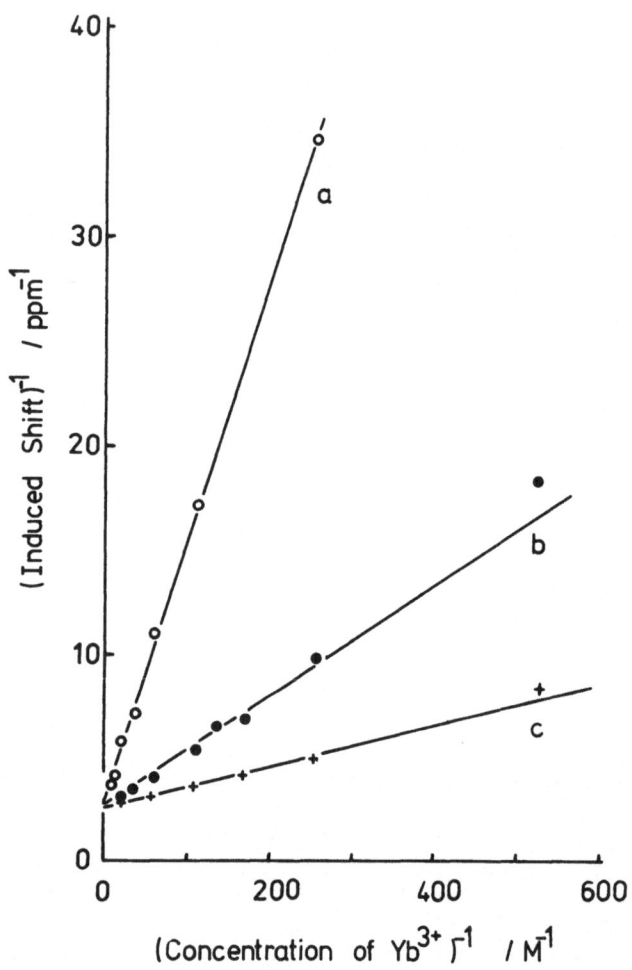

Fig. 7. Plots of the reciprocal of the induced shift of thr 51 (CH$_3$) (corrected for the linear shift, see Figure 1) against the reciprocal of the concentration of Yb^{3+}. (a) pH 4.3; (b) pH 5.3; (c) pH 5.8.

The data from the pH titrations themselves provide more information. In order to interpret these results it is nessary to consider a specific type of behaviour.

Consider the binding of a lanthanide ion (M^{3+}) to a singly ionised group in the protein molecule (L$^-$), and assume that no binding to the protonated form LH occurs. Let the association constant for M^{3+} be K_2, and the pK value of the protein group be pK_1. The relevant processes are therefore

$$L^- + H^+ \rightleftharpoons LH$$

and

$$L^- + M^{3+} \rightleftharpoons LM^{2+}$$

Now $1/K_1 = [LH]/[L^-][H^+]$, and a plot of pH ($-\log [H^+]$) against log $([L^-]/[LH])$ is linear and can be used to determine the pK_1 value. Also, when $[L^-] = [LH]$, $K_1 = [H^+]$ and at this point $pK_1 = pH$.

By definition, $K_2 = [LM^{2+}]/[L^-][M^{3+}]$.

Thus, $K_2 = ([LM^{2+}]/[LH][M^{3+}])[H^+]/K_1$. Provided that $[M^{3+}] \simeq [M^{3+}] + [LM^{2+}]$ over the pH range in question as will be the case with an excess of M^{3+} in solution, a plot of pH against log $([LM^{2+}]/[LH])$ will be linear. By analogy with the previous paragraph, when $[LH] = [LM^{2+}]$, $K_2 = ([H^+]/[M^{3+}])/K_1$. For a pH value at least one unit below the pK_1 value of LH, $[LH] \simeq [LH] + [L^-]$ and the condition that $[LH] = [LM^{2+}]$ is merely that for 50% of the protein being bound to M^{3+}. For a given $[M^{3+}]$, this point will be at a pH value denoted pH (50% bound) and

$$pK_1 - pH \text{ (50\% bound)} = \log [M^{3+}] + \log K_2.$$

This treatment can now be used to characterise the strong binding site of lysozyme, if it is a valid model for lysozyme. It can be seen that in order to define K_2, it is necessary to know pK_1, and also the pH value

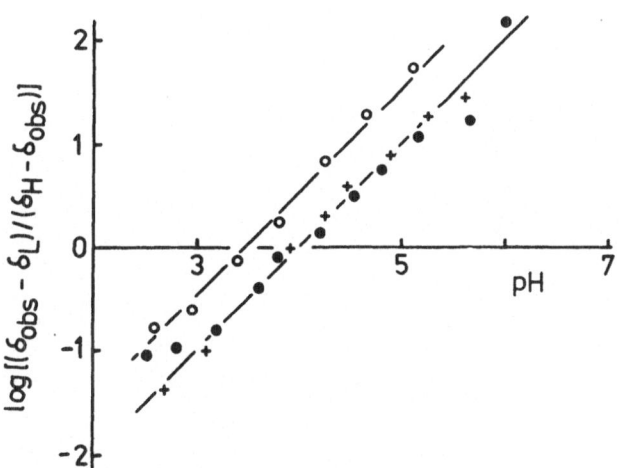

Fig. 8. Plots of log $((\delta_{obs} - \delta_L)/(\delta_H - \delta_{obs}))$ against pH for paramagnetic lanthanides. Open circles, tyr 53 (ortho protons) with 0.05 M Eu^{3+}; closed circles, trp 108 (C(2)H) with 0.05 M Pr^{3+}; crosses, thr 51 (CH_3) with 0.05 M Yb^{3+}. The slopes are drawn to be 1.00.

at which 50% of the ligand (lysozyme) is bound to a lanthanide at the major site.

The paramagnetic shift data were treated as follows. We can write when M^{3+} is in large excess,

$$[LM^{2+}]/[LH] = (\delta_{obs} - \delta_L)/(\delta_H - \delta_{obs})$$

where δ_L is the paramagnetic induced shift for a given resonance at low pH (no M^{3+} bound), δ_H is the shift at high pH (L^- fully bound) and δ_{obs} is the shift at intermediate pH values. In calculating the paramagnetic induced shifts, correction was made for the pH independent shift (see above). Then, $\log ((\delta_{obs} - \delta_L)/(\delta_H - \delta_{obs}))$ was plotted against pH. This is shown for three resonances in Figure 8. The plots are linear, with slopes

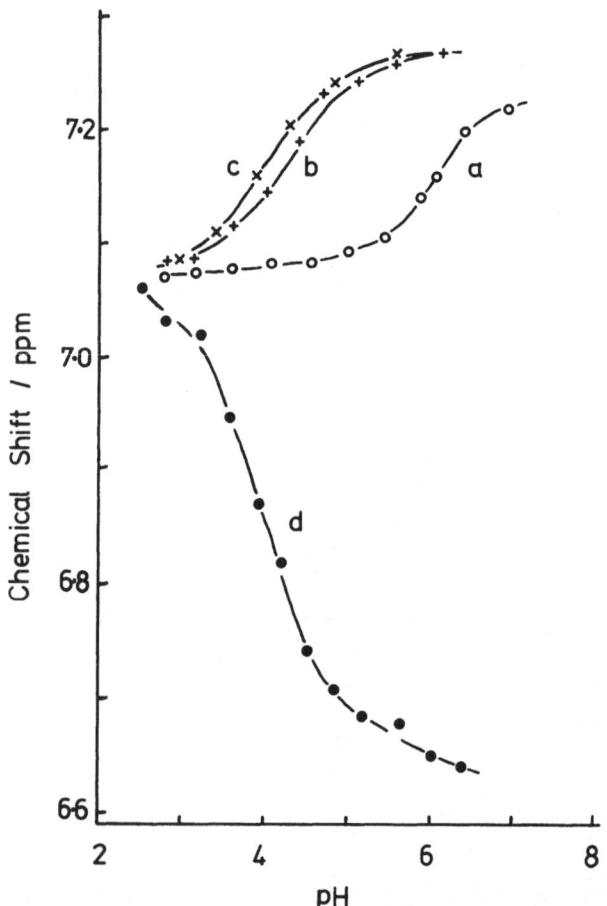

Fig. 9. pH dependence of the chemical shift value of trp 108 (C(2)H) in the presence of (a) 0.4 M KCl; (b) 0.05 M La^{3+}; (c) 0.05 M Lu^{3+}; (d) 0.05 M Pr^{3+}.

equal to unity from at least pH 3.5 to 6.0. This shows that only one group with a pK value above 3.5 is involved in binding the lanthanide. The pH (50% bound) value is readily determined from the plots. Had the slope of the plots not been unity, the initial assumptions in the method would have been invalid.

In order to dertermine pK_1, we are able to make use of a fortunate observation. The resonance of the C(2)H and N(1)H protons of trp 108 are strongly affected by pH, even in the absence of metal ions. In Figure 9 the titration of the C(2)H proton of trp 108 with pH is shown. The chemical shift values for the resonances at high pH (δ_H), and at low pH (δ_L) and at intermediate pH values (δ_{obs}) were measured. Writing $[L^-]/[LH] = (\delta_{obs}-\delta_L)/(\delta_H-\delta_{obs})$, pH was plotted against log $((\delta_{obs}-\delta_L)/(\delta_H-\delta_{obs}))$ as shown in Figure 10. This plot has a slope of 1, and indicates that the observed shifts arise from a single ionisation $L^- + H^+ \rightleftarrows LH$.

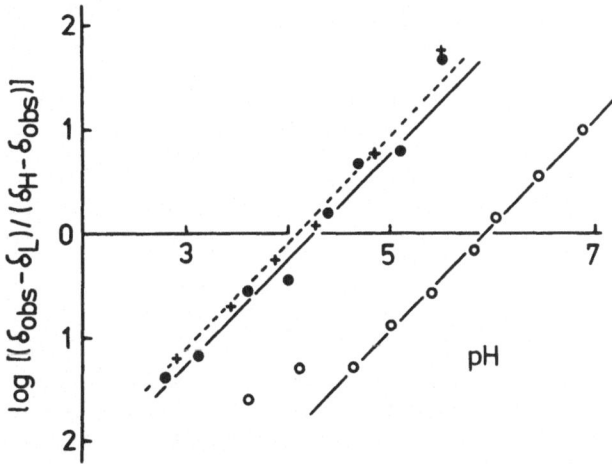

Fig. 10. Plots of log $((\delta_{obs}-\delta_L)/(\delta_H-\delta_{obs}))$ against pH for trp 108 (C(2)H). Open circles, 0.4 M KCl; closed circles, 0.05 M La^{3+}; crosses, 0.05 M Lu^{3+}. The slopes are drawn to be 1.00.

The pK value of the ionisation is 6.0 ± 0.2 (μ = 0.4). This pK value is known to be that of glu 35, from chemical studies, and the pH dependence of the trp 108 resonances appears to arise from an interaction between glu 35 when protonated, and the indole ring of trp 108 (Campbell et al., 1975b, c). It is now necessary to show that glu 35 is in fact the group with which the lanthanides are interacting. This has been done as follows. In the presence of 50mM La^{3+}, the pH titration of the trp 108 resonances is different, for the apparent pK value is lower. This strongly suggests that the La^{3+} is competing with H$^+$ for glu 35, i.e. is binding to glu 35. In this case, we can write

$$[LM^{2+}]/[LH] = (\delta_{obs}-\delta_L)/(\delta_H-\delta_{obs})$$

and a logarithmic plot of this function against pH is again linear
(Figure 10), with a slope of 1. The apparent pK value is therefore the
pH at which 50% of the La^{3+} is bound to glu 35, and is 4.2 ± 0.1. This
pH(50%) value at this lanthanide concentration is very close to that
obtained for the paramagnetic shift data on all the resonances affected
by the strong binding site.

TABLE I
Association constants[a] for different lanthanides[b]

Lanthanide	pH(50% bound)[c] at 0.05 M	$K^{[d]}$ (M^{-1})
La	4.2	1.3×10^3
Pr	4.0	2.1×10^3
Eu	3.5	6.7×10^3
Yb	4.0	2.1×10^3
Lu	4.1	1.7×10^3

[a] at 54 °C, μ = 0.40.

[b] the binding constant for Ca^{2+} has also been calculated.
At 0.5M $CaCl_2$, pH(50% bound) = 5.0, giving K = 20 M^{-1}.

[c] values ± 0.1.

[d] for binding to a group of pK = 6.0.

All the data are summarised in Table I, where the values of K_2 can now
be given. We can now summarise the results of this analysis. Only one
group of pK value above 3.5 is involved in the major binding site for
lanthanides. This group is glu 35 (pK = 6.0). Any other group involved
in binding must have a pK value of less than 3.5. By analogy with the
crystallographic data, asp 52 might be expected to be a binding group.
The pK value of this group is 3.5 ± 0.5, and it is therefore conceivable
that it is involved. There is however no firm evidence for this. We note
however that the titration curves for the trp 108 resonances show a slight
inflexion at 3.5 ± 0.5 which may be attributed to the effect of the
ionisation of asp 52 (Campbell et al., 1975).

Having established that the strong binding site involves glu 35,
it is resonable to suppose that the binding position in solution is similar to that is the crystal, where glu 35 was also identified. The data
for this weak binding sites will merely be summarised here, as we shall
not consider them further.

5.2. *The Minor Binding Sites*

Many of the resonances observed to be affected by the binding at weak
sites have been assigned. The binding groups must be close to these
affected resonances as the induced shifts fall off as $1/r^3$. By examination of the X-ray structure it was possible to deduce the likely binding

groups and these are summarised in Table II. In fact, it is likely that
weak binding occurs to other exposed carboxylic acid groups, but these
have not been identified. The approximate pK values of the binding groups,
and the approximate values for the binding constants are also given.

TABLE II
Summary of Lanthanide binding sites

Site	Binding groups and pK values at μ=0.4	Approximate K (M^{-1})
A	glu 35 (6.0) [asp 52 (3.5)]	10^3
B	asp 101 (ca. 4.5)	10
C	glu 7 (ca. 2.0)	<10
D	asp 85 (ca. 4.0)	<10

6. MEASUREMENT OF SHIFT AND RELAXATION RATIOS

The experimental titration described above may now be used to determine
the ratios of shift and relaxation effects on different nuclei which
arise from binding at the major binding site. These data define the
conformation of the ligand (protein) which binds the metal. In this
section the methods of calculation are described.

6.1. *Shift Ratios*

The data from the titrations at fixed pK values may now be treated as
follows. The required shift ratios are of the form Δ_i/Δ_j. The experimental measurements are observed shifts, $\delta_i = f\Delta_i$, and $\delta_i/\delta_j = \Delta_i/\Delta_j$.
Consider the case where shifts of nuclei arise from binding not only
at the major site, but also from binding at a minor site. Here,

$$\delta_i = f\Delta_i + f'\Delta_i'$$

and

$$\delta_i/\delta_j = (f\Delta_i + f'\Delta_i')/(f\Delta_j + f'\Delta_j')$$

Here, f' represents the fraction of lysozyme with a lanthanide bound
to the minor site, and Δ_i' and Δ_j' are the fully bound shifts which result
from this binding. Provided that the binding sites are independent, and
that the binding constants at the major and minor sites are sufficiently
different (say more than a factor of 10 as found for lysozyme), at low
concentrations of lanthanides f≫ f'. At infinite concentration of
lanthanides, f = f' = 1.0.

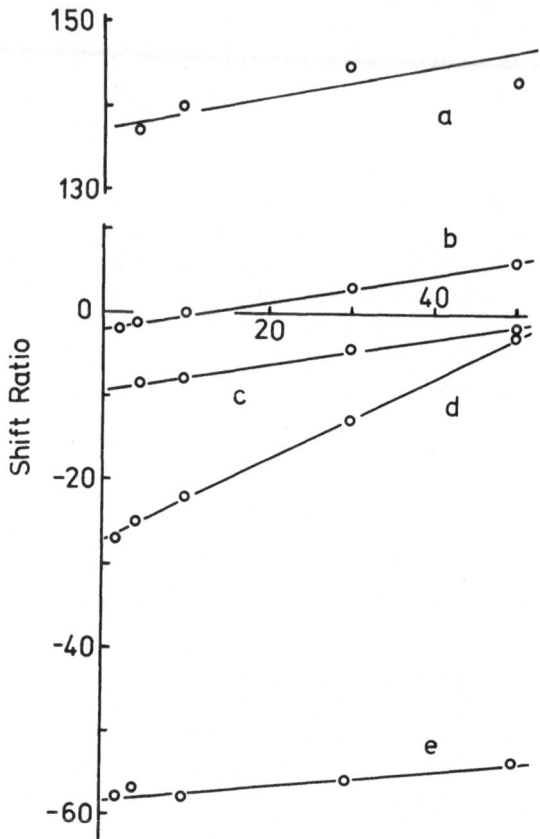

Fig. 11. Ratios of shifts, relative to trp 108 (C(2)H) = 100, with Pr^{3+} at pH 6.0. (a) tyr 53 (ortho protons); (b) met 12 (CH_3); (c) trp 63 (C(5)H or C(6)H); (d) ile 98 (γCH_3); (e) ala 107 (CH_3).

Thus, calculation of δ_i/δ_j at a fixed ligand concentration and a wide range of lanthanide concentrations, followed by extrapolation of the values to zero lanthanide concentrations, gives a close approximation to Δ_i/Δ_j. This procedure also removes the effects of the linear contribution to the observed titration curves. Some data are plotted in Figure 11. The extrapolation is essentially linear, as found in small molecule systems (Barry et al., 1971, 1974).

    Shift ratios can also be calculated from the pH titrations described above. However, it is necessary to ensure again that the effects of binding at the major site only are measured. Thus, the paramagnetic shifts (corrected for the effects of La^{3+} or Lu^{3+}) were calculated for different resonances at each pH value. These shifts of one nucleus (δ_i) at differ-

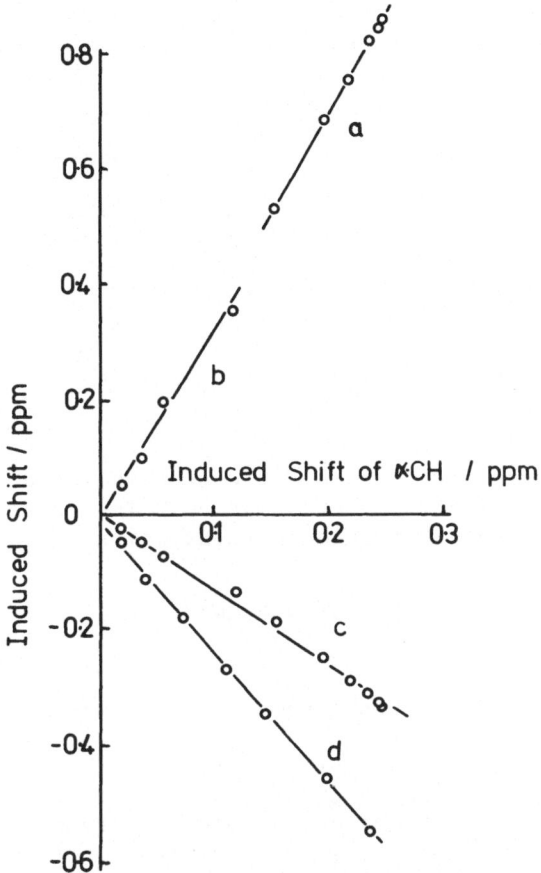

Fig. 12. Shifts induced by 0.05M Pr^{3+} in the resonances of
(a) tyr 53 (meta protons); (b) thr 51 (CH_3); (c) ala 107
(CH_3); (d) trp 108 (C(2)H); plotted against the shift induced
in another resonance (unassigned α-CH at 5.81 ppm). Each point
represents a different pH value, and the relative slopes are
the shift ratios.

ent pH values were plotted against shifts of another nuclei (δ_j) as
shown in Figure 12. Only if the two nuclei have the same pH dependence
(i.e. arise from binding at the same site) will these plots be linear.
The slope of the plot if linear gives $\delta_i/\delta_j = \Delta_i/\Delta_j$. In particular it
is noted that the pH independent contribution to the shift is eliminated
from the shift ratio. Table III lists some shift ratios determined by
the two methods. Close agreement is found for those resonances affected
by the major binding site above. However, unlike the first method, this
pH method does not readily allow calculation of shift ratios for resonances affected by more than one site.
    The linearity of the plots of Figure 12 allows a further conclusion
to be drawn. This is that the shift ratios are independent of pH over

TABLE III
Shift ratios[a] determined by different methods

Residue	Lanthanide used	Shift Ratio[b]		
		A	B	C
tyr 53(m-protons)	Pr	159	160	164
	Eu	135	146	156
	Yb	120	140	123
ile 98 (γ-CH$_3$)	Pr	-35	c	-15
	Eu	-28	c	-22
	Yb	-33	c	-15

[a] relative to tyr (o-protons).

[b] Method A is extrapolation to zero lanthanide concentration at constant pH (6.0). Method B is the pH titration method at 50mM lanthanide. Method C is the use of a constant total lanthanide concentration of 24 mM at pH 5.3.

[c] Not applicable because of the effects of weak binding.

the range pH 3.5—6.0. This shows that the same mode of binding occurs over the whole pH range.

Finally, in this section, a more rapid but less satisfactory method of obtaining the shift ratios may be mentioned. Here, a solution containing a certain concentration (say 25 mM) of paramagnetic lanthanide is titrated into a solution containing the same concentration of diamagnetic La^{3+} or Lu^{3+}. The total concentration of lanthanide remains constant throughout the titration, and essentially linear plots of induced shift against the concentration of paramagnetic lanthanide are obtained (see Campbell et al, 1973b). No further correction for the diamagnetic contribution to the shift is needed, but no indication about the effects of weak binding is obtained. The observed shift ratios are however close to those obtained by the other methods (Table III), except when the shifts are not solely due to binding at the major site.

Thus, by a combination of experiments, the shift ratios for binding at the major site were obtained for the different resonances studied.

6.2. Gd^{3+} Broadening Ratios

In order to obtain quantitative measurements of the relaxation induced by Gd^{3+} in different resonances, difference spectroscopy was used as described previously (Campbell et al., 1973a). Spectra were run of solutions containing different concentrations of Gd^{3+}, and these spectra subtracted in turn from the spectrum of the solution recorded before the addition of Gd^{3+}. Only resonances which have been strongly relaxed (broadened) by Gd^{3+} appear in the difference spectra. An example of a difference spectrum is given in Figure 13. Under the conditions used

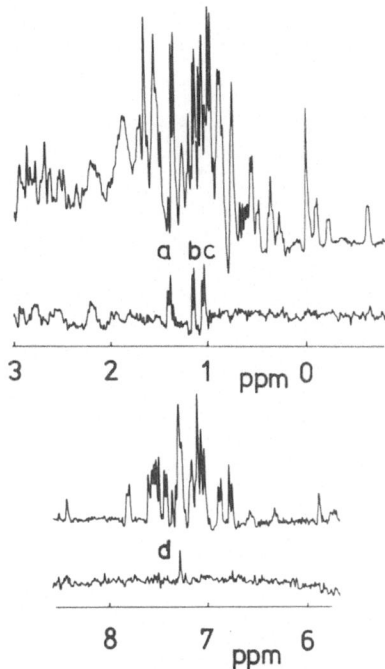

Fig. 13. Gd^{3+} difference spectrum of lysozyme (5 mM lysozyme, 24 mM La^{3+}, pH 5.3, difference between spectra with and without 1.1×10^{-5} M Gd^{3+}) showing that the resonances of the CH_3 groups of ala 110 (a) and val 109 (b) and (c), and of the C(2)H proton of trp 108 (d) are the most broadened resonances. Above the difference spectrum is shown the spectrum of lysozyme (obtained using the convolution difference procedure, Campbell et al., 1973a) in the absence of Gd^{3+}.

here, a plot of $1/I_0$ vs $1/[Gd^{3+}]$ is linear (Campbell et al., 1973a) with a slope proportioned to T_{2M}. I_0 is the height of a peak in the difference spectrum, and the method allows relative values of T_{2M} to be obtained for different resonances. The analysis assumes that the initial linewidths of all resonances are the same, and this has been shown to be a good approximation for most resonances in lysozyme (Campbell et al., 1976). Some of the data are plotted in Figure 14. Corrections for weak binding sites were not made in this analysis, and this will be mentioned below.

7. THE CONFORMATIONAL ANALYSIS

Up to this point, no mention of the protein conformation has been made, as the methods of analysing the lanthanide data have been of prime importance. Here we shall indicate some of the results of the conformational analysis. First, the relaxation data will be considered.

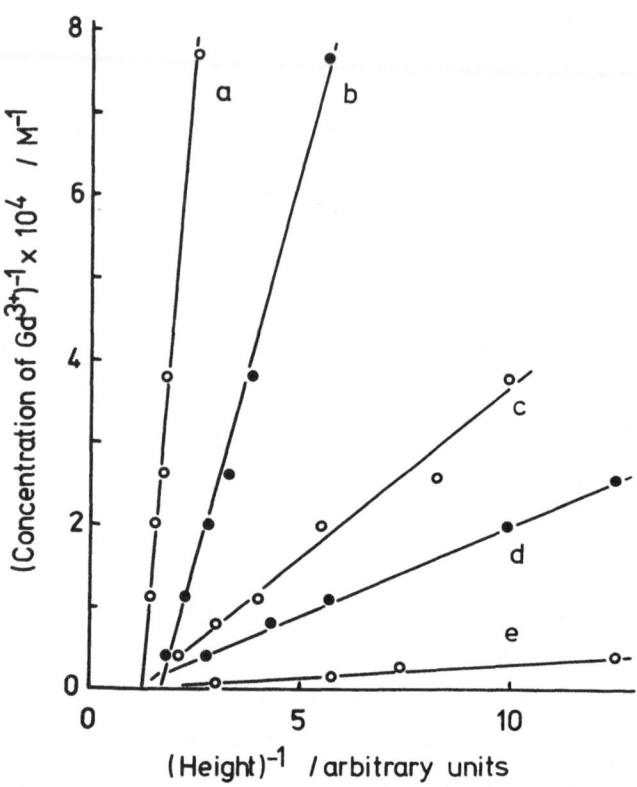

Fig. 14. Plots of the reciprocals of the heights of peaks
in difference spectra (such as Figure 13) against the reciprocal of the concentration of Gd^{3+}. (a) val 109 (CH_3); (b)
trp 108 (C(2)H); (c) thr 51 (CH_3); (d) leu 56 (CH_3); (e) leu
17 (CH_3).

7.1. *The Gd^{3+} Broadening Ratios*

From the Gd^{3+} broadening ($1/T_{2M}$) ratios, the relative effective distances
to assigned groups were measured. The word effective allows for possible
effects of motion on these values as discussed below. Then relative distances to some groups were calculated from the X-ray structure, taking
the position of the bound metal from the X-ray data (Blake and Rabstein,
1970). The high resolution X-ray diffraction study was of the protein
in the absence of bound metal ions. We discuss below that the conformation in the presence of metal ions is little altered. Now, in calculating relative distances to compare with the nmr data, allowance was made
for two types of motion which has been observed to occur in solution.
First, methyl groups rotate rapidly about the C—CH_3 bond. An average
of r^6 was calculated on the basis of this, and used to obtain an effective distance. Secondly, aromatic tyrosine residues have been observed
to rotate rapidly about the C_β–C_γ bond (Campbell et al., 1975d) and an

effective distance was calculated for tyrosine residues on the basis of this motion.

TABLE IV
Observed and calculated relative effective distances[a]

Proton		Relative distances[b]	
		Calculated[c]	Observed
val 109	(CH$_3$)	0.53	0.60
val 109	(CH$_3$)	0.73	0.60
trp 108	(C(2)H)	0.73	0.65
ala 110	(CH$_3$)	0.81	0.62
thr 51	(CH$_3$)	0.84	0.93
tyr 53	(o-protons)	1.00	1.00
ala 107	(CH$_3$)	1.08	0.94
leu 56	(CH$_3$)	1.18	1.01
ile 98	(γ-CH$_3$)	1.19	1.05
leu 56	(CH$_3$)	1.36	1.08
met 105	(CH$_3$)	1.38	1.20
leu 17	(CH$_3$)	2.00	1.30

[a] relative to tyr 53 (o-protons) = 1.00. The calculated effective distance is in fact 10.0 Å to these protons.

[b] see comments in text for deviations at large effective distances.

[c] the coordinates of the bound ion are x = 9.89, y = 22.73 and z = 21.79 in the coordinate scheme defined in Imoto et al. (1972).

In table IV, the calculated and measured effective distances to various residues are given. The agreement between the two sets of data is reasonable, but it is noticeable that at longer distances (more than ca. 12 Å) the measured effective distances become shorter than the calculated X-ray structure effective distances. This is undoubtedly due to the increasing importance of the weaker binding sites as the perturbations from the major site become small. No discrepancies between the experimental and calculated distances are so large for any assigned residue that they are outside the present limitations of the calculations. We shall return to this point below.

7.2. *The Paramagnetic Shift Data*

The shift ratios measured with different lanthanides have been shown (Campbell et al., 1975b) to be sufficiently similar to assume axial symmetry of the bound lanthanide ions, (with the exception of Tm^{3+} and Er^{3+}) by analogy with results from small molecule systems. Similarly, it may be assumed that the complexes are essentially homologous, and that the shifts are solely pseudocontact in origin. The most accurate set of shift ratios has been determined for Pr^{3+} (Table V). Then, using the X-ray coordinates, values of $(3 \cos^2\theta - 1)/r^3$ were calculated for the different groups, assuming a direction of the symmetry axis, defined by two angles α and β (see Barry et al., 1973a, b). These were converted into ratios, and allowance was again made for CH_3 and tyrosine motion. Then, α and β were varied over all space, calculating the effective ratios at each value of α and β. For only a small range of values of α and β, the calculated effective ratios were close to the experimental shift ratios (Campbell et al., 1975b). Shifts which were downfield with Pr^{3+}, corresponded to values of $(3 \cos^2 \theta - 1)$ which were positive, in agreement with other studies of small molecules in aqueous solution (eg. Barry et al., 1974).

TABLE V
Observed and calculated shift ratios[a]

P	Proton		Observed shift ratios	Calculated shift ratio[c] $\alpha = 100^o$ $\beta = 70^o$	$\alpha = 110^o$ $\beta = 100^o$
tyr	20	(o-)	0	-3	-6
tyr	23	(o-)	-14	-10	-3
tyr	53	(o-)	100	100	100
tyr	53	(m-)	160	144	93
trp	63	(C(2)H)	53	42	-20
trp	108	(C(2)H)	-117	-137	-124
trp	108	(N(1)H)	-45	-21	-84
met	12		-2	-6	-12
thr	51		151	111	166
leu	56[d]		-9	-19	-35
leu	56[d]		0	-1	-23
ile	98γ		-32	-23	-35
ile	98δ		0	14	-29
met	105		-14	-25	-11
ala	107[d]		-68	-49	-41
val	109[d]		-109	65	631
val	109[d]		73	175	271
ala	110		59	-80	73

[a] relative to tyr 53 (o-) = 100.

[b] the first group are aromatic protons, the second are CH_3 groups.

[c] α and β refer to rotation angle about the crystallographic x and z axes. $\alpha = 0$ is the z direction.

[d] distinction between assignments of the two CH_3 groups of each residue cannot be made.

In Table V are listed calculated ratios for two values of α and β. The variation of these calculated values with α and β can be seen from this Table. Overall, the agreement with the experimental ratios is good. The largest discrepancy is of val 109. This is not surprising, as this residue cannot be clearly defined in the X-ray electron density map, suggesting that its position cannot be defined clearly even in the crystal.

7.3. *The Overall Conformation and Conformational Mobility*

The level of agreement between the experimental conformational parameters and the X-ray structure shows that there is indeed close similarity between the solution and crystal structures. However, we have been defining a time-averaged structure. In order to make the comparison, the effects of conformational mobility of methyl groups and even tyrosine residues have had to be considered. Allowance has not been made for the motion of other groups, and it is likely that this exists. In order to make a close comparison of the protein in the two states (solution and crystal) this mobility must be defined.

The detection of the existence of mobility of groups in a protein is a departure from the apparent rigidity of the molecule when the X-ray structure is viewed. In a separate paper we describe evidence for extensive and rapid independent mobility of groups (Campbell et al., 1976). The motion of individual groups can be studied in detail now that the assignment of many resonances has been made.

7.4. *Conformational Changes on Binding Lanthanides*

In this analysis, it has been assumed that no conformational change occurs in the protein when a lanthanide ion binds. This appears to be generally true, because of the level of agreement achieved in the analysis above (the X-ray structure was determined for the metal-free protein). Nmr can provide information directly concerning the extent and rate of protein conformational changes (see for example Dobson and Williams, 1975).

The binding of the diamagnetic La^{3+} or Lu^{3+} ions to lysozyme causes small shifts of resonances in the active site (Campbell et al., 1975b). That these shifts are small shows that any induced conformational change is small. A small rearrangement of the structure must however take place to produce these shifts. The most obvious change is that described earlier concerning the interaction between glu 35 and trp 108. The carboxylic group of glu 35 must move away from trp 108 (where it rests when protonated) into free solution for binding to the lanthanide, just as it moves when glu 35 becomes ionised (Campbell et al., 1975b, c). The lack of exchange broadening in the spectrum shows that this movement is fast ($> 10^3$ s^{-1}).

8. CONCLUSIONS

This paper has been concerned with the binding of lanthanide cations to lysozyme in solution. The lanthanides bind to several sites in the mol-

ecule, although to only site is the binding strong. This major binding
site involves glu 35 (pK value 6.0) but no other ionisable group with
a pK value above 3.5. There is only indirect evidence for the involvement of asp 52 in the binding, and the pK value of this group is low
(3.5±0.5). However no significant binding to asp 52 is possible except
that involving glu 35 in its ionised form.

The binding of lanthanide cations to lysozyme causes only a small
conformational change. This change is fast, and probably involves merely
a reorientation of binding groups to accomodate the metal ion.

The shift and broadening induced by the binding of paramagnetic
lanthanides at the major site has been measured for a number of assigned
resonances. By interpretation of these data the protein conformation and
orientation of groups around the metal binding site have been shown to
be similar in the solution and crystalline states. However, it was found
that groups in the protein are not rigidly held in space, but are undergoing indepent motion. In this way the solution conformation of lysozyme
differs from the static model of the protein constructed from X-ray diffraction studies.

As far as the nature of the binding of the lanthanide ions to the
protein is concerned, the following points can be made. First, all binding of significant strength is to carboxylate groups rather than to other functional groups. Secondly, only one strong binding site exists,
and this is presumable due to the binding between two carboxylate groups
(asp 52 and glu 35), whilst weak binding occurs to single carboxylate
groups. Thirdly, no major conformational change takes place on binding,
which has allowed in this case the lanthanide cations to be used as
probes of the structure of the native protein.

ACKNOWLEDGEMENTS

This is a contribution from the Oxford Enzyme Group which is supported
by the Science Research Council. The work is also supported by the Medical Research Council.

REFERENCES

Barry, C.D., Glasel, J.A., North, A.C.T., Williams, R.J.P., and Xavier,
   A.V.: 1971, Nature 232, 236.
Barry, C.D., Dobson, C.M., Ford, L.O., Sweigart, D.A., and Williams,
   R.J.P.: 1973a, in Sievers, R.E. (ed.), Nuclear Magnetic Resonance
   Shift Reagents, Academic Press, New York, 173.
Barry, C.D., Hill, H.A.O., Sadler, P.J., and Williams, R.J.P.: 1973b,
   Ann. N.Y. Acad. Sci. 206, 247.
Barry, C.D., Glasel, J.A., Williams, R.J.P., and Xavier, A.V.: 1974,
   J. Molec. Biol. 84, 471.
Blake, C.C.F., Johnson, L.N., Mair, G.A., North, A.C.T., Philips, D.C.,
   and Sarma, V.R.: 1967, Proc. Roy. Soc. B167, 378.
Blake, C.C.F. and Rabstein, M.A.: 1970, unpublished data.
Bleaney, B.: 1972, J. Mag. Resonance 8, 91.

Bloembergen, N.: 1957, J. Chem. Phys. 27, 595.
Campbell, I.D., Dobson, C.M., Williams, R.J.P., and Xavier, A.V.: 1973a, J. Mag. Resonance 11, 172.
Campbell, I.D., Dobson, C.M., and Williams, R.J.P.: 1973b, Ann. N.Y. Acad. Sci. 222, 163.
Campbell, I.D., Dobson, C.M., and Williams, R.J.P.: 1975a, Proc. Roy. Soc. A345, 23.
Campbell, I.D., Dobson, C.M., and Williams, R.J.P.: 1975b, Proc. Roy. Soc. A345, 41.
Campbell, I.D., Dobson, C.M., and Williams, R.J.P.: 1975c, Proc. Roy. Soc. B189, 485.
Campbell, I.D., Dobson, C.M., and Williams, R.J.P.: 1975d, Proc. Roy. Soc. B189, 503.
Campbell, I.D., Dobson, C.M., and Williams, R.J.P.: 1976, in press.
Dobson, C.M. and Levine, B.A.: 1976, in press.
Dobson, C.M. and Williams, R.J.P.: 1975, FEBS Letters 56, 362.
Dwek, R.A., Richards, R.E., Morallee, K.G., Nieboer, E., Williams, R.J.P., and Xavier, A.V.: 1971, Eur. J. Biochem. 21, 204.
Imoto, T., Johnson, L.N., North, A.C.T., Philips, D.C., and Rupley, J.A.: 1972, in Boyer, P.D. (ed.), The Enzymes, vol. VII (3rd ed.), Academic Press, New York, p. 666.
Kurachi, K., Sieker, L.C., and Jensen, L.H.: 1975, J. Biol. Chem. 250, 7663.
Jones, R., Dwek, R.A., and Forsén, S.: 1974, Eur. J. Biochem. 47, 271.
Levine, B.A. and Williams, R.J.P.: 1975, Proc. Roy. Soc. A345, 5.
Osserman, E.F., Canfield, R.E., and Beychok, S. (eds.): 1974, Lysozyme, Academic Press, New York.
Perkins, S.J.: 1975, personal communication.
Philips, D.C.: 1967, Proc. Natl. Acad. Sci. U.S.A. 57, 484.
Secenski, I.I. and Lienard, G.E.: 1974, J. Biol. Chem. 249, 2932.
Solomon, I.: 1955, Phys. Rev. 99, 559.

DISCUSSION

Ward: In your talk, you mentioned that your data indicate that a tyrosine side chain in the active site is undergoing complete rotation in solution. You also mention that other aromatic side groups are rotating. Can you specify which other side chains in the protein are spinning in a similar manner?

Is there a correlation between those that are rotating in solution and those that have a higher average B factor in the X-ray structure analysis?

Dobson: The nmr data indicate that all three tyrosine residues in lysozyme are spinning or flipping about the C_β–C_γ bonds. This motion of tyrosine and phenylalanine residues has been detected in several proteins. There is no direct evidence for complete rotation of other side chains, for it is only in the case of the symmetric phenyl ring that such evidence can be readily obtained. However, relaxation time measurements (Campbell et al., 1976) show that all groups in the proteins possess some internal mobility, but this could be oscillation rather than complete rotation about bonds in some cases. Correlations between B factors and motion have not been examined. However, it has been observed that groups which are on the surface are more mobile in solution than internal groups, and are often not well defined by the X-ray diffraction data.

Truter: The hydrogen atoms in the methyl groups were not located by X-rays so the rotation of these groups does not present a conflict of evidence. The spinning of the phenyl groups is reminiscent of the controversy about cyclooctatetraene $Fe(CO)_3$, which shows one ^1H nmr resonance suggesting equivalence of δ carbon atoms but a structure with Fe coordinated to 4 out of 8 carbon atoms and a non-coplanar molecule; the explanation is that the nmr time scale is long compared with X-rays, electromagnetic radiation.

Dobson: I would like to comment further on the difference between the X-ray and nmr methods. The X-ray method measures an average electron density for a large number of molecules, over a long period of tim. Consider the aromatic ring of a tyrosine in lysozyme. If it has a preferred conformation in the molecule, this conformation will be detected by the X-ray method. Rotation about the C_β—C_γ bond by $180°$ would give the same preferred conformation if the time spent in this rotation is short compared to the lifetime in the preferred conformation, the X-ray method will not detect the fact that the rotation is occurring. In fact, whether or not rotation occurs in the crystal cannot be stated. However, the nmr method detects the motion directly because, for example, the two ortho protons can be shown to experience equivalent environment, which can aris only by complete C_β—C_α rotation or flipping. This must occur rapidly compared the nmr time scale (which for the case of tyr 53 is ca. 10^{-4} s). Thus, in solution we known this motion occurs. It is not known whether or not motion occurs in the crystal. The example given by Dr Truter is simil.

Co(III)-ATP COMPLEXES AS AFFINITY LABELING REAGENTS OF MYOSIN AND
COUPLING FACTOR-1 ATPases

M.M. WERBER
Polymer Dept., Weizmann Institute of Science, Rehovot, Israel

A. DANCHIN
*Dept. de Biologie Moléculaire, Institut Pasteur and Institut
de Biologie Physico-chimique, Paris, France*

and

Y. HOCHMAN, C. CARMELI and A. LANIR
*Dept. of Biochemistry, The George S. Wise Center for Life
Sciences, Tel-Aviv University, Tel-Aviv, Israel*

1. INTRODUCTION

In most ATPases the substrate is found in the form of a complex of ATP,
with a metal ion, such as Ca^{++} or Mg^{++}. We have therefore attempted to
replace these labile metal ions by Co^{3+} which, by virtue of the inertness
of its complexes, could remain attached to its ligands.
    Recently, inert Co(III) complexes of nucleotides have been used
as affinity labeling reagents of regulatory and active sites of enzymes
(Danchin and Buc, 1973; Werber et al., 1974). In particular, a Co(III)-
ATP complex was shown to bind specifically and irreversibly to the ATPase
site of myosin (Werber et al., 1974). It was therefore of interest to
find out if this reagent, Co(phen)-ATP is of general applicability for
ATPases, and possibly kinases. In this study Co(III)-ATP complexes were
used as affinity labeling reagents of two ATPases: myosin (from rabbit)
and coupling factor-1 from chloroplasts-(CF_1). Although the inactivation
patterns observed were different, stoichiometric labeling of nucleotide
binding sites was obtained in both cases.

2. EXPERIMENTAL SECTION

2.1. *Synthesis of Co(III)-Nucleotide Complexes*

Co(III)-ATP complexes were synthesized in aqueous solution at 10 mM
concentration of the components (with a 10% excess of Na_2ATP). The solution pH was brought to 10 at 0° and an aliquot of H_2O_2 was added, so
that the final concentration in H_2O_2 was 40 mM. The pH was kept at 10
with NaOH until almost no pH decrease was observed (about 20 min). The

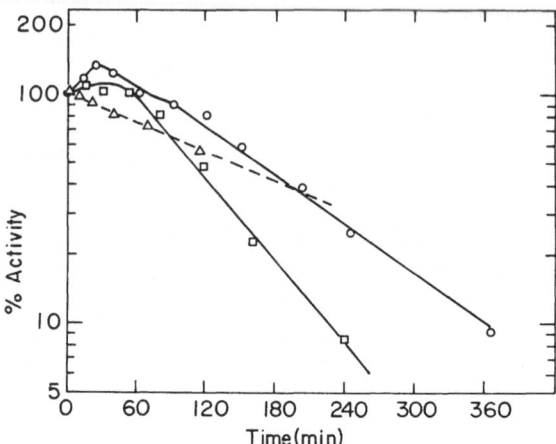

Fig. 1. Time course of Co-(phen)-ATP labeling of myosin and subfragment-1 at $0°$, as monitored by Ca^{2+}-ATPase activity. Initial concentrations: Co-(phen)-ATP, 9.6×10^{-4} M; enzymes, 6.3×10^{-6} M. O - O, - Myosin at low ionic strength (0.15M KCl); □ - □, - Myosin at high ionic strength (0.60M KCl); △   △, - subfragment-1 at low ionic strength (0.15M KCl).

brown solution was then precipitated twice in ethanol (2:1 v/v) at $-20°$, in order to remove the uncomplexed reactants, dried under a stream of nitrogen and redissolved in water (pH 10). The complexes were kept frozen (at $-18°$).

2.2. *Enzyme Preparations and Assays*

Myosin from rabbit white muscle and its subfragments were prepared as previously described (see Werber et al., 1972). Ca^{2+}-ATPase activity was determined as described previously (Werber et al., 1974). Coupling factor from chloroplasts (CF_1) was prepared from lettuce leaves and was trypsin-activated essentially according to Lien and Racker (1971). The conditions for the determination of Ca^{2+}-ATPase activity were as in Hochman et al. (1976). The affinity labeling of both enzymes and the reactivation with dithiothreitol (DTT) were performed under essentially the same conditions as in Werber et al. (1974), except that the labeling of CF_1 was carried out at $15°$. In the case of myosin and its subfragments, the stoichiometry was determined after dialysis of excess reagent from the cobalt content evaluated from the absorbance of the complex obtained in the presence of DTT (Danchin and Buc, 1973).

In the case of CF_1, the stoichiometry was determined from the radio-activity bound to the enzyme after its incubation with Co-(phen)-[^{32}P]-ATP (see below and Figure 3).

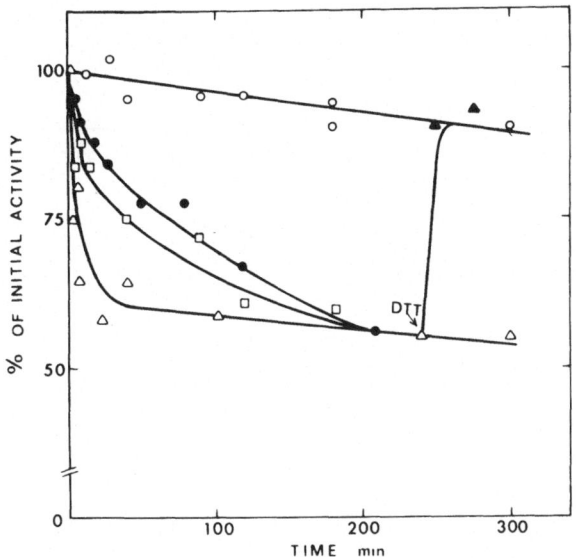

Fig. 2.   Time course of CF_1-ATPase inactivation in the presence
of various concentrations of Co(III)-(phen)-ATP: ($\bullet$ - $\bullet$) 8.8 mM;
($\square$ - $\square$) 4.5 mM; ($\triangle$ - $\triangle$) 2.5 mM; (O - O) no complex; ($\blacktriangle$ - $\blacktriangle$)
after addition of DTT. Trypsin-activated CF_1 was incubated with
the Co(III) complex at $15°$. At various times 1 μl aliquots
were assayed for ATPase activity by following the decompositon
of highly labeled [^{32}P]-ATP at 37 $°$C as described in Hochman
et al. (1976). The reaction mixture contained 10 μg CF_1, 40 mM
HEPES buffer, pH 8 and 5.4 mM CaATP in 1.5 ml.

3. RESULTS AND DISCUSSION

The time course of the Co-(phen)-ATP labeling of myosin, at two ionic
strengths, and of subfragment-1, is shown in Figure 1. Whereas sub-
fragment-1 as well as heavy meromyosin (Werber et al., 1974) become
inactivated in a monotonous single process, myosin first undergoes an
enhancement of activity as a result of the labeling process. The unusual
pattern of labeling can be interpreted in terms of an uncoupling of the
anti-cooperativity of the sites in myosin. Thus, when one of the sites
is labeled, the activity of the other one is enhanced. However, the
extent of this enhancement is greater at low ionic strength, where
myosin is in the aggregated state, suggesting that a regulatory role
might be associated to filament formation by myosin in muscle (Oplatka
et al., 1974). The time course of the labeling of CF_1 at 3 concentrations
of Co-(phen)-ATP is presented in Figure 2. The inactivation rates depend
on the complex concentration, but in all cases at least 40% of the activity is lost at the end of the labeling process under the conditions
of the assay.
        The original activities of Co-(phen)-ATP-labeled myosin and CF_1
can be restored by DTT treatment (Table I). The stoichiometry of the

TABLE I
Inactivation and reactivation (by DTT) of labeled ATPases [a]

	Myosin		Coupling Factor 1		
Labeling reagent	Co(III)-(phen)-ATP		No treatment or Co(II)-ATP	Co(III)-ATP	Co(III)-(phen)-ATP
Labeling time (h)	0.75	3.5	4	4	4
Residual activity %	150[b]	10	96	70	60
Activity after DTT treatment %	99	95	90	70	90

[a] Ca^{2+} ATPase activities as described above

[b] Maximum of activity in the first phase of the labeling

TABLE II
Stoichiometry of the labeling of myosin (and its subfragments) and of CF_1 by Co(phen)-ATP

	Myosin meromyosin	Heavy meromyosin	Subfragment-1	Coupling factor 1
n_1 [a]	1.2+0.2			
n_2 [a]	2.0+0.2	1.8+0.1	1.1+0.1	1.3; 2.0[b]

[a] n-moles of Co(III) complex per mole of protein at the maximum (n_1) and at the end of the labeling (n_2), i.e. when residual activity is lowest.

[b] After 1 h of labeling with about 50% residual activity, see also Figure 3.

labeling of the ATPases is shown in Table II. Whereas myosin and heavy meromyosin, which are both double-headed, are labeled with 2 moles of complex per mole of enzyme, only one mole of complex is attached per mole of the monoheaded subfragment-1. At the peak of the enhancement phase about one mole of complex is retained per mole of myosin, in agreement with the kinetic scheme previously proposed (Werber et al., 1974) for myosin labeling.

In the case of CF_1 about 2 moles of Co-(phen)-ATP per mole of enzyme remain tightly bound after 90 min incubation with 100-fold excess of complex. The labeled CF_1 was freed from unbound complex by chromatography on Sephadex G-50 (Figure 3). It was assumed that the radioactivity represented binding of the complex since it has been established that Co-(phen)-ATP cannot serve as a substrate for the ATPases. This is probably due to the inertness of the nucleotide in these complexes.

That Co-(phen)-ATP is indeed an affinity label of the two ATPases is shown by the fact that the complexes compete with the substrate for

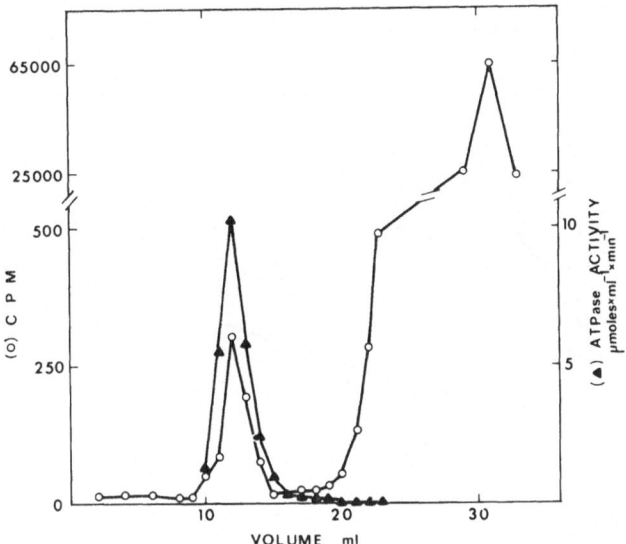

Fig. 3.   Binding of Co-(phen)-[^{32}P]-ATP to CF$_1$. 12 mM of
complex were incubated with CF$_1$ for 90 min. The labeled
enzyme was separated from excess reagent on a Sephadex G-50
column (1.5x30 cm) equilibrated with 0.1 M NaCl, 40 mM HEPES,
pH 8. Fractions of 1 or 2 ml were collected. Each fraction
was assayed for CF$_1$ concentration, CF$_1$-ATPase activity and
for radioactivity.

TABLE III
Protection against labeling of heavy meromyosin by substrates
and analogs

Substrate	Concentration (mM)	Protection[a]
MgATP	2	95
MgADP	2	100
MnADP	2	62
MgPP[b]	5	62

[a] Protection is defined as $(v_p-v_0)/(100-v_0)$, where v_p and v_0
are the % of activity in the presence and absence of analog,
respectively.

[b] $[Mg^{2+}] = 2$ mM.

the same site. Thus, a competitive inhibition of CF$_1$ by Co-(phen)-ATP
was demonstrated (Figure 4) as in the case of myosin (Werber et al.,
1974). Furthermore, substrate protection against inactivation by Co-
(phen)-ATP could be shown in the case of heavy meromyosin (Table III).

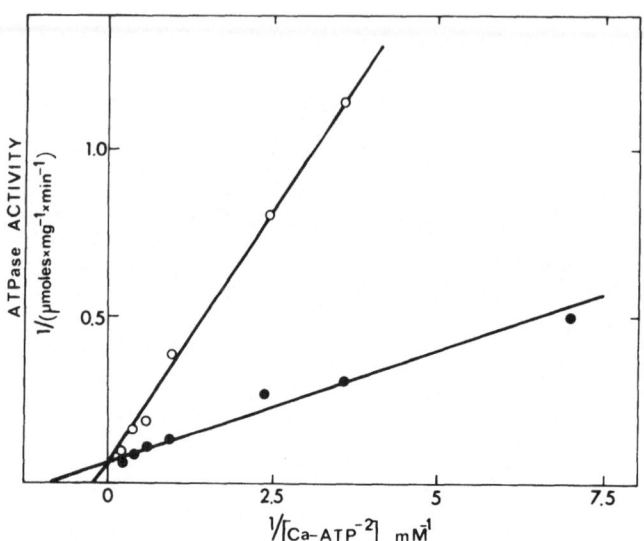

Fig. 4. Dependence of the rate of CF_1 ATPase activity on the concentration of CaATP. CF_1 (3 mg ml^{-1}) was incubated with 12 mM Co(III)-(phen)-ATP for 90 min. Aliquots were assayed for ATPase activity as indicated under Figure 2. (● - ●) Native CF_1; (O - O) labeled CF_1.

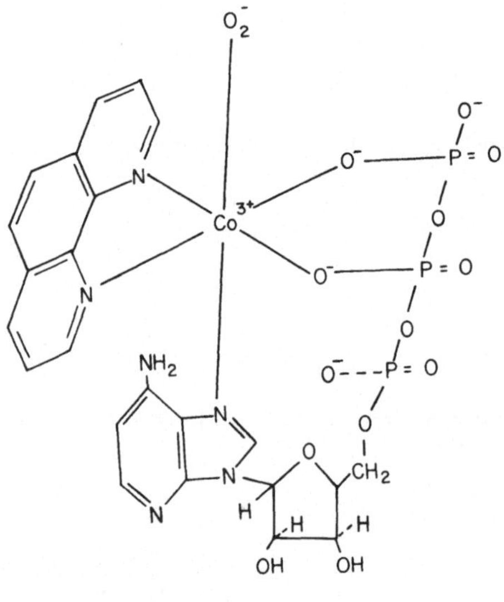

Fig. 5.   Proposed structure of Co-(phen)-ATP.

It has been suggested (Danchin, 1973) that the Co(III) affinity labels might be negative prints of the sites to which they can rigidly attach themselves. We believe that Co-(phen)-ATP (Figure 5) binds to these proteins by virtue of an exchange between the non-essential O_2^- and a donor ligand of the enzymes. The full characterization of the structure of this complex (Danchin and Werber, to be published) might shed light on the conformation of the substrate binding sites of these ATPases.

It thus appears that these reagents are useful tools for the elucidation of the details of the interaction of enzymes with nucleotide-metal ion complexes and work with other ATPases and with kinases is in progress.

REFERENCES

Danchin, A.: 1973, Biological Macromolecules Labeling with Covalent Complexes of Magnesium Analogs: I. The Cobaltic Co(III) Ion, Biochimie 55, 17-27.
Danchin, A. and Buc, H.: 1973, Affinity Labeling of the Adenosine 5'-Monophosphate Binding Site of Rabbit Muscle Glycogen Phosphorylase b with an Adenosine 5'-Monophosphate Cobalt(III) Complex, J. Biol. Chem. 248, 3241-3247.
Danchin, A. and Werber, M.M.: 1976, Structure of a Cobaltic-Nucleotide Affinity Label of the Substrate Binding Sites of ATPases, to be published.
Hochman, Y., Lanir, A. and Carmeli, C.: 1976, Relations Between Divalent Cation Binding and ATPase Activity in Coupling Factor from Chloroplast, FEBS letters 61, 255-259.
Lien, S. and Racker, E.: 1971, Preparation and Assay of Chloroplast Coupling Factor CF$_1$ in A. San Pietro (ed.) Methods in Enzymology, Academic Press, New York and London, vol. XXIII A, 547-555.
Oplatka, A., Werber, M.M., and Danchin, A.: 1974, Specific Interaction of Cobaltic Complexes with Myosin, FEBS letters 47, 7-10.
Werber, M.M., Szent-Györgyi, A.G., and Fasman, G.D.: 1972, Fluorescence Studies on Heavy Meromyosin-Substrate Interaction, Biochemistry 11, 2872-2883.
Werber, M.M., Oplatka, A., and Danchin, A.: Affinity Labeling of a Rabbit Muscle Myosin with a Cobalt(III)-Adenosine Triphosphate Complex, Biochemistry 13, 2683-2688.

DISCUSSION

Haynes: Can you compare the binding site specificity of Co(Phe)AMP, Co(Phe)ATP and MgATP on the basis of their K_i or K_m values?

Werber: Co(phen)ATP has a K_i value of about 10^{-4} M toward myosin and of about 2×10^{-5} M toward coupling factor 1. The K_m for MgATP in the case of myosin is about 10^{-7} M and that for CaATP, which was the substrate in our experiments is around 10^{-6} M. As for coupling factor 1 the substrate is CaATP with a K_m of about 10^{-3} M.

Kretsinger: (1) In your DTT (dithiothreitol) experiment was the

DTT added before or after formation of the Co-(phen)-ATP complex?
(2) Regarding your interpretation of cooperativity did you do a control
experiment using SF-1 myosin fragments?

Werber: (1) The DTT was added after the enzyme was labeled with
Co(phen)ATP and caused return of the ATP-ase activity to its original
value. (2) We performed both control experiments with the monoheaded
subfragment-1 and with the double-headed heavy-meromyosin. In both cases
there was a single process of decrease in activity. Only in the case of
myosin did the time course of labeling pass through an enhancement of
activity (higher than the original activity). This could occur only
if in the native myosin one site is depressing the activity of the
other. This effect of anticooperativity is being uncoupled by the
labeling of one site per molecule of myosin.

Zundel: Why do you study the Co complex and not the Mn complex? The
Mn complex should be similar to MgATP.

Werber: We study Co(III) complexes since they are inert (slow
exchanging) and therefore the whole complex remains rigidly attached
to the site where it binds. Mn(II) does not form inert complexes.

In addition, Co(III) has a very similar ionic radius (about 0.65 Å)
in many complexes to that of Mg(II).

Caughey: Can you please comment upon the basis for the assignment
of a superoxide structure as the exchangeable ligand bound to Co(III)
and do you envision a special role for the superoxide as ligand
as distinct from other possible ligands? Has it been established that
O_2^- is a usually facile ligand in these compounds?

Werber: The assignment of a superoxide ligand as the non exchangeable
ligand in Co(III)-ATP complexes is based on e.s.r. studies as well as
on studies of CN^--promoted displacement of O_2^- from the complex
(Danchin, A. and Werber, M.M., to be published).

The special role of O_2^- is that it is much more easily displaced
from the complex by a suitably located ligand group in the protein,
thus enabling the Co(III) complex to become rigidly attached to it.

CRYSTALLOGRAPHIC STUDIES OF VALINOMYCIN AND A23187

G. DAVID SMITH and WILLIAM L. DUAX
*Medical Foundation of Buffalo, 73 High Street, Buffalo,
New York 14203, U.S.A.*

1. INTRODUCTION

Ionophores are a class of antibiotics which are capable of transporting
ions across synthetic and biological membranes. Great differences in molecular composition are observed since these antibiotics may be linear
or cyclic compounds and may be either polyethers or polypeptides. Ion
specificity is also found to show a considerable variation.

These antibiotics may also be classified according to the mechanism
by which ions are transported. Valinomycin and the polyether antibiotics
are thought to transport ions by a carrier mechanism in which the ionophore forms a complex with the ion on one side of the membrane, trans-
ports it through the membrane, and finally releases the ion on the other
side of the membrane. This mechanism of ion transport may be detected
by measuring the change in conductance upon freezing the membrane, during which a decrease in the conductance of approximately 10^5 is observed.
Gramicidin A and other antibiotics which transport ions by a channel
mechanism exhibit no appreciable change in conductance when the membrane
is frozen (Krasne et al., 1971).

The carrier ionophores may undergo considerable conformational
changes as complexation takes place. Polar groups of the uncomplexed antibiotic must be in exposed positions on the surface of the molecule in
order to initiate the complexation process. During this process, the
metal ion and the polar atoms which are bound to it must move towards
the center of the complex and away from the surface. Once complexation
is complete, the exterior surface of the complex must be primarily hydrophobic so that passage through the membrane is possible.

From a precise knowledge of the structures of the complexed and
uncomplexed modifications of these ionophores, it should be possible to
propose mechanisms by which the complexation and the release of ions
may take place and to further explain membrane transport phenomena.

The structures of several ionophores have been determined by X-ray
crystallographic techniques in both the complexed and uncomplexed modifications. The structure of the sodium ion complex of the macrotetra-
lide, nonactin (Dobler and Phizackerely, 1974), shown in Figure 1a, has
been found to differ in conformation from that of the uncomplexed form

*B. Pullman and N. Goldblum (eds.), Metal-Ligand Interactions in Organic
Chemistry and Biochemistry, part I, 291-315. All Rights Reserved.
Copyright © 1977 by D. Reidel Publishing Company, Dordrecht-Holland.*

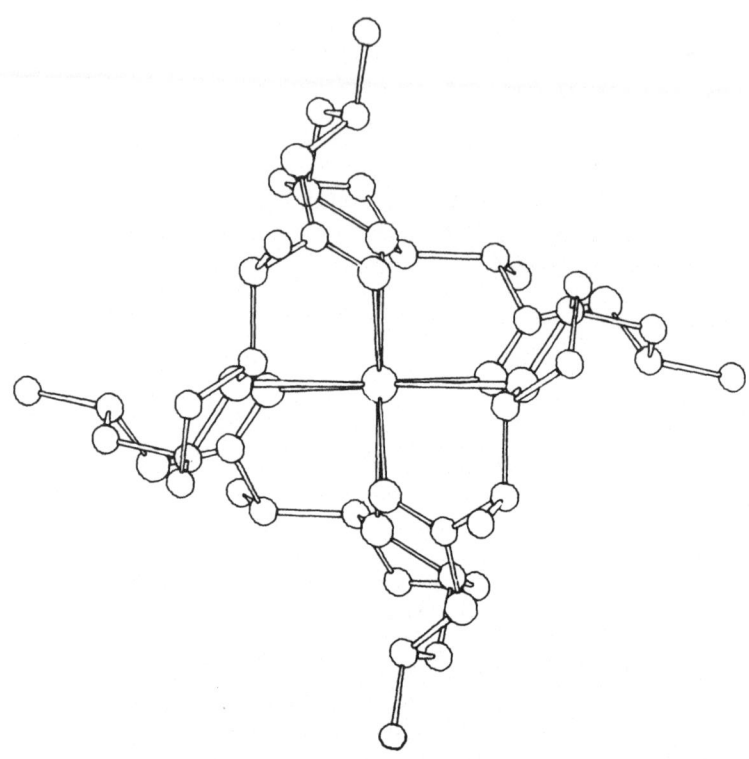

Fig. la.   Structure of the sodium ion complex of nonactin
(Dobler and Phizackerely, 1974).

(Dobler, 1972) (Figure lb) primarily by an alteration of eight torsion
angles by approximately 120 $^{\circ}$. The complexed form of tetranactin (Iitaka
et al., 1972) is quite similar to that of complexed nonactin, but surprisingly enough, the structures of the uncomplexed modifications
(Nawata et al., 1974) (Figure lc) are quite different. While the nonactin molecule resembles a torus with $\bar{4}$ symmetry, the tetranactin molecule
is elongated and has only two-fold symmetry.
     The structures of both the complexed (Pinkerton and Steinrauf, 1970)
and free (Lutz et al., 1971) acid form of monensin, a linear polyether
antibiotic, have been determined. Monensin coordinates metal ions by
wrapping itself about the cation and forming two hydrogen bonds between
the carboxyl group and the two hydroxy groups at the other end of the
molecule as seen in Figure 2. The carboxyl group does not serve as a
ligand to the metal ion in this case, although other ionophores do utilize one of the carboxyl oxygen atoms as a ligand. Although there is a
great difference in the number and arrangement of the hydrogen bonds,
the torsion angles of the free acid do not differ by more than 17 $^{\circ}$ from
that of the complexed form.
     The discussion which follows will be limited to the structures,
the conformations, and the mechanisms of ion complexation of valinomycin,
a cyclic depsipeptide, and of A23187, a linear polyether.

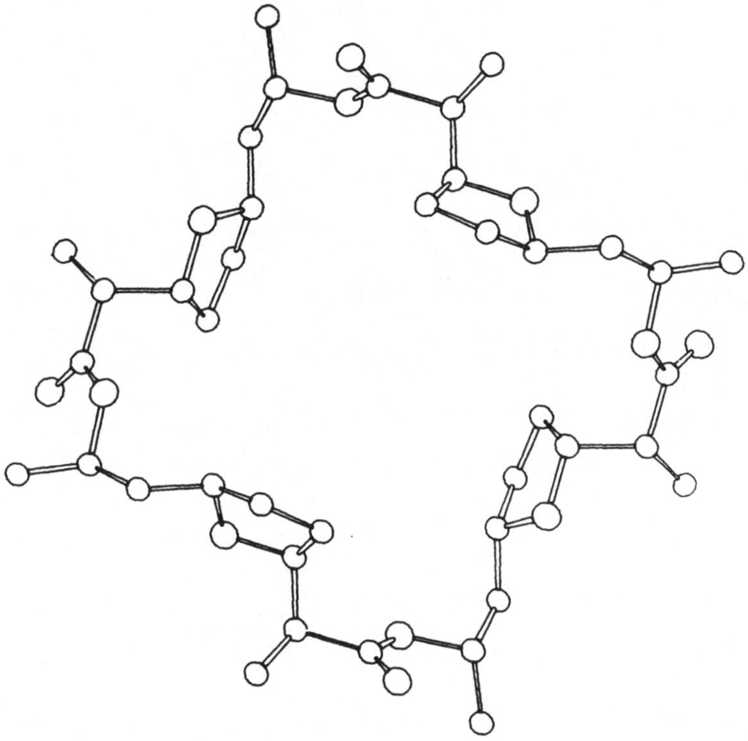

Fig. 1b.   Structure of uncomplexed nonactin (Dobler, 1972).

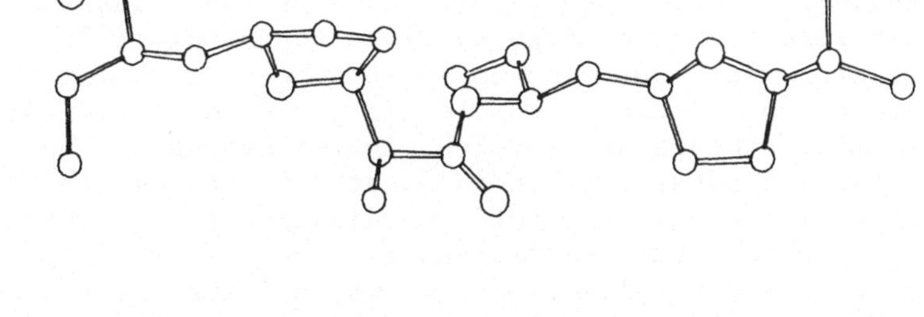

Fig. 1c.   Structure of uncomplexed tetranactin (Nawata et al.,
1974).

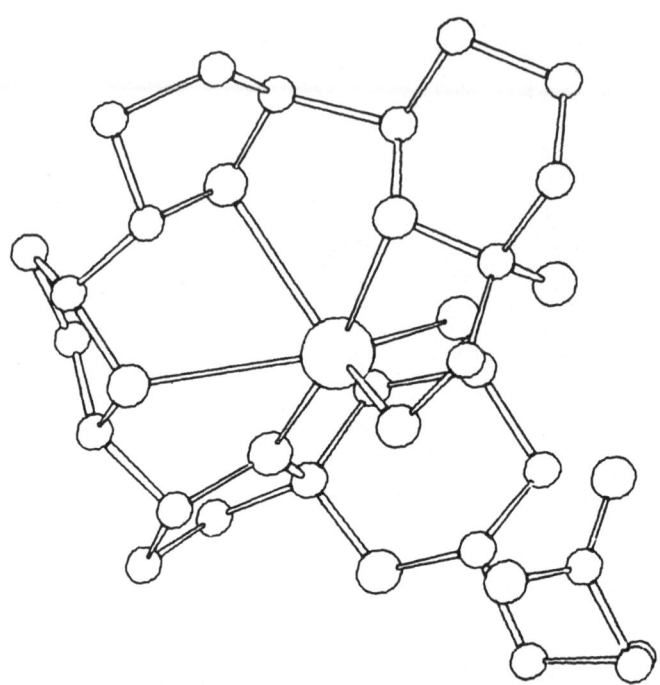

Fig. 2. Structure of the complexed form of monensin (Pinkerton and Steinrauf, 1970).

2. VALINOMYCIN

Valinomycin is a naturally occuring antibiotic which exhibits a high selectivity for potassium ion and is capable of transporting this ion across both natural and synthetic membranes (Tosteson et al., 1967 and Wipf et al., 1970). This ionophore has been the subject of a considerable number of investigations in order to determine the conformation of both the complexed and uncomplexed forms in a variety of solvents and to ascertain the mechanism of complexation.

The primary structure of valinomycin, a cyclic dodecadepsipeptide, is shown in Figure 3. The sequence (L-Val-D-HyIv-D-Val-L-Lac) is repeated three times producing a molecule which may have three-fold symmetry. Preliminary results of a crystallographic study on the potassium complex of valinomycin were first reported by Pinkerton et al. (1969). Although these investigators were not able to locate the side chains unambiguously, their results showed that the molecule possesses three-fold symmetry in which all amide protons are involved in 1—4 hydrogen bonds, illustrated in Figure 4. The potassium ion is located in the center of the doughnut shaped molecule and has six-fold coordination to the amino carbonyls. More recently, complete structural results have been reported for the potassium ion complex of valinomycin, shown in Figure 5, and are in good agreement with the earlier work (Neupert-Laves and Dobler, 1975). Infrared, nmr and Raman studies agree that this conformation also exists in

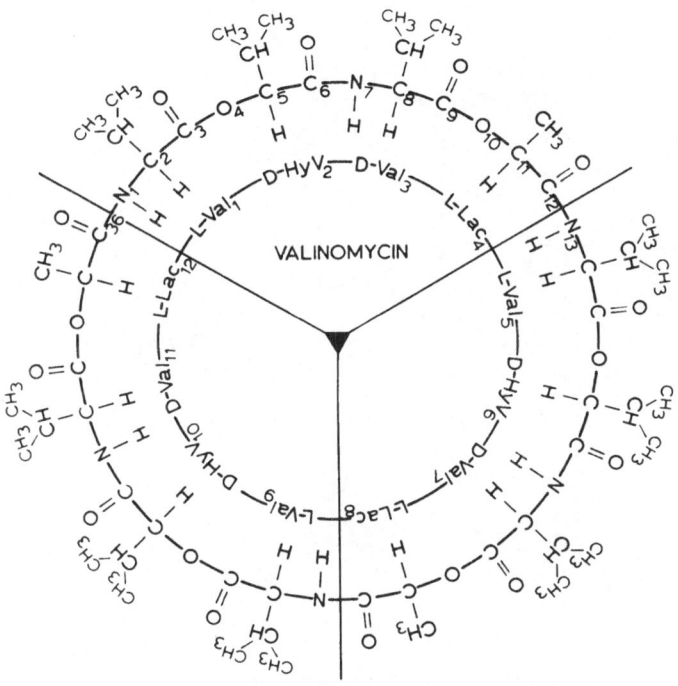

Fig. 3.   The primary structure of valinomycin.

solution (Rothschild et al., 1973; Urry and Kumar, 1974 and Patel and
Tonelli, 1973).

The question involving the conformation of uncomplexed valinomycin
has been considerably more difficult to resolve. IR, nmr and ORD results
have indicated that the hydrogen bonding scheme is primarily a function
of solvent polarity. Thus, in polar solvents, an open conformation has
been proposed in which there is no intramolecular hydrogen bonding while
a conformation similar to that of the potassium ion complex has been
proposed for non-polar solvent systems. For solvent polarities between
these two extremes, a hydrogen bonding scheme has been proposed which
involves only the protons of the D-Val or of the L-Val residues. However,
all of these models have required the presence of three-fold symmetry
(Patel and Tonelli, 1973).

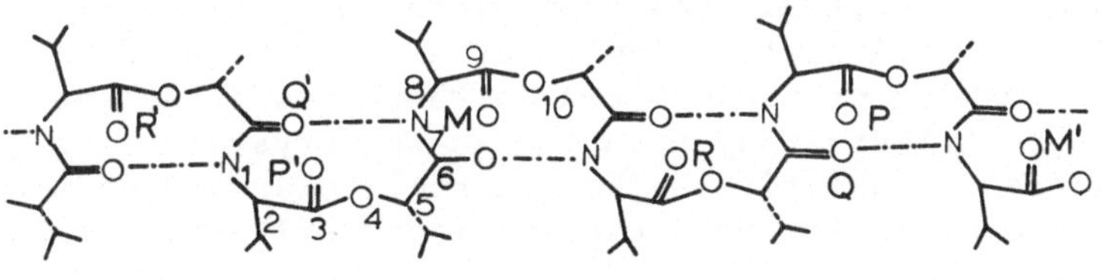

Fig. 4.   The hydrogen bonding scheme in the potassium ion
complex of valinomycin.

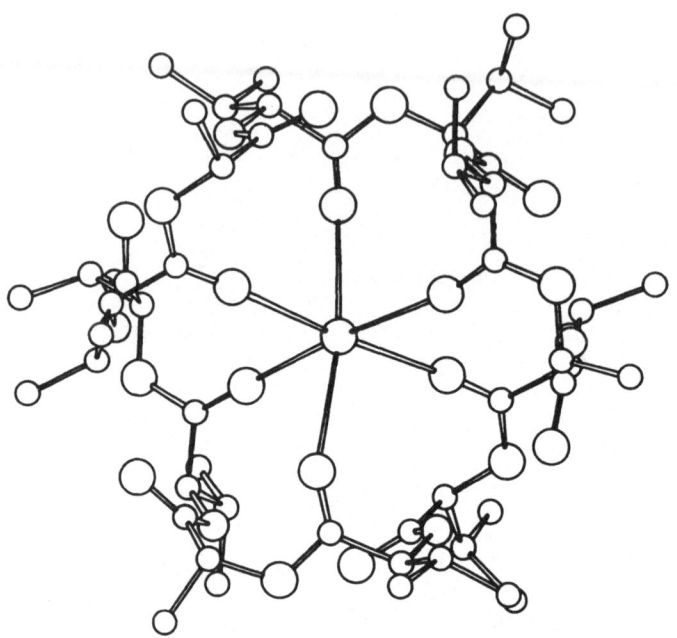

Fig. 5.   The structure of the potassium complex of valinomycin
(Neupert–Laves and Dobler, 1975).

The X-ray crystal structure of uncomplexed valinomycin, crystallized
from n-octane, was first reported by Duax et al. (1972) and is illustrated in Figure 6. The observed conformation was quite unlike anything
which had been proposed based on solution studies. Three-fold symmetry
has been replaced by a pseudo-center of inversion and two different kinds
of hydrogen bonds are present. In addition to four beta type hydrogen
bonds, 1–5 hydrogen bonds are observed between residues L-Val(9) and
L-Val(5) and between D-Val(3) and D-Val(11) and are illustrated in Figure 7. This latter type of hydrogen bonding has been used to explain
the solid state Raman spectra of uncomplexed valinomycin (Rothschild
et al., 1973). The net effect of these 1—5 hydrogen bonds is to flatten
the doughnut shaped conformation of the complexed form and to position
several carbonyl oxygen atoms at the surface of the molecule.
      Recently, two independent determinations of a triclinic crystalline
form of valinomycin, crystallized from different solvent systems, have
been reported in the literature (Smith et al., 1975 and Karle, 1975).
The gross conformation of valinomycin in these two determinations is
identical to the earlier study even though the crystals were grown from
solvents of different polarities. No intermolecular hydrogen bonding is
observed and several of the more exposed isopropyl groups are disordered
in all three studies.
      The interpretation of the spectral data for uncomplexed valinomycin
in a variety of solvents has always required the presence of three-fold
symmetry, not observed in any of the crystallographic studies. However,

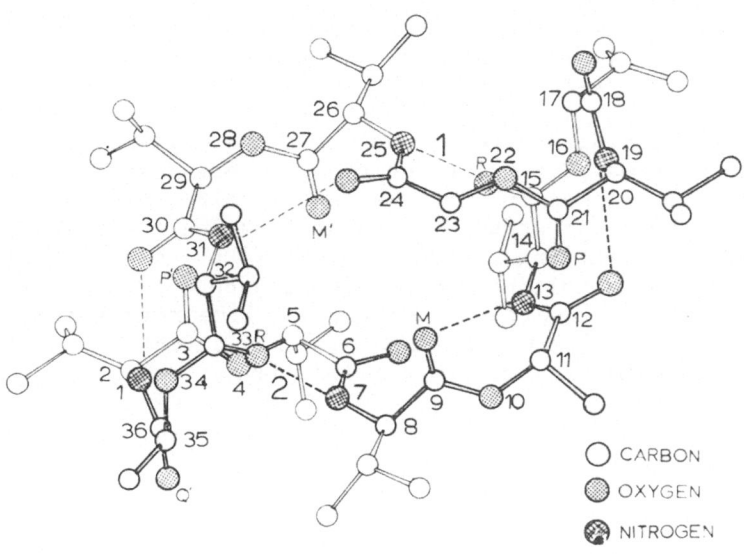

Fig. 6.   The structure of uncomplexed valinomycin (Duax et al.,
1972).

the pair of crystallographically observed 1—5 hydrogen bonds, related
by a pseudo-center of inversion, may be formed between three pairs of
different residues. By altering torsion angles of two residues, neither
of which is involved in accepting or donating a proton, by less than
90 °, an amide proton is redirected towards a carbonyl oxygen four residues away, thus producing the familar beta turn. If this interconversion
of hydrogen bonding is fast on the nmr time scale, three-fold symmetry
would in fact be observed. Since the 1—5 hydrogen bonds are inconsistent
with three-fold symmetry, it is easy to understand why their presence
has been overlooked.

The three crystallographic studies on valinomycin have yielded structural parameters for five independent valinomycin molecules. In addition

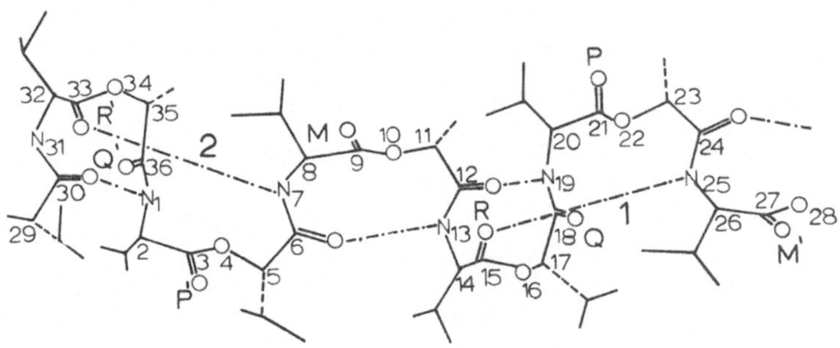

Fig. 7.   The hydrogen bonding scheme in uncomplexed valinomycin
(Duax et al., 1972).

to the original determination of the monoclinic form (molecule A), (Duax et al., 1972), the triclinic crystals that have been the subject of two studies contain two crystallographically independent molecules in the unit cell (molecule B1 and B2 (Smith et al., 1975); molecule C1 and C2 (Karle, 1975)). As stated before, the gross conformation of all five molecules is the same. Careful scrutiny, however, does reveal the presence of minor differences in conformation. A half-normal probility plot is a method of comparing structural parameters and their associated estimated standard deviations of one molecule with those of another (DeCamp 1973). Ideally, the resulting plot would produce a straight line with a slope of unity and an intercept of zero. Non-linearity indicates differences in conformation while deviation of the slope from unity probably indicates incorrect estimation of the standard deviations of the structural

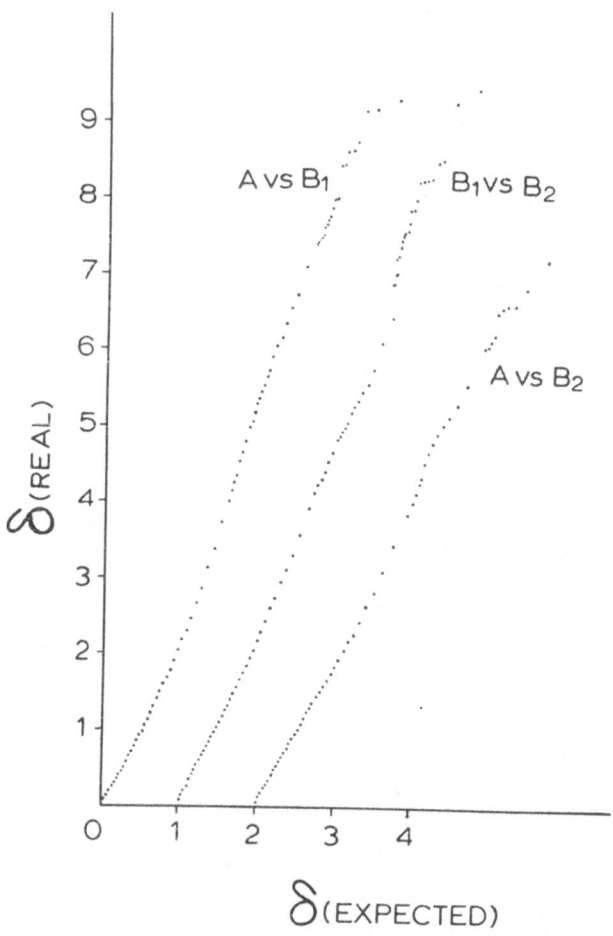

Fig. 8a. Half-normal probability plots comparing molecules A, B1 and B2 (Smith et al., 1975). Two of the plots have been shifted by 1.0 and 2.0 in δ (expected).

parameters. When modifications A, B1 and B2 were compared in this fashion, it was found that all three plots show some deviation from linearity, particularly those involving molecule B1 (Figure 8a). This indicated that there are minor differences in the conformations of all three molecules but that A and B2 have the most in common. Careful examination of the individual torsion angles for each molecule also reveal these same trends. When this same technique is applied to the two independent determinations of the triclinic form it is found that B2 and C1 are identical. Small differences are observed between B1 and C1 but the plots involving C2 are very clearly non-linear. These plots are reproduced in Figure 8b. Several important conclusions may be drawn from the results of these half-normal probability plots. In addition to demonstrating that molecules B2 and C1 are identical in all respects, the standard deviations of the structural parameters are shown to be underestimated by a factor of only 1.2, the slope of the least-squares straight line through the points. At the same time, the non-linear behavior of the plot comparing either molecule B1 or C2 indicates that, compared to the other three, there are minor but real differences in conformation amounting to small concerted changes throughout the molecule. Finally, the fact that three out of five conformations are identical suggests that this is probably the minimum energy form.

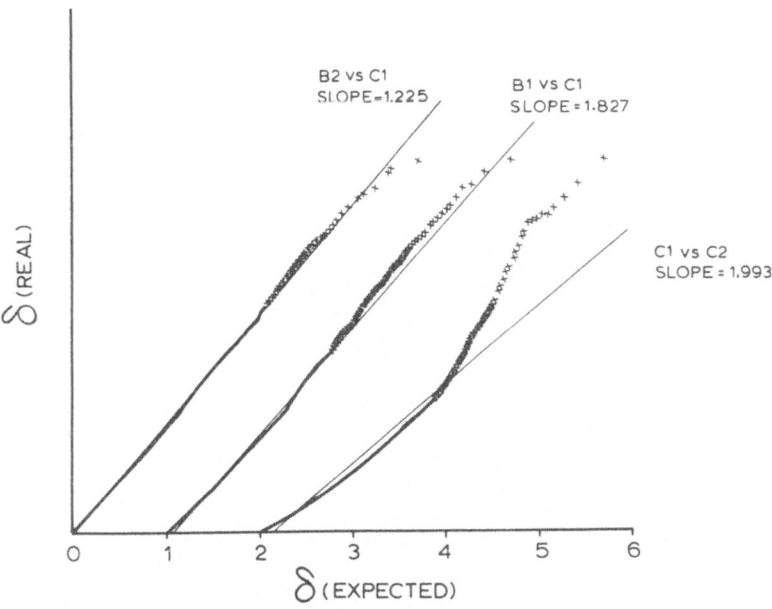

Fig. 8b. Half-normal probability plots comparing molecules B1 and B2 (Smith et al., 1975) to molecules C1 and C2 (Karle, 1875). δ (real) has an arbitrary scale for each of the three plots.

The conformation observed in the solid state does suggest a mechanism
by which valinomycin may coordinate potassium ion. There are two carbony
oxygen atoms which are in exposed positions at the surface of the molecu
and can initiate complexation to a potassium ion (Figure 9a). Once this

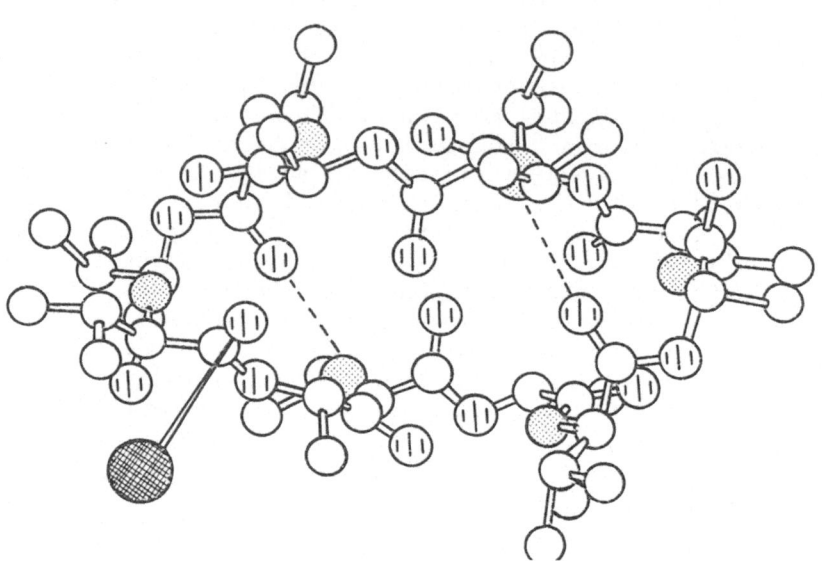

Fig. 9a — f.   The complexation process of valinomycin.

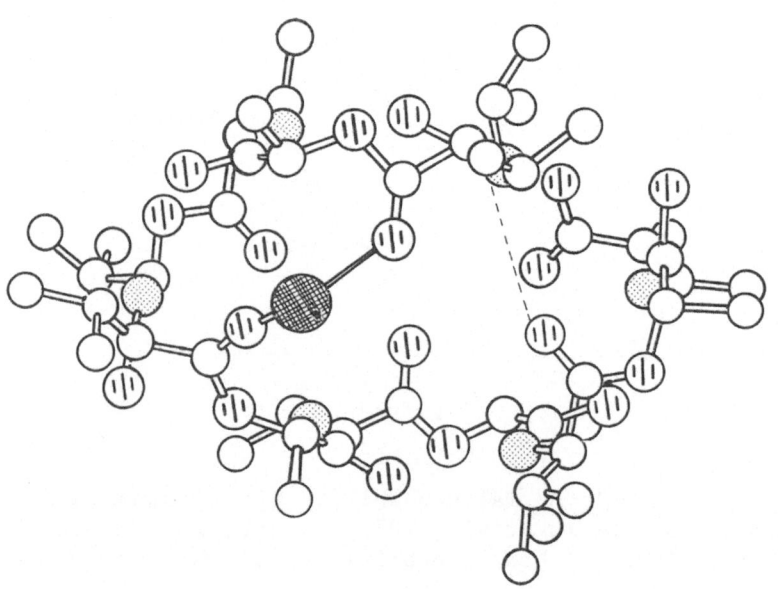

Fig. 9b.

loose complex is formed, the 1—5 hydrogen bonds can be easily broken and these carbonyl oxygen atoms are then free to also form bonds to potassium (Figures 9b, c). As the ion is brought into full coordination, the molecule rounds out and the appropriate free carbonyls are brought into position to hydrogen bond to the free protons (Figures 9d, 9e) giving rise to the observed structure of the potassium ion complex (Figure 9f). For the most part, this mechanism requires minor displacements of

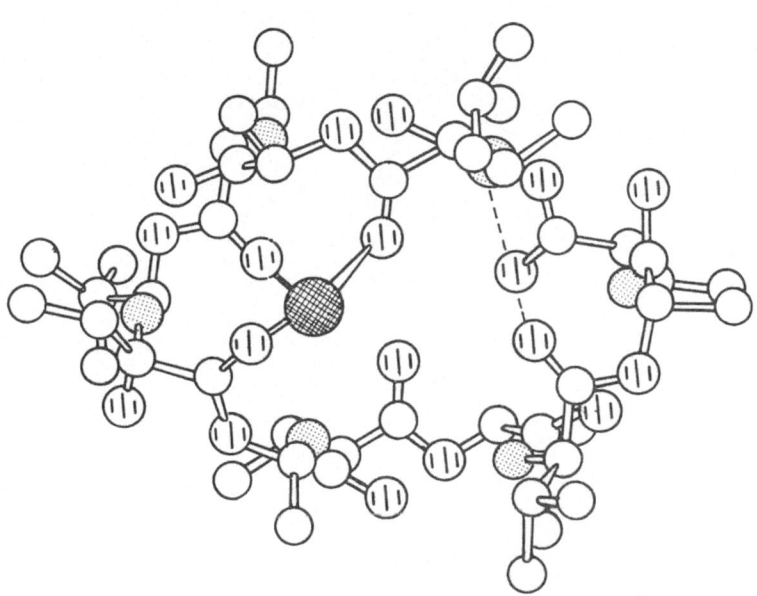

Fig. 9c.

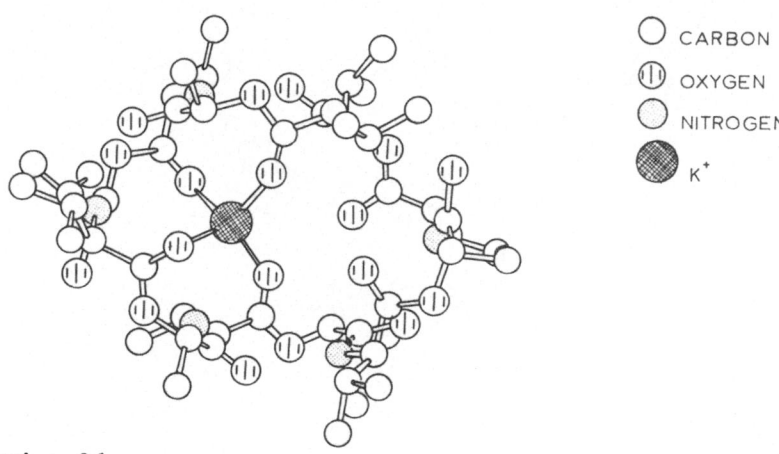

Fig. 9d.

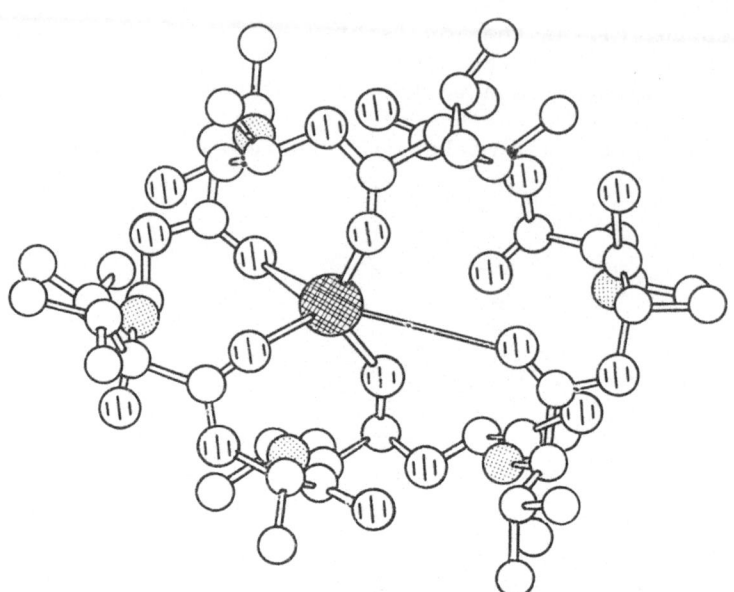

Fig. 9e.

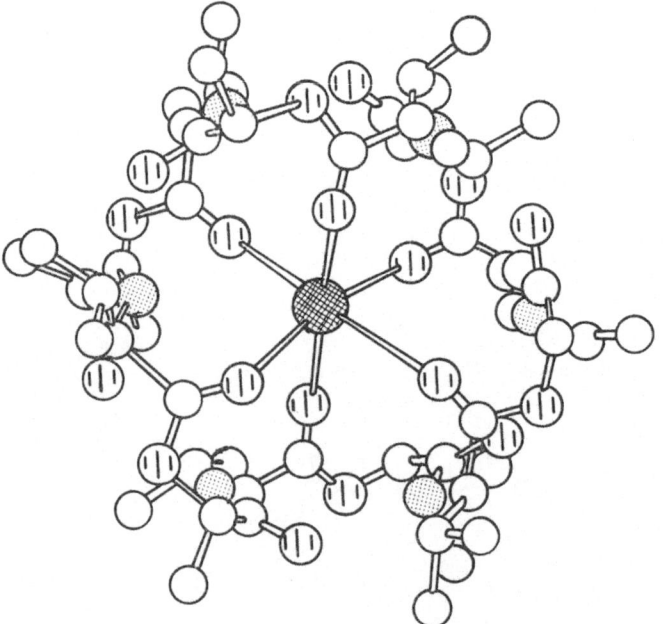

Fig. 9f.

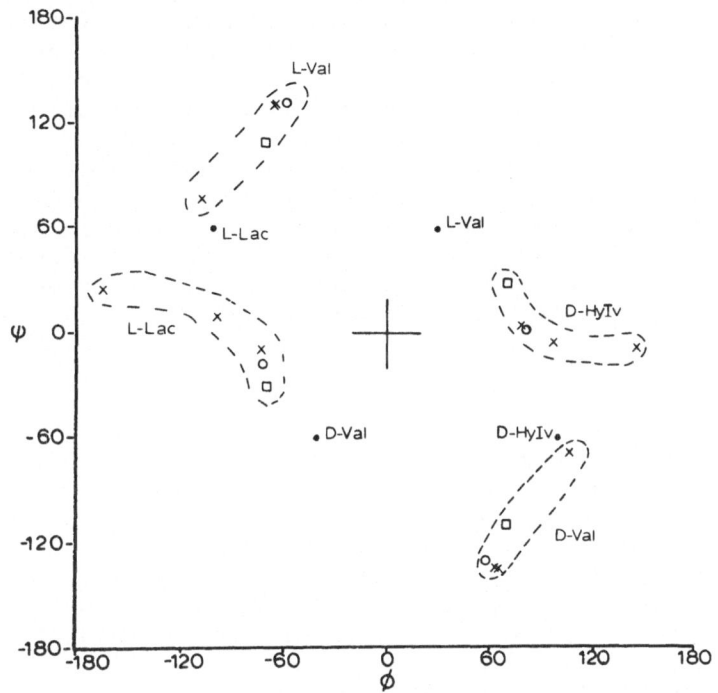

Fig. 10. ϕ-ψ Plot for valinomycin; (×) torsion angles observed
in the solid state for uncomplexed valinomycin (Smith et al.,
1975), (o) torsion angles observed in the solid state for the
potassium ion complex of valinomycin (Neupert–Laves and Dobler,
1975), (■) calculated angles from nmr data for the potassium
complex of valinomycin in methanol (Patel and Tonelli, 1973),
and (•) calculated angles for uncomplexed valinomycin in a
dioxane–water solvent system (Patel and Tonelli, 1973).

all atoms involved. The conformational changes illustrated in Figure 9
are similar to those that would occur in the rapid interconversion of
1–4 to 1–5 hydrogen bonds in the uncomplexed modification.

The average torsion angles of the uncomplexed and complexed
modifications of valinomycin are plotted on a Ramachandran plot in
Figure 10 along with the values predicted from spectral studies. Two of
the points for each amino acid residue are quite close to that observed
in the crystalline potassium ion complex and are consistent with the
values one would expect for a 1–4 bond. The remaining point corresponds
to the conformation of an amino acid involved in a 1–5 hydrogen bond.
The hydroxy acid residues fall into three distinct conformational groups.

Quantum mechanical calculations have shown that there are three
low energy conformations which valinomycin can adopt. Two of these involve the 1–4 hydrogen bonding scheme while the third is a combination
of 1–4 and 1–5 hydrogen bonds such as that observed in the solid state
(Maigret and Pullman, 1975).

Since the conformations of three independent polymorphs of valinomycin
in the solid state have been found to be identical despite variations
in the crystallization conditions, crystal packing, the presence of minor disorder in the isopropyl groups, and the presence of occluded solvent in the monoclinic lattice, it is extremely unlikely that this conformation is primarily a result of crystal packing. It seems reasonable
to assume instead that in non-polar solvents a non-negligible amount of
this conformer must be present, even though it has not yet been detected. Because of the competition for hydrogen bonding, this is likely
to be a less populated form in polar solvents. The minor differences in
conformation which have been observed in the solid state may be interpreted as local minima in the potential energy surface.

3. ANTIBIOTIC A23187

The antibiotic A23187, a monocarboxylic acid, exhibits a high specificity for divalent cations, particularly calcium and magnesium. Various
biological processes which require the presence of calcium ion have
been stimulated through the use of this ionophore (Grenier et al., 1974
and Karl et al., 1975). Because of its fluorescent properties, it has
been used extensively as a probe for divalent cations in membranes and
for the determination of the mode of action of ionophore mediated transport (Case et al., 1974). Although it is known that a 1:2 complex is
produced between the metal ion and the anionic form of the ionophore,
little work has been done to determine the mechanism by which complexation takes place.

    The chemical structure of A23187 is shown on Figure 11. The molecule
consists of a substituted benzoxazole, an α-ketopyrrole, and a spiro

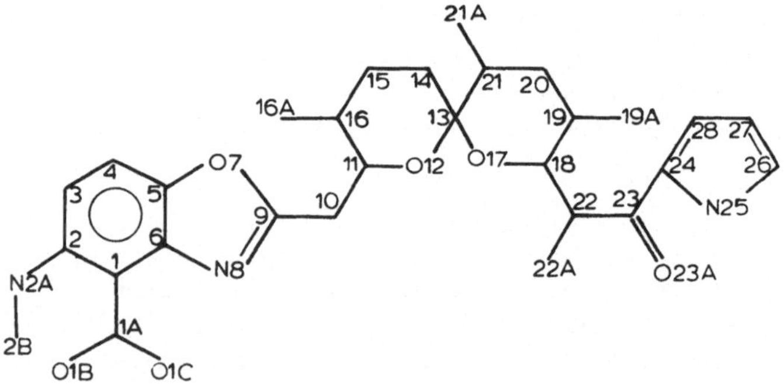

Fig. 11.   Primary structure of A23187.

ring system. The crystal structure of the free acid has been reported
and is illustrated in Figure 12 (Chaney et al., 1974). One of the three
intramolecular hydrogen bonds, illustrated as dotted lines in Figure 12,
occurs between the pyrrole nitrogen and a carboxyl oxygen and serves to
hold the ends of the molecule together. This kind of head to tail hydro-

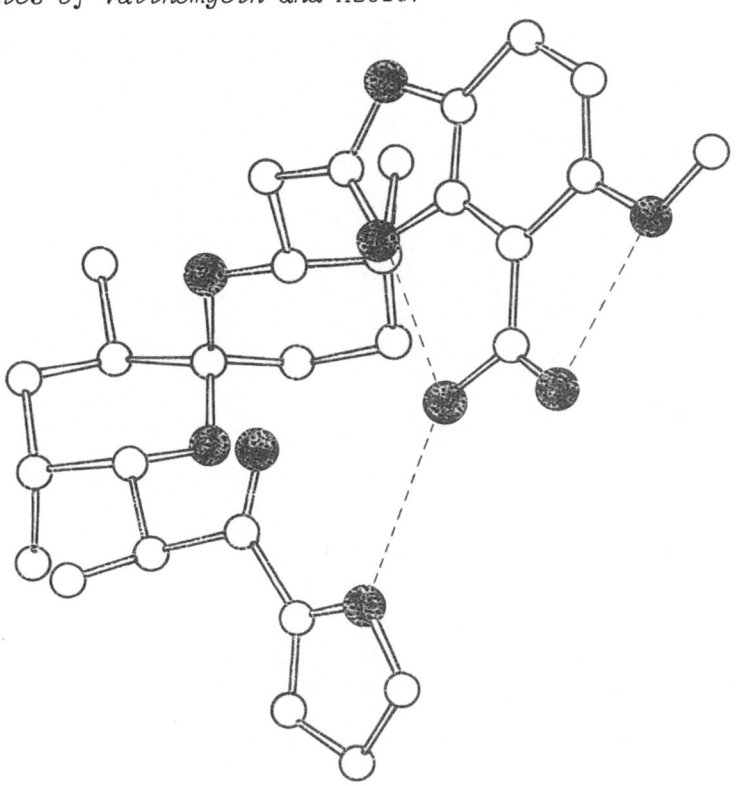

Fig. 12. The structure of the free acid form of A23187 (Chaney et al., 1974).

gen bonding is often observed in the crystal structures of uncomplexed polyether antibiotics.

Very recently two different structures of a calcium ion complex with A23187 have been reported and are shown in Figure 13 (Chaney et al., 1976; and Smith and Duax, 1976). In both cases, two ions of A23187 are coordinated to a calcium ion through bonds from a carboxyl oxygen, a carbonyl oxygen and the nitrogen of the benzoxazole ring system. A pseudo two-fold axis passes through the calcium ion of both structures and relates one A23187 ion of the complex to the other. The observed hydrogen bonding pattern within each complex is also the same. Head to tail hydrogen bonds are formed between the nitrogen of the pyrrole ring of one anion and the carboxyl oxygen of the other A23187 anion. The nitrogen of the secondary amino group is hydrogen bonded to an oxygen of the carboxyl group, a feature also observed in the free acid form.

Despite all of these remarkable similarities, the calcium in one determination has octahedral coordination (Chaney et al., 1976) but in the second case has *seven-fold* coordination (Smith and Duax, 1976).

The seventh coordination site is occupied by a water molecule and lies almost on the pseudo two-fold axis. The coordination of the carbonyl and carboxyl oxygen atoms is approximately square planar about the cal-

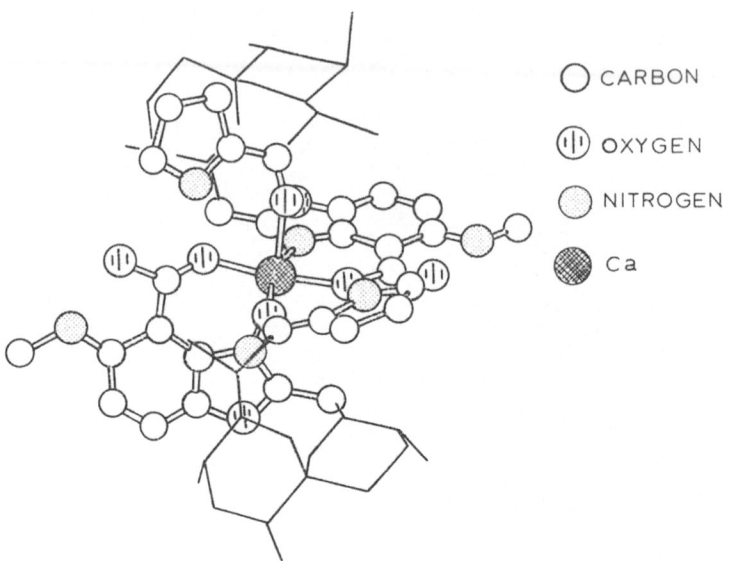

Fig. 13a.   The structure of the six coordinate form of the calcium complex of A23187 (Chaney et al., 1976).

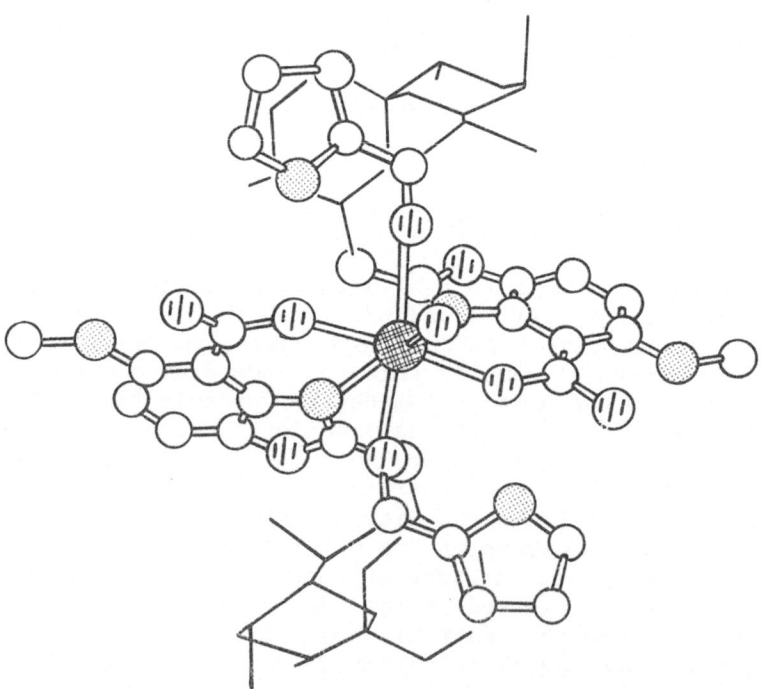

Fig. 13b.   The structure of the seven coordinate form of the calcium complex of A23187 (Smith and Duax, 1976).

cium ion. The benzoxazole nitrogen atoms, together with the calcium ion, lie in a plane nearly perpendicular to the plane of the four oxygen atoms and almost bisect the carbonyl oxygen-calcium-carboxyl oxygen angles of each ionophore. The water molecule occupies the remaining apical position and is in a very exposed position at the surface of the complex. This coordination about calcium is illustrated in Figure 13b.

One very interesting feature of this structure is the overall shape of the complex. The side of the complex containing the calcium ion and the water molecule bound to it has, in exposed positions, atoms capable of donating or accepting a hydrogen bond. If the presence of the water molecule is ignored, the calcium is in an exposed position on a side of the complex that is rather flat.

Not unexpectedly, distances from calcium to the various atoms which are bonded to it are longer in every instance in the seven coordinated structure than in the six-fold case and are listed in Table I. The short-

TABLE I
Bond distances to calcium in the six coordinate (Chaney et al., 1976) and the seven coordinate (Smith and Duax, 1976) complexes

	7-coordinate form	6-coordinate form
Ca - O(1C1)	2.27 Å	2.01 Å
Ca - O(1C2)	2.28	1.92
Ca - N(81)	2.69	2.21
Ca - N(82)	2.58	2.22
Ca - O(23A1)	2.37	2.10
Ca - O(23A2)	2.38	2.02
Ca - O(1S)	2.38	—

est distances observed for both structures are to the carboxyl oxygen atoms while the longest involve the distances to the nitrogen atoms. Nitrogen-calcium distances in the seven coordinated structure are quite long. The nitrogen-calcium-nitrogen angle is 77°; the nitrogen-nitrogen distance is 3.28 Å, the same as is observed in the octahedral coordination. Thus, the formation of a shorter bond to calcium from these two atoms would also produce a stronger repulsive interaction between the two nitrogens. The averages of the calcium-ligand distances in the six and seven coordinated structure are 2.08 and 2.42 Å respectively. It is interesting to note that the ratio of these distances is equal to the ratio of the coordination numbers.

A manganese complex (Chaney et al., 1976) and a magnesium complex* of A23187 have also been crystallized and found to be isomorphous with the six-fold coordinated calcium complex.

Because of the rigidity of the spiro ring system, the benzoxazole ring and the pyrrole ring, there are only four bonds about which free ro-

---

* Work in progress at the Medical Foundation of Buffalo.

TABLE II
Torsion angles involving the sp^3 carbon atoms adjacent to the spiro ring system for the complexed (Chaney et al., 1976; Smith and Duax, 1976) and free acid forms (Chaney et al., 1974)

| | | Seven coordinate form | | Free acid form | Six coordinate form | |
		Ion (1)	Ion (2)		Ion (1)	Ion (2)
X_1	O(7)-C(9)-C(10)-C(11)	-84°	-96°	132°	-97	-100
X_2	N(8)-C(9)-C(10)-C(11)	90	80	-46	73	77
X_3	C(9)-C(10)-C(11)-O(12)	51	56	174	66	57
X_4	C(9)-C(10)-C(11)-C(16)	176	178	-65	-171	-179
X_5	O(17)-C(18)-C(22)-C(23)	62	57	66	62	73
X_6	C(19)-C(18)-C(22)-C(23)	-178	179	-172	-171	-171
X_7	C(18)-C(22)-C(23)-C(24)	-116	-96	-158	-103	-111

tation can produce conformational changes. Torsion angles for these
bonds are listed in Table II for the free acid form and for both anions
of the six and seven coordinate forms. While there are only small differences in the torsion angles of the two complexed forms, major changes
are observed when either complexed form is compared to the structure of
the free acid. The effect of these torsion angle changes on the overall
conformation is illustrated in Figure 14 where one ion of the complex is

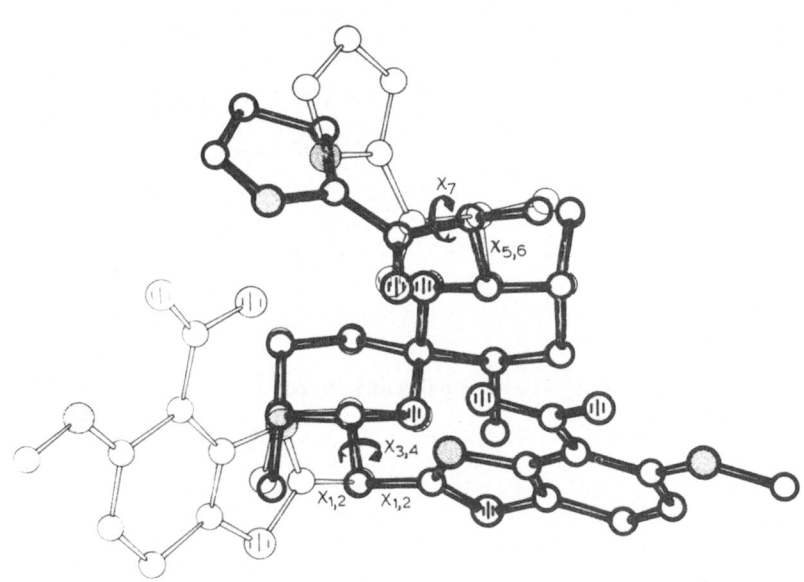

Fig. 14.   Superposition of one A23187 (dark lines) in the seven
coordinate form (Smith and Duax, 1976) on one molecule of the
free acid form (Chaney et al., 1974). The spiro ring systems
in both drawings have identical orientations.

superimposed on that of the free acid; in both cases, the orientation of
the spiro ring system is identical.
      The structure of the seven coordinate complex with the water molecule
exposed at the surface suggests that this structure is an intermediate
in the complexation process and that the six coordinate form is the structure that would be present during transport through a membrane.
      Accordingly, a mechanism of complexation can be proposed based upon
the observed conformation of all three molecules and is illustrated in
Figure 15. A hydrated calcium ion approaches the A23187 ion and forms a
weak bond to the carboxyl oxygen atom, simultaneously breaking the intra-
ionic head to tail hydrogen bond. Following the rupture of this hydrogen
bond, the anion may now undergo conformational changes. The calcium can
now form a weak bond to the nitrogen of the benzoxazole ring system. Further conformational changes of the anion move the spiro ring system away
from the position which will be occupied by calcium and allow the metal

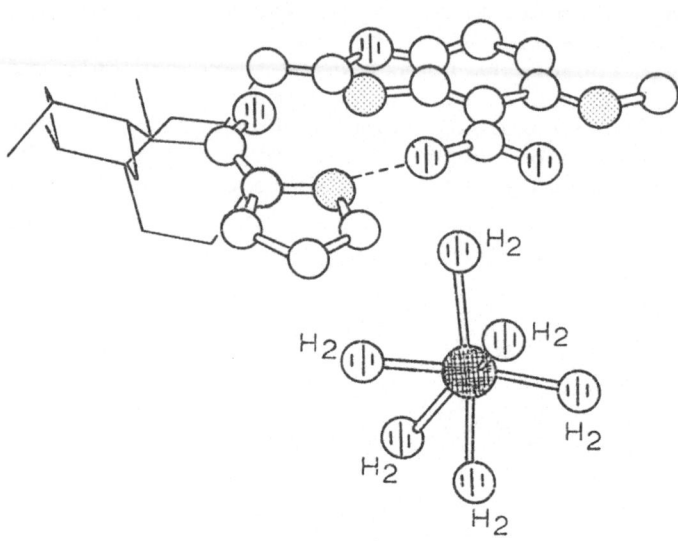

Fig. 15a — e.   The complexation of calcium ion by a single
A23187 anion.

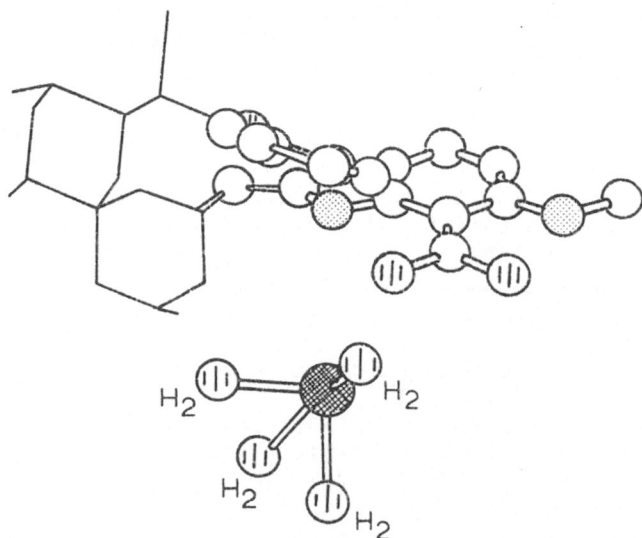

Fig. 15b.

to coordinate to the carbonyl oxygen atom. Unlike valinomycin, where the
conformational changes may occur uniformly and produce no short interatomic contacts, the torsion angle χ_3 may not undergo much change at the
beginning of this process or an exceedingly short interatomic distance i
produced between the carbonyl oxygen and the nitrogen of the benzoxazole
ring. Complexation of the second A23187 ion to the intermediate form pro-

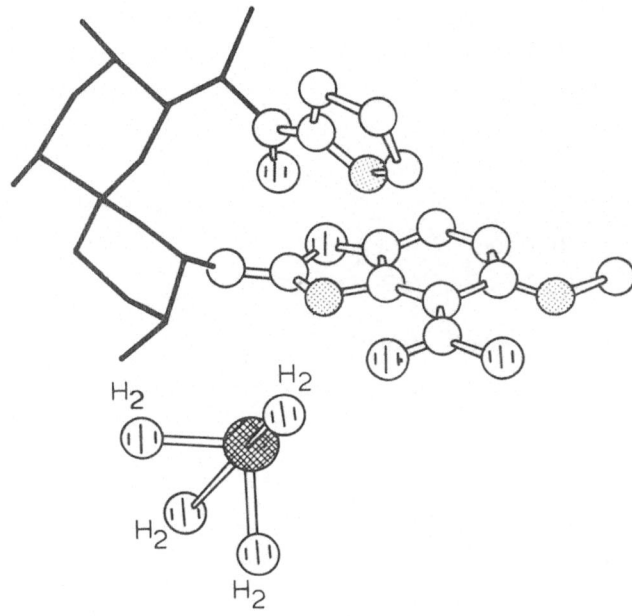

Fig. 15c.

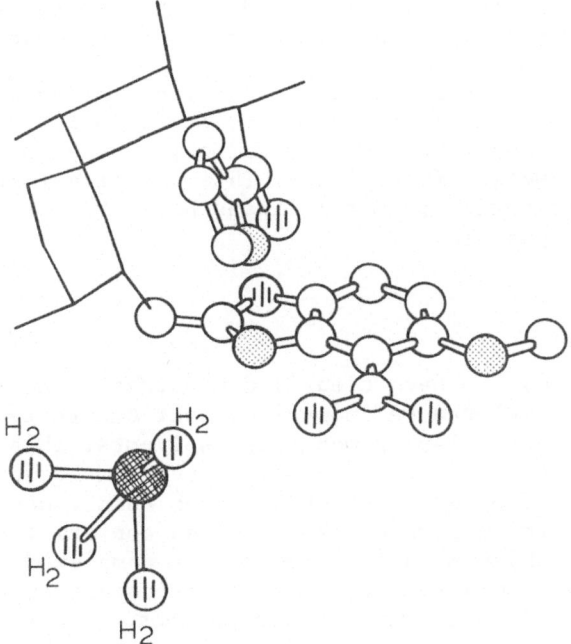

Fig. 15d.

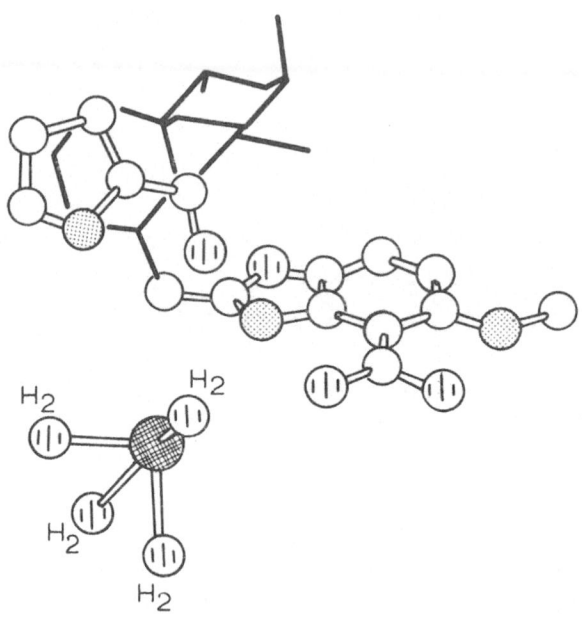

Fig. 15e.

ceeds in exactly the same fashion. At the end of this latter step, the
interionic hydrogen bonds are formed, giving rise to the observed seven
coordinated complexed form.

The transformation of the seven to the six coordinated structure
is easily accomplished by loss of the water molecule followed by a rotation of one A23187 ion about the calcium carboxyl oxygen bond as well
as a contraction of the metal ligand bonds. This rotational process
does not require the alteration of metal ligand bonds or the disruption
of the hydrogen bonding pattern.

4. CONCLUSIONS

X-ray crystallographic results have clearly demonstrated the existence
of a conformation of uncomplexed valinomycin as yet undetected by solution spectroscopy in either polar or non-polar solvents. This discrepancy can be explained, at least in a non-polar environment, by a fast
permutation of the two different kinds of hydrogen bonds, observed in
the solid state, around the molecule. The observed conformation also
suggests a mechanism by which potassium ion may be complexed. Further
studies are currently underway to determine the structure of a valino-
mycin.modification crystallized from a polar solvent.

A possible mechanism for complexation is postulated based upon the
crystallographically observed structures of the free and complexed forms
of A23187. The seven coordinate complex with calcium partially hydrated

provides a stable intermediate in which calcium and several hydrophilic groups of each A23187 anion are located on the surface of the molecule and ideally situated to interact with a polar surface. The conversion of the seven coordinated to the six coordinated complex, involved in active transport, has been shown to be accomplished by a simple rotation about one bond. The existence of the partially hydrated intermediate suggests that similar intermediates might also be crystallized for other ionophores whose structures have already been determined. Work of this nature can provide valuable insights into the mechanism of complexation as well as membrane ionophore interactions.

ACKNOWLEDGEMENT

The authors wish to express their gratitude to Miss M. Tugac, Miss G. Del Bel and Mr A. Erman for their assistance in the preparation of the illustrations, to Miss D. Hefner, Mrs V. Kamysz, Mrs E. Robel and Mrs B. Giacchi for the typing of the manuscript and to Dr J. Griffin for the stimulating discussions concerning this study. Some of the illustrations were prepared from structural data stored in the NIH PROPHET system, an NIH interactive computer network. This work has been supported by Grant No. GM-19684, awarded by the National Institute of General Medical Sciences, DHEW, and by Grant No. LM-02353, awarded by the National Library of Medicine, DHEW.

REFERENCES

Case, G.D., Vanderkooi, J.M., and Scarpa, A.: 1974, Physical Properties of Biological Membranes Determined by the Fluorescence of the Calcium Ionophore A23187, Arch. Biochem. Biophys. 162, 174-185.
Chaney, M.O., Demarco, P.V., Jones, N.D., and Occolowitz, J.L.: 1974, The Structure of A23187, a Divalent Cation Ionphore, J. Am. Chem. Soc. 96, 1932-1933.
Chaney, M.O., Jones, N.D., and Debono, M.: 1976, The Structure of the Calcium Complex of A23187, a Divalent Cation Ionophore, J. Antibiotics 29, 424-427.
DeCamp, W.H.: 1973, Probability Plot Comparison of Crystallographically Independent Molecules, Acta Cryst. A29, 148-150.
Dobler, M.: 1972, The Crystal Structure of Nonactin, Helv. Chim. Acta 55, 1371-1384.
Dobler, M. and Phizackerely, R.P.: 1974, The Crystal Structure of the NaNCS Complex of Nonactin, Helv. Chim. Acta 57, 664-674.
Duax, W.L., Hauptman, H., Weeks, C.M., and Norton, D.A.: 1972, Valinomycin Crystal Structure Determination by Direct Methods, Science 176, No. 4037, 911-914.
Grenier, G., Van Sande, J., Glick, D., and Dumont, J.W.: 1974, Effect of Ionophore A23187 on Thyroid Secretion, FEBS Letters 49, 96-99.
Iitaka, Y., Sakamaki, T., and Nawata, Y.: 1972, Molecular Structures of Tetranactin and Its Alkali Metal Ion Complexes, Chem. Lett. (Tokyo), 1225-1230.

Karl, R.C., Zawalich, W.S., Ferrendelli, J.A., and Matochinski, F.M.: 1975, The Role of Ca^{+2} and Cyclic Adenosine 3':5'-Monophosphate in Insulin Release Induced in Vitro by the Divalent Cation Ionophore A23187, J. Biol. Chem. 250, 4575-4579.

Karle, I.L.: 1975, Conformation of Valinomycin in a Triclinic Crystal Form, J. Amer. Chem. Soc. 97, 4379-4386.

Krasne, S., Eisenman, G., and Szabo, G.: 1971, Freezing and Melting of Lipid Bilayers and the Mode of Action of Nonactin, Valinomycin, and Gramicidin, Science 174, 412-415.

Lutz, W.K., Winkler, F.K., and Dunitz, J.D.: 1971, Crystal Structure of the Antibiotic Monensin. Similarities and Differences between Free Acid and Metal Complex, Helv. Chim. Acta 54, 1103-1108.

Maigret, B. and Pullman, B.: 1975, Etudes quantiques sur la conformation de la valinomycine et de ses éléments constitutifs, Theoret. Chim. Acta (Berl.) 37, 17-36.

Nawata, Y., Sakamaki, T., and Iitaka, Y.: 1974, The Crystal and Molecular Structures of Tetranactin, Acta Cryst. B30, 1047-1953.

Neupert-Laves, K. and Dobler, M.: 1975, The Crystal Structure of a K^+ Complex of Valinomycin, Helv. Chim. Acta 58, 432-442.

Patel, D.J. and Tonelli, A.E.: 1973, Solvent-Dependent Conformations of Valinomycin in Solution, Biochem. 12, 486-495.

Pinkerton, M., Steinrauf, L.K., and Dawkins, P.: 1969, The Molecular Structure and Some Transport Properties of Valinomycin, Biochem. Biophy. Res. Comm. 35, 512-518.

Pinkerton, M. and Steinrauf, L.K.: 1970, Molecular Structure of Monovalent Metal Cation Complexes of Monensin, J. Mol. Biol. 49, 533-546.

Rothschild, K.J., Asher, I.M., Anastassakis, E., and Stanley, H.E.: 1973, Raman Spectroscopic Evidence for Two Conformations of Uncomplexed Valinomycin in the Solid State, Science 182, 384-386.

Smith, G.D. and Duax, W.L.: 1976, The Crystal and Molecular Structure of the Calcium Ionophore A23187, J. Amer. Chem. Soc. 98, 1578 - 1580.

Smith, G.D., Duax, W.L., Langs, D.A., DeTitta, G.T., Edmonds, J.W., Rohrer, D.C., and Weeks, C.M.: 1975, The Crystal and Molecular Structur of the Triclinic and Monoclinic Forms of Valinomycin, $C_{54}H_{90}N_6O_{18}$, J. Amer. Chem. Soc. 97, 7242-7247.

Tosteson, D.C., Cook, P., Andreoli, T., and Tieffenberg, M.: 1967, The Effect of Valinomycin on Potassium and Sodium Permeability of HK and LK Sheep Red Cells, J. Gen. Physiol. 50, 2513-2525.

Urry, D.W. and Kumar, N.G.: 1974, Affirmation of Critical Proton Magnetic Resonance Data on the Solution Conformation of the Valinomycin-Potassiu Ion Complex, Biochem. 13, 1829-1831.

Wipf, von H.-K., Olivier, A., and Simon. W.: 1970, Mechanisms and Selektivität des Alkali-Ionentransportes in Modell-Membranen in Gegenwart des Antibioticums Valinomycin, Helv. Chim. Acta 53, 1605-1608.

DISCUSSION

Simon: In the stepwise procedure of bringing K^+ into the coordination sphere of valinomycin you neglected the solvation shell of K^+. In order to keep the activation energy of the process low, these solvent molecules have to be replaced one by one by the coordinating carbonyl groups of valinomycin. Is this consistent with your model?

Duax: The hydration sphere of potassium was omitted from the figures illustrating the valinomycin complexation in order to simplify them. We fully agree that the sequential replacement of the water molecules coordinated to potassium by the carbonyl groups lining the inside of the uncomplexed molecule is the most energetically plausible way for the complexation to proceed.

Saenger: For the complex formation between Ca^{2+} and A23187 you proposed a 1 : 1 complex followed by the addition of one more A23187 molecule. Is this stepwise aggregation based on experimental results?

Duax: The model is highly speculative. Because two body collisions are generally considered to be more likely to occur than three body collisions and because the observed uncomplexed molecule does not have a dimeric form which might suggest a simple introduction of Ca^{++} into a dimer in the manner proposed for the valinomycin K^+ complex, we elected to propose sequential formation of a 1 : 1 followed by a 2 : 1 complex and proceeded to modify the molecular conformation accordingly. Alternate paths could probably be developed for this part of the process. In fact it might be possible to crystallize a dimer or A23187 from less polar solvents. However, I think it is safe to say that the seven-coordinate complex to six-coordinate complex transition suggested as relevant at a water-lipid interface may be easier to accept at face value.

A. Pullman: The reason for the lengthening of the distances in the 7-fold complex is simply the repulsion between the ligands. When the ligands are small enough, the cation tends to accommodate them so as to form a complex similar to the hexacoordinated complex that it forms with water molecules. When a seventh ligand tries to be accommodated, all the others retire to a larger distance on account of the repulsion. The seven-coordinated complex may form if the extra binding energy so gained overcompensates the loss of binding energy undergone by moving the other ligands farther out.

EFFECTS OF CATIONS OF GROUPS IA AND IIA ON CROWN ETHERS

MARY R. TRUTER
*Molecular Structures Department, Rothamsted Experimental
Station, Harpenden, Herts AL5 2JQ, U.K.*

ABSTRACT. Changes in conformation of macrocyclic polyether (crown) molecules on complex formation with alkali and alkaline earth metal cations
are described. An empirical relation (Brown and Shannon) for metal-
oxygen distance gives approximate electrostatic attraction terms while
four different relations are compared for calculations of conformational
energies. Repulsive interactions between cations and some methylene
hydrogen atoms probably account for the stoicheiometry 1:1 or 1:2 of the
complexes of benzo-15-crown-5.

1. INTRODUCTION

For cations of groups IA and IIA the most favourable ligands might be
expected to have two or more oxygen or nitrogen donor atoms arranged
so that little configurational entropy is lost on chelate complex
formation. In fact such molecules as 1, 10-phenanthroline and phenacyl
kojate, I of Figure 1, do form complexes with alkali metals (Truter,
1973, and references therein) without change of conformation (Philips
and Truter, 1975a) but the complexes have low stabilities in solution.
    Many of the compounds which form alkali metal complexes stable in
solution do so with change of conformation, e.g. valinomycin, nonactin
and Pedersen's macrocyclic ethers (Pedersen, 1967). It is with the last
compounds that I shall be concerned here. They are neutral and multidentate and form complexes which may be positively charged or neutral,
depending upon the remaining coordination to the metal. Crystal structure analysis has shown that there are no significant changes in bond
lengths and bond angles between free molecules and those in various
complexes. There are changes in conformation and we have attemted to
calculate the conformational energies and to design ligands on conformational principles. I shall first describe the changes found in macrocyclic ethers in order of ring size, then give the results of two empirical approaches.

*B. Pullman and N. Goldblum (eds.), Metal-Ligand Interactions in Organic
Chemistry and Biochemistry, part 1, 317-335. All Rights Reserved.*
Copyright © 1977 by D. Reidel Publishing Company, Dordrecht-Holland.

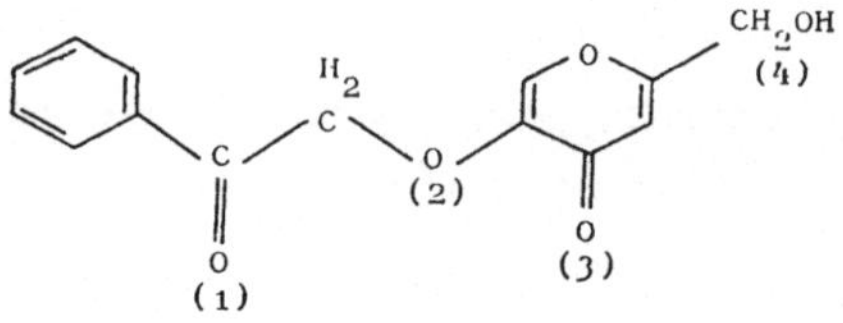

Fig. 1a.

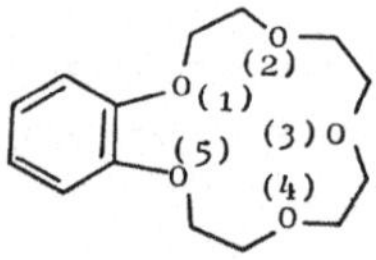

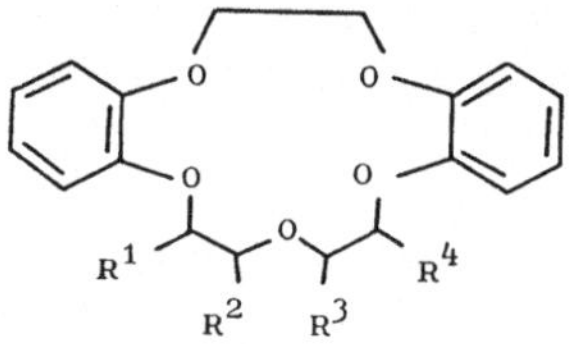

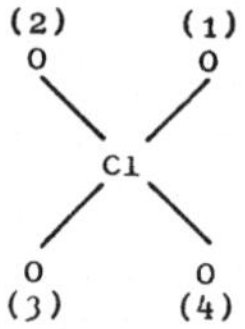

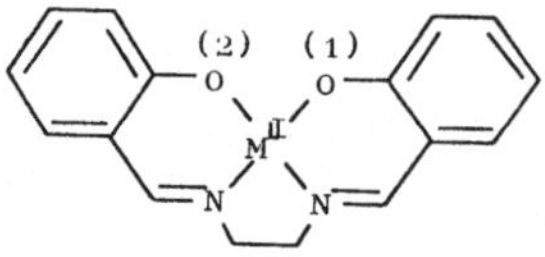

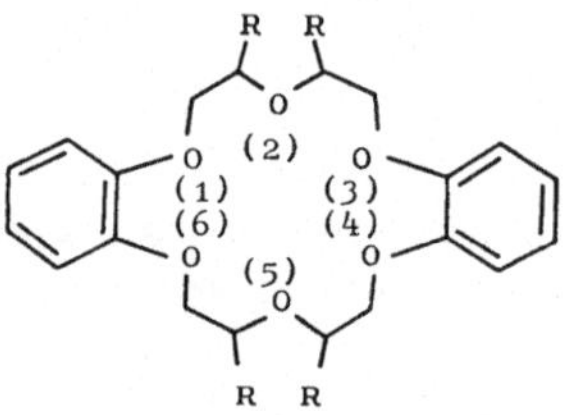

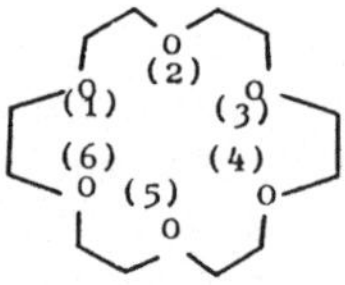

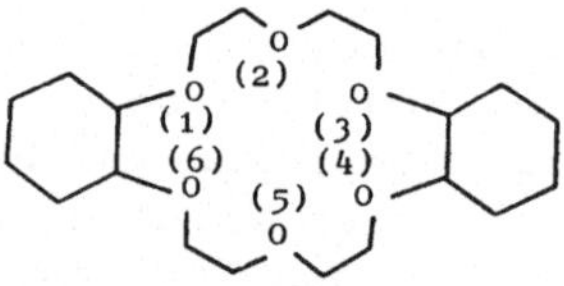

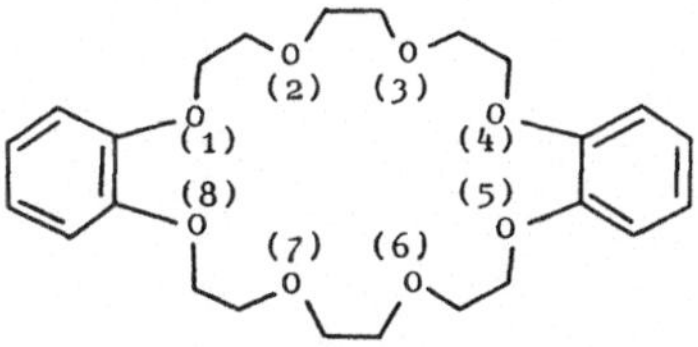

X

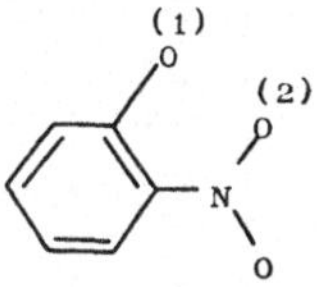

XI

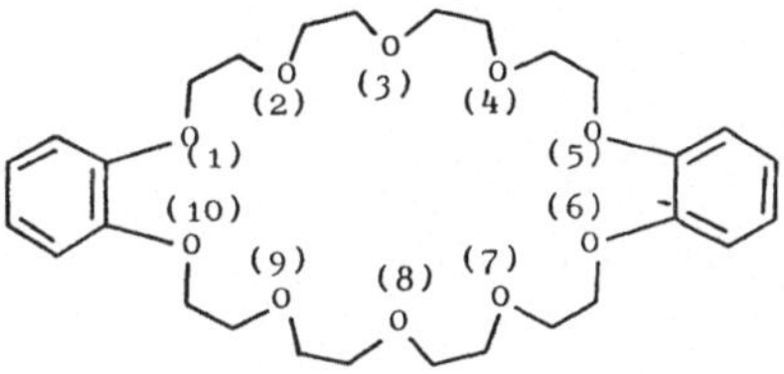

XII

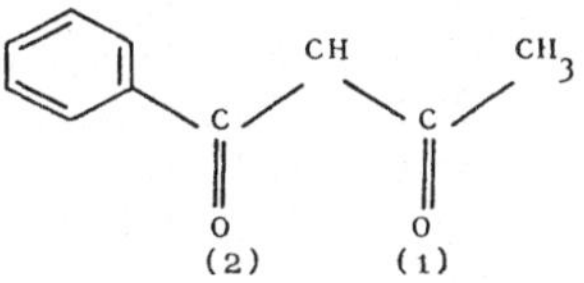

XIII

HOCH₂ CH₂ OH
(3)   (4)

XIV

Fig. 1b.

TABLE I

Torsion angles in benzo-15-crown-5, II, in various complexes with cations, each taken to be above the plane of the paper. The oxygen atoms are designated as in Figure 1.

complex bond	$[Na,II,H_2O]^+ I^-$	$[K(II)_2]^+ I^-$	$[Ca,II,(NCS)_2 MeOH]^o$	$[Ca,II(NCS)_2 H_2O]$
O(5)—C—C—O(1)	3	1	4	7
C—C—O(1)—C	147	175	172	173
C—O(1)—C—C	173	-160	-152	-158
O(1)—C—C—O(2)	50	-60	-50	-54
C—C—O(2)—C	-177	167	-177	-178
C—O(2)—C—C	171	-89	-178	180
O(2)—C—C—O(3)	-62	-61	51	53
C—C—O(3)—C	173	-179	82	83
C—O(3)—C—C	-160	-143	-168	-163
O(3)—C—C—O(4)	58	22	57	52
C—C—O(4)—C	-178	116	175	172
C—O(4)—C—C	173	-173	-170	-172
O(4)—C—C—O(5)	-56	52	-54	-53
C—C—O(5)—C	-177	150	173	176
C—O(5)—C—C	-161	-160	-124	-108

2. 15-CROWN-5 RINGS

The first example is the molecule, benzo-15-crown-5, II of Figure 1, this usually gives a 1:1 complex with sodium salts and always a 1:2 with potassium. The differences between the two conformations of the completed molecules are shown in Table I as torsion angles. All the aliphatic carbon atoms are on one side of the plane through the five oxygen atoms and the potassium ion is at 1.67 Å on the other side of this plane (Mallinson and Truter, 1972), while in the 1:1 sodium complex the cation is 0.75 Å from the plane of the oxygen atoms with four aliphatic carbon atoms on the same side (0.07 to 0.33 Å) and four on the opposite side (0.02 to 0.46 Å) of this plane (Bush and Truter, 1972a). Thus the difference in conformation allows two ligand molecules to sandwich the potassium ion. Our first calculation of purely repulsion energies was from the formula.

$$C.E. = \Sigma \exp(4.0 - 4.0 \ R_{ij}/R_o)$$

with R_{ij} as the interatomic distance, maximum 7.0 Å, and R_o the sum of the van der Waal's radii of the atoms i and j. The values for II in the sodium and potassium complexes were similar and this, coupled with experience in making complexes (Poonia and Truter, 1973) that the stoicheiometry of potassium ones was always $K(II)_2$, suggested two approaches to further complex formation.

Firstly, with the aid of space filling models we designed a molecule having a 15-crown-5 ring but with steric hindrance to the conformation with all carbons on one side of the plane through the oxygen atoms. Substitution by two instead of one benzene ring somewhat reduces the flexibility of the molecule and addition of methyl groups as shown in III, particularly at R^1 and R^4, gives ready formation of 1:2 complex with potassium when the *meso* isomer is used but very small effect for the racemic *dl* isomer. While this is in the predicted direction the selectivity is against potassium rather than for sodium and measurements of the change in emf in methanol solution of MX (X=Cl, Br) follow the same rank order for both cations (Parsons, 1975).

Secondly, we made a sandwich of sodium by using a large, and non-coordinating anion. We chose first the tetraphenylborate (Parsons et al., 1975) which also had the merit that hydrogen bonding from coordinated water to it was unlikely; this gave the desired 1:2 complex Na BPh$_4$ (benzo-15-crown-5)$_2$ as deduced from the infra red spectrum in the finger print region. A sandwich was also synthesised with perchlorate, IV, as the anion, the stoicheiometry 1:1 or 1:2 depending upon the reaction medium and the concentration.

While the perchlorate ion is much-quoted as a non-complexing anion there is ample evidence for it acting as a uni and as a bidentate ligand. In fact we have found it one of the better anions for the isolation of complexes (Wingfield and Parsons, 1976a). This is, at least in part, because the lattice energy is in the lower range compared with the alkali-metal halides. It has the practical advantage that the IR spectrum in the 630 cm^{-1} region is diagnostic. Elemental analysis and spectroscopic evidence indicates that sodium perchlorate forms both NaClO$_4$

(benzo-15-crown-5)$_2$ containing the sandwich form of the crown and NaClO$_4$
(benzo-15-crown-5) containing the same crown conformation as [Na(benzo-
15-crown-5)H$_2$O]$^+$I$^-$ and coordinated perchlorate. Crystal structure analysis of the latter confirms (Owen, 1976) the conformation of the crown
with the sodium 0.78 Å from the plane of the oxygen atoms towards a
chelating perchlorate anion IV at Na—O 2.5 Å. This entity, a coordinated sodium perchlorate ion pair has been found previously with the
additional ligand atoms being oxygen atoms from two salicylaldimine
ethylenediamine copper complex molecules (V, M^{II} = Cu) (Milburn et al.,
1974). In that case also the Na—O (perchlorate) distance (2.55 Å was
longer than Na—O (neutral) 2.35, 2.37 Å. The significance of cation-
oxygen distances is discussed later.

   Although the calcium ion is nearly the same size as sodium, it
usually has a larger coordination number and in the cyclic ether system
this effect persists, for example the benzo-15-crown-5 complex of
Ca(NCS)$_2$ is actually 8 coordinated (Owen and Wingfield, 1976) there
being one additional molecule of solvent; two thiocyanate nitrogen atoms
and one oxygen from solvent, water or methanol, are on one side of the
calcium ion and the five oxygens of II are on the other; in contrast
with both sodium and potassium complexes these oxygens are not coplanar
and the symmetry of the molecule is much lower, as shown by the torsion
angles in Table I. The larger barium and strontium ions, like potassium,
yield 1:2 complexes with similar IR spectra.

3. 18-CROWN-6 RINGS

Very subtle effects of the ligand molecules may change the stoicheiometry
of the complexes. For example the isomeric tetramethyl dibenzo-18-crown-6
compounds, called F and G, Figure 1, VI, form 1:1 and 1:2 complexes
respectively with CsNCS (Parsons, 1975b), the meso form, F, provides
six oxygen atoms for the complex and the charge is neutralised by two
nearly equally shared thiocyanate anions (Mallinson, 1975a); a similar
arrangement is found in [CsNCS(18-crown-6)]$_2$ (Dobler and Phizackerley,
1974a). Isomer G is optically active and it is the racemate which we
used for complex formation so the Cs ion is 12 coordinated being sand-
wiched between one d and one l molecule (Mallinson, 1975a). I might
venture to predict that an asymmetric synthesis would lead to a 1:1
complex were it not for the fact that dibenzo-18-crown-6, VII, itself
forms a 1:2 complex (Pedersen, 1970, and preliminary X-ray investigation)
Comparison of uncomplexed isomer F, (Mallinson, 1975b), with its complex
and with dibenzo-18-crown-6, VII, shows changes in the torsion angles,
Table II. Complexes of VI(F) and VI(G) and VII show similar patterns in
the torsion angles, 0° at the benzene rings, gauche at the other O—C—C—O
bonds and trans for the others.

   For the unsubstituted 18-crown-6 ring, VIII, which is more flexible
than the dibenzo, or even the dicyclohexyl, derivatives the conformation
varies with the cation; in particular, sodium is coordinated by 5 approximately coplanar ring oxygen atoms with the sixth forming one apex of a
pentagonal bipyramid and a water molecule at the other apex (Dobler et
al., 1974). This distortion of the molecule gives closer Na—O distances

TABLE II
Torsion angles in 18-crown-6 rings

	VII	VI(F)	VIII	M(VII)	[Cs(VI(F)NCS]₂	Cs(VI(G))₂	Na(VIII)H₂O	K(VIII)	[Cs(VIII)NCS]₂
O(6)—C—C—O(1)	-5	1	75	-1	2	-3	61	70	68
C—C—O(1)—C	155	-177	-155	-178	179	-178	-171	-176	-177
C—O(1)—C—C	-74	172	166	179	180	158	-177	-177	179
O(1)—C—C—O(2)	-77	-69	-68	66	69	61	-59	-65	-63
C—C—O(2)—C	103	84	176	175	175	-173	-173	-178	-173
C—O(2)—C—C	176	100	175	-180	177	-159	-174	171	177
O(2)—C—C—O(3)	164	-170	175	-62	-69	-69	52	65	61
C—C—O(3)—C	-159	-174	170	-167	-165	-143	71	-179	172
C—O(3)—C—C	174	175	80	173	180	150	-172	-178	174
O(3)—C—C—O(4)	5	-1	-75	-1.0	-1	-3	63	-70	-66
C—C—O(4)—C	-155	177	155	-180	-169	-178	-176	176	-176
C—O(4)—C—C	74	-172	-166	174	167	158	77	177	-172
O(4)—C—C—O(5)	77	69	68	61	72	61	47	65	65
C—C—O(5)—C	-103	-84	-176	177	162	-173	115	178	173
C—O(5)—C—C	-176	-100	-175	-179	172	-159	-74	-171	-179
O(5)—C—C—O(6)	-164	170	-175	-65	-61	-69	-59	-65	-66
C—C—O(6)—C	159	174	-170	-177	-177	-143	167	179	180
C—O(6)—C—C	-174	-175	-80	-179	179	150	173	178	-178

than those which would be available in a planar hexagonal arrangement
and allows for a usual feature in sodium complexes, a coordinated water
molecule which is hydrogen bonded to the anion.

With other cations the oxygen atoms are coplanar, the potassium
ion is central in the ring (Seiler et al., 1974) and the caesium ion
1.44 Å above the plane of the oxygen atoms; there is no significant
difference between the torsion angles, gauche for O–C–C–O and trans for
the others.

Hydrogenation of VII leads to two isomers of dicyclohexyl-18-crown-6,
IX. One isomer yields a sodium complex in which the cation is 8-coordi-
nated by the six coplanar oxygen atoms of the ring, with torsion angles
similar to those of K(VIII) in Table II, and water molecules at the
apices (Mercer and Truter, 1973a).

4. 24-CROWN-8 RINGS

Dibenzo-24-crown-8, is a molecule of great versatility, it may enclose
two univalent cations (but not divalent ones) or only one cation. There
are two polymorphic forms of the free molecule; one, obtained by unsuc-
cessful attempts to form complexes with small cations, has quite a different infrared spectrum from the normal, obtained by crystallisation
from a pure solution (Wingfield, 1975). It is for the latter that the
crystal structure has been determined (Hanson et al., 1976). Two complexes, both containing two cations have also been studied crystallographically (KNCS)$_2$ dibenzo-24-crown-8, (Mercer and Truter, 1973b),
and (Nao–nitrophenolate)$_2$dibenzo-24-crown-8 (Hughes, 1975). Torsion
angles, Table III, are quite different in all three forms and analysis
of the results on the uncomplexed molecule indicates that packing may be
an important influence.

In the potassium complex the complexed entity can be seen as a
dimeric ion pair SCN⟨K⟩NCS with each potassium ion being coordinated
by 5 ligand oxygen atoms, two of these atoms being shared by the two
potassium ions. In the sodium complex the dimeric ion pair consists of
chelated sodium ions sharing the phenolic oxygen atoms XI, and sodium
ion is coordinated by only three of the ligand oxygen atoms with a fourth
making a long contact.

5. 30-CROWN-10-RINGS

Dimensions are available for dibenzo-30-crown-10, XII, and for the complex in which it wraps round one potassium ion, as with the other crown
compounds there is a change in some conformation angles, Table IV (Bush
and Truter, 1972b).

TABLE III
Torsion angles in dibenzo-24-crown-8, X, and its complexes

	X	(KNCS)$_2$X	(Na XI)$_2$X
O(8)—C—C—O(1)	1	0	2
C—C—O(1)—C	-172	-175	-177
C—O(1)—C—C	-175	165	-180
O(1)—C—C—O(2)	-78	67	-61
C—C—O(2)—C	144	94	-157
C—O(2)—C—C	166	84	-180
O(2)—C—C—O(3)	-84	61	117
C—C—O(3)—C	177	95	-180
C—O(3)—C—C	85	169	-157
O(3)—C—C—O(4)	-78	68	-61
C—C—O(4)—C	-177	165	-180
C—O(4)—C—C	-175	-174	-177
O(4)—C—C—O(5)	-1	0	2
C—C—O(5)—C	172	175	-171
C—O(5)—C—C	175	-165	-175
O(5)—C—C—O(6)	78	-67	-60
C—C—O(6)—C	-144	-94	-176
C—O(6)—C—C	-166	-84	179
O(6)—C—C—O(7)	84	-61	139
C—C—O(7)—C	-177	-95	179
C—O(7)—C—C	-85	-169	-176
O(7)—C—C—O(8)	78	-68	-60
C—C—O(8)—C	177	-165	-175
C—O(8)—C—C	175	174	-171

TABLE IV
Torsion angles in dibenzo-30-crown-10, XII, and its complex
with potassium in K(XII)I. (Bush and Truter, 1972b)

	XII	[K(XII)]$^+$
O(10)—C—C—O(1)	-2	-4
C—C—O(1)—C	-174	-174
C—O(1)—C—C	165	175
O(1)—C—C—O(2)	63	62
C—C—O(2)—C	-177	79
C—O(2)—C—C	175	-176
O(2)—C—C—O(3)	72	62
C—C—O(3)—C	-177	85
C—O(3)—C—C	180	-169
O(3)—C—C—O(4)	70	69
C—C—O(4)—C	169	172
C—O(4)—C—C	-95	62
O(4)—C—C—O(5)	67	44
C—C—O(5)—C	-148	178
C—O(5)—C—C	154	147

[*TABLE IV (Continued)*]

	XII	$[K(XII)]^+$
for O(5)—C—C—O(6)	2	-4
to C—O(10)—C—C	-154	147
repeat with sign	changed	same

TABLE V
Bond strength, S, according to the empirical relation $S = S_0 (R/R_0)^{-N}$.

Cation in 5 coordination
(1-phenyl butane-1,3-dionato)(ethylene glycol)sodium
                                                    (Bright et al., 1971)

          Na(XIII)(XIV)

Na XIII O(1) 0.244; XIII O(2) 0.213; XIII O(1') 0.204; XIV O(3) 0.226;
XIV O(4') 0.217 ΣS = 1.104

Cation in 6 coordination
sodium iodide and phenacyl kojate   (Phillips and Truter, 1975)

          $[Na(I)(H_2O)_2]^+$

Na I O(1) 0.167; I O(2) 0.209; I O(3) 0.186. I O(4') 0.127; H_2O 0.124
and 0.236 ΣS = 1.049
sodium iodide and benzo-15-crown-5   (Bush and Truter, 1972a)

          $[Na(II)H_2O]^+$

Na II O(1) 0.200; II O(2) 0.182; II O(3) 0.207; II O(4) 0.175; II O(5)
0.199; H_2O 0.246 ΣS = 1.209
bis [NN'ethylene bis(salicylideneiminato)copper(II)perchlorato sodium
                                                    (Milburn et al., 1974)

          $[Na(V, M=Cu)(IV)]^0$

Na IV O(1) O(2) 0.132; V O(1) O(1') 0.209; V O(2) O(2') 0.200
ΣS = 0.541 x 2 = 1.082
sodium tetraphenylborate and NN'ethylene bis(salicylideneiminato) cobalt
solvated with tetrahydrofuran   (Floriani et al., 1973)

          $[Na(V, M=Co)(THF)_2]^+$

Na THF 0.209 0.200; V O(1) 0.178; V O(2) 0.178; V O(1') 0.190; V O(2')
0.155 ΣS = 1.110
potassium valinomycin triiodide   (Neupert-Laves et al., 1975)
K O(1) 0.130; O(2) 0.163; O(3) 0.126; O(4) 0.152; O(5) 0.148; O(6)
0.145 ΣS = 0.86

[*TABLE V (Continued)*]

Bond strength, S, according to the empirical relation $S = S_0 \ (R/R_0)^{-N}$

Cation in 6 or 7 coordination
(sodium o-nitrophenolate)$_2$dibenzo-24-crown-8   (Hughes, 1975)

(Na XI)$_2$X

Na XI O(1) 0.238; XI O(2) 0.184; XI O(1') 0.225; X O(1) 0.159; X O(2) 0.113; X O(8) 0.137; X O(7) 0.057 ΣS = 1.099

Cation in 7 coordination
aquo sodium (18-crown-6) thiocyanate   (Dobler et al., 1974)

[Na VIII H$_2$O]$^+$

Na VIII O(1) 0.115; VIII O(2) 0.113; VIII O(3) 0.157; VIII O(4) 0.133; VIII O(5) 0.165; O(6) 0.111; H$_2$O 0.224 ΣS = 1.017

(potassium thiocyanate)$_2$dibenzo-24-crown-8   (Mercer et al., 1973b)

[(K NCS)$_2$X]

K X O(1) 0.150; X O(2) 0.127; X O(3) 0.097; X O(7) 0.111; X O(8) 0.138
ΣS = 0.623   K–NCS 2.887 and 2.852 Å

Cation in 8 coordination
sodium bromide (dibenzo-18-crown-6) dihydrate   (Bush et al., 1971)

Molecule A   [Na (VII)(H$_2$O)$_2$]$^+$

Na VII O(1) 0.109; VII O(2) 0.104; VII O(3) 0.083; VII O(4) 0.075; VII O(5) 0.094; VII O(6) 0.111; H$_2$O 0.254 0.230 ΣS = 1.060

Molecule B   [Na (VII) H$_2$O Br]^o

Na VII O(1) 0.098; VII O(2) 0.066; VII O(3) 0.070; VII O(4) 0.094; VII O(5) 0.135; VII O(6) 0.102; H$_2$O 0.209 ΣS = 0.773
Na–Br = 2.82 Å
sodium bromide (dicyclohexyl-18-crown-6)dihydrate   (Mercer et al., 1973a)

[Na (IX)(H$_2$O)$_2$]$^+$

Na IX O(1) O(4) 0.101; IX O(2) O(5) 0.100; IX O(3) O(6) 0.054; H$_2$O 0.211 ΣS = 0.469 x 2 = 0.938
potassium thiocyanate (18-crown-6)   (Seiler et al., 1974)

[K(VIII)]$^+$

K VIII O(1) O(4) 0.140; VIII O(2) O(5) 0.130; VIII O(3) O(6) 0.125
ΣS = 0.395 x 2 = 0.790   K–NCS 2 at 3.19 Å

[*TABLE V (Continued)*]
Bond strength, S, according to the empirical relation $S = S_0 (R/R_0)^{-N}$.

---

calcium thiocyanate (benzo–15–crown–5) methanol   (Owen et al., 1976)

$Ca(NCS)_2$ (II) CH_3OH

Ca II O(1) 0.226; II O(2) 0.207; II O(3) 0.210; II O(4) 0.224; II O(5)
0.207; H_2O 0.310 $\Sigma S = 1.384$ Ca—NCS 2.485 and 2.402 Å
calcium thiocyanate (benzo–15–crown–5) hydrate   (Owen et al., 1976)

$Ca(NCS)_2(II)H_2O$

Ca II O(1) 0.216; II O(2) 0.211; II O(3) 0.241; II O(4) 0.258; II O(5)
0.178; H_2O 0.295 $\Sigma S = 1.399$ Ca—NCS 2.422 and 2.424 Å
sodium chloride and phenacyl kojate   (Phillips, 1976)

$[Na(I)_2]^+$

Na I O(1) O(1') 0.130; I O(2) O(2') 0.102; I O(3) O(3') 0.107; I O(4'')
O(4''') 0.124 $\Sigma S = 0.463$ x 2 = 0.925
potassium iodide and phenacyl kojate   (Hughes et al., 1974)

$[K(I)_2]^+$

K I O(1) 0.138; I O(2) 0.107; I O(3) 0.126; I O(4'') 0.130; I O(1')
0.137; I O(2') 0.124; I O(3') 0.145; I O(4''') 0.148 $\Sigma S = 1.055$
sodium thiocyanate and nonactin   (Dobler and Phizackerly, 1974)

Na $>$O 0.080 and 0.088; O=C 0.170 and 0.188 $\Sigma S = 0.526$ x 2 = 1.052
potassium thiocyanate and nonactin   (Dobler et al., 1969)

K $>$O 0.115 and 0.130; O=C 0.130 and 0.147 $\Sigma S = 0.522$ x 2 = 1.044

Cation in 10 coordination
potassium iodide (benzo–15–crown–5)$_2$ (Mallinson et al., 1972)

$[K(II)_2]^+$

K II O(1) O(1') 0.126; II O(2) O(2') 0.110; II O(3) O(3') 0.120;
II O(4) O(4') 0.101; II O(5) O(5') 0.138 $\Sigma S = 0.595$ x 2 = 1.190
potassium iodide (dibenzo–30–crown–10)   (Bush et al., 1972b)

$[K(XII)]^+$

K XII O(1) O(6) 0.112; XII O(2) O(7) 0.121; XII O(3) O(8) 0.115; XII
O(4) O(9) 0.121; XII O(5) O(10) 0.105 $\Sigma S = 0.574$ x 2 = 1.148

Primed designations refer to atoms in molecules other than those
displayed in Figure 1.

6. METAL–OXYGEN DISTANCES

We have found the simple empirical approach of Brown and Shannon (1973) to the metal-oxygen distances in our compounds and related ones to give intuitively reasonable results. From about 30 structures for each cation they evaluated constants R_0, S_0 and N in the empirical relation $S = S_0 (R/R_0)^{-N}$ by a least-squares procedure, the criterion being that for a cation surrounded only by oxygen atoms the observed M^{n+}–O distances, R, should yield a sum of the bond strengths, S, in valency units equal to the charge n+. For sodium the values are $S_0 = 0.166$, $R_0 = 2.449$ Å and N = 5.6, for potassium $S_0 = 0.125$, $R_0 = 2.833$ and N = 5.0 and for calcium $S_0 = 0.125$, $R_0 = 2.468$ and N = 6.0. Table V shows the values of S obtained for the complexes whose torsion angles are given in Tables I — IV with some other discrete complexes for comparison.

For sodium compounds the relation works well, ΣS, does not vary systematically with coordination number (the empirical values of R_0 and S_0 could be seen as indicating a normal coordination number of 6 with Na–O = 2.449 Å). For our complexes, the value of S shows that the one water molecule of $[Na(benzo-15-crown-5)H_2O]^+$ has a bond strength equal to the two from the perchlorate anion. The relatively high electrostatic interaction to sodium suggests an enhanced hydrogen-bonding for the water. Another feature of interest is the reasonable value for the complex $[(Na$ o-nitrophenolate$)_2(dibenzo-24-crown-8)]$ having two atoms in the crown ether apparently not coordinated. Further insight is obtained for those systems in which the coordinating atoms are not all oxygen, the ion pairs being to Br in $[NaBr(dibenzo-18-crown-6)H_2O]$ for example, ΣS for oxygen leaves a Na$^+$...Br$^-$ of 0.227 valence units and the distance found, 2.82 Å, is intermediate between 2.983 Å in NaBr (S=0.166) and 2.502 Å in Na$^+$Br$^-$ (gas) (S=1.0).

In the calcium compounds of known dimensions, i.e. Ca(NCS)$_2$(II) solvent, the values of ΣS in Table V show a contribution of 1.1 valence unit from the macrocyclic ether, 0.3 from solvent leaving 0.3 per Ca–NCS contact. There is a subtle balance of ion pairing and solvation. For the larger divalent cations, Ba^{2+} and Sr^{2+}, the complexes with II are of the sandwich type, $[M(II)_2]^{2+}$, and the valency unit effect of the two ligands can be visualised.

For potassium the results are less satisfactory. No simple adjustment of the parameters gives the same value of ΣS for the complexes with dibenzo-30-crown-10, XII, nonactin and valinomycin in which the cation is completely enclosed by 10, 8 and 6 oxygen atoms respectively. For 10 coordination in $[K(II)_2]^+$ a similar effect is obtained which can also be seen as an indication of shorter than expected K–O distances; in the sandwich the shortest O–O contacts are 3.4 Å leading originally to a suggestion that this was a van der Waals' contact with potassium filling the space, not necessarily interacting. The persistence of the 1:2 stoicheiometry whatever the anion, suggests that there is some source of stability.

In the two complexes having potassium-thiocyanate interaction the results are qualitatively reasonable. For 18-crown-6, VIII, $\Sigma S = 0.79$ so that only 0.21 valence unit remain to be shared between the thiocyanate anions along the axial direction; in fact the anions are disordered with the N (or S) distance corresponding to 0.105 valence units at 3.19 Å. In

TABLE VI
Equation and parameters

(1) C.E. $= \Sigma \exp(a - a\, r_{(ij)}/r'_{(ij)})$ (Rees and Smith, 1975)

(2) C.E. $= \Sigma\, B\, r_{ij}^{-9} - A\, r_{ij}^{-6}$ (Hopfinger, 1973)

(3) C.E. $= \Sigma\, B'r_{ij}^{-12} - A'r_{ij}^{-6}$ (Coiro et al., 1974)

(4) C.E. $= \Sigma\, a\, \exp(-b\, r_{ij}) - \dfrac{C\, r_{ij}^{-6}}{r_{ij}^{d}}$

atom		(1)a	(1)r'_{ij}	(2)A	(2)B	(3)A'	(3)B'	(4)a	(4)b	(4)c	(4)d
i	j				($\times 10^3$)		($\times 10^4$)	($\times 10^3$)			
H	H	4.0	2.40	47	0.43	70.0	0.62	6.6	4.08	49.2	0
H	C	4.0	2.70	128	2.08	127.4	3.743	44.8	2.04	125.0	6
H	O	4.0	2.60	124	1.66	123.9	2.503	42.0	2.04	132.7	6
C	C	4.0	3.00	370	9.70	372.5	28.58	301.2	0	327.2	12
C	O	4.0	2.90	367	8.18	367.2	20.52	278.7	0	342.3	12
O	O	4.0	2.80	367	6.88	367.2	14.49	259.0	0	258.0	12
		arbitrary units		K cal/mole		K cal/mole		K cal/mole			

[*TABLE VI (Continued)*]

	(1)	(2)	(3)	(4)
[Na,II,H$_2$O]$^+$I$^-$	10.7	-10.0	-13.6	12.0
[K(II)$_2$]$^+$I$^-$	10.7	- 8.9	- 9.2	15.3
[Ca II(NCS)$_2$MeOH]^o	11.2	-10.2	-13.4	13.9
[Ca II(NCS)$_2$H$_2$O]^o	11.3	-10.3	-13.6	14.1
VII	15.5	-13.7	-16.1	18.5
M VII	14.1	-11.6	-14.2	19.1
VI(F)	20.7	-20.3	-25.5	20.1
[Cs VI(F)NCS]$_2$	18.0	-16.0	-19.2	20.4
Cs(VI(G))$_2$	13.1	-12.2	-13.8	13.5
VIII	12.5	-12.6	-18.5	12.0
Na(VIII)H$_2$O	12.9	-13.3	-19.8	11.3
K(VIII)	11.2	-11.0	-16.8	13.1
[Cs(VIII)NCS]$_2$	11.4	-11.1	-17.0	12.9
X	19.6	-17.8	-22.5	22.0
(K NCS)$_2$X	19.2	-14.9	-17.2	30.4
(Na XI)$_2$X	18.4	-13.7	-15.4	33.5
XII	26.1	-25.0	-32.8	28.0
[K(XII)]$^+$	27.5	-31.2	-41.8	14.5

the complex of SCN⟍K⟋NCS with dibenzo-24-crown-8, X, the oxygens provide

$\Sigma S = 0.623$ per potassium leaving a mean of 0.188 per K-N contact of mean 2.87 Å.

7. MOLECULAR MECHANICS

The macrocyclic ethers seem suitable for treament by molecular mechanics accepting the restriction that results for only one kind of molecule in different conformations may be comparable. From the wealth of equations and parameters available we have chosen four, one allowing for repulsive interactions only (1); two modifications of the Lennard-Jones potential (2) and (3); and one more general form (4). The last approximates to the Lennard-Jones 6-12 for bonds excluding hydrogen. As this is the atom most frequently involved in non-bonded interactions, it is unfortunate that there are doubts about the correct potentials and even a suggestion (Hopfinger, 1973) that different values are required according to the type of atoms to which the hydrogen is bonded. The equations and parameters used are set out in Table VI together with the total CE's for the compounds with torsion angles in Tables I-IV (Owen, 1976).
    Equation (1) allows only for repulsion and Equation (4) with the given parameters results in net repulsion. Equations (2) and (3) give a net attraction. Equations (2), (3) and (4) give the same relative conformational energies except for the caesium complexes of VI, isomers F and G, for which Equations (2) and (3) show a more favourable value for isomer F and Equation (4) a more favourable one for isomer G. Examination of the values for each interaction confirm the basic problem of this approach e.g. one O...O in VIII gives -0.165 and +0.285 kcal mole-1 for Equations (3) and (4) respectively. A problem with some of the complexes considered here is that although the bond lengths and angles are not significantly different as determined experimentally, some non-bonded distances may differ by enough to give 0.5 to 1 kcal mole-1 difference in interaction energy; this effect is manifest by the catechol oxygen atoms which apparently range from 2.52 Å to 2.60 Å and give a particularly large repulsive term in Equation (4). (Obviously these can be omitted but they serve to illustrate the point).
    Having made the reservations we can consider the useful implications. Firstly, there does appear to be some penalty in conformational energy in the sandwich form of benzo-15-crown-5, II. If the empirical relation of Brown and Shannon is regarded as a term indicating electrostatic attraction between the cation and the oxygen atoms it is more favourable in the 1:1 sodium complex (0.96 units) than in the 1:2 potassium complex (0.60 units per molecule of II). Another explanation must be sought and it seems probable that this is in the cation...hydrogen repulsions. In the [Na II H$_2$O]$^+$ ion the conformation gives four C-H bonds pointing approximately parallel to the normal to the plane of the five oxygen atoms, along which lies Na$^+$ at 0.75 Å and H$_2$O at a further 2.285 Å; the Na...H distances are 3.15, 3.04, 3.25 and 3.05 Å compared with a sum of the van der Waals' radii of 3.0 Å . The mean Na-O (ether) is 2.39 Å;

if a potassium ion were situated far enough along the normal to give an average K—O (ether) distance of 2.75 Å the K...H interactions would be at 2.94, 3.17, 3.06 and 3.27 Å compared with a sum of the van der Waals' radii of 3.35 Å.

For the 18-crown-6 rings the 1:1 or 1:2 stoicheiometry shown by caesium compounds cannot be attributed to an analogous effect. When these rings are in the complexing conformation with all oxygen atoms equidistant from a point, the minimum distance of a hydrogen atom to the normal through that point is approximately equal to the sum of the van der Waals' radii of caesium and hydrogen, 3.7 Å. Nor do possible permutations of the methyl group positions in VI(F) and VI(G) explain the slightly different conformations found in the 1:1 and 1:2 complexes. (The conformational energy calculations are incomplete because some hydrogen atoms on methyl groups could not be located (Mallinson, 1975) but such hydrogen atoms are unlikely to be restrained by a close environment). Table VI shows that the unsubstituted ring, VIII, has a more favourable conformational energy in the unsymmetrical form, which partly wraps the sodium ion, than in the one with approximately $\bar{3}$ m symmetry in the potassium and caesium complexes.

When the crown ether completely wraps around the cation, as in K XII, there is a gain of conformational energy on complex formation. The binuclear complexes of X show losses of conformational energy; at present we are investigating 1:1 complexes of X and tentatively predict a configuration corresponding to a gain in conformational energy.

ACKNOWLEDGEMENT

I am grateful to my colleagues, particular Dr. J.D. Owen, for some of their unpublished results.

REFERENCES

1. Bright, D., Milburn, G.H.W., and Truter, M.R.: 1971, J. Chem. Soc. (A), 1582.
2. Brown, I.D. and Shannon, R.D.: 1973, Acta Cryst. A29, 266—282.
3. Bush, M.A. and Truter, M.R.: 1971, J. Chem. Soc. (B), 1440.
4. Bush, M.A. and Truter, M.R.: 1972a, J. Chem. Soc. (Perkin II), 341.
5. Bush, M.A. and Truter, M.R.: 1972b, J. Chem. Soc. (Perkin II), 345.
6. Coiro, V.M., Mazza, F., and Mignucci, G.: 1974, Acta Cryst. B30, 2607.
7. Dobler, M., Dunitz, J.D., Kilbourn, B.T.: 1969. Helv. Chim. Acta 52, 2573—2583.
8. Dobler, M., Dunitz, J.D., and Seiler, P.: 1974, Acta Cryst. B30, 2741—2743.
9. Dobler, M. and Phizackerley, R.P.: 1974a, Acta Cryst. B30, 2748—2750.
10. Dobler, M. and Phizackerley, R.P.: 1974b, Helv. Chim. Acta 57, 664—674.
11. Dunitz, J.D. and Seiler, P.: 1974, Acta Cryst. B30, 2570.
12. Fenton, D.E.: 1973, J. Chem. Soc. (Dalton), 1380.

13. Hanson, I.R., Hughes, D.L., and Truter, M.R.: 1976, J. Chem. Soc. (Perkin II), in press.
14. Hopfinger, A.J.: 1973, Conformational Properties of Macromolecules, Academic Press, New York, p. 47.
15. Hughes, D.L.: 1975, J. Chem. Soc. (Perkin II), 2374
16. Hughes, D.L., Phillips, S.E.V., and Truter, M.R.: 1974, J. Chem. Soc (Dalton), 907.
17. Janes, N.F.: 1975, personal communication.
18. Mallinson, P.R.: 1975a, J. Chem. Soc. (Perkin II), 261.
19. Mallinson, P.R.: 1975b, J. Chem. Soc. (Perkin II), 266.
20. Mallinson, P.R. and Truter, M.R.: 1972, J. Chem. Soc. (Perkin II), 1818.
21. Mercer, M. and Truter, M.R.: 1973a, J. Chem. Soc. (Dalton), 2215.
22. Mercer, M. and Truter, M.R.: 1973b, J. Chem. Soc. (Dalton), 2469.
23. Milburn, H., Truter, M.R., and Vickery, B.L.: 1974, J. Chem. Soc. (Dalton), 841—846.
24. Neupert-Laves, K. and Dobler, M.: 1975. Helv. Chim. Acta 58, 432—442.
25. Owen, J.D. and Wingfield, J.N.: 1976, J. Chem. Soc. (Chem. Comm.), in press.
26. Parsons, D.G.: 1975a, personal communication.
27. Parsons, D.G.: 1975b, J. Chem. Soc. (Perkin I), 245.
28. Parsons, D.G., Truter, M.R., and Wingfield, J.N.: 1975, Inorg. Chim. Acta 14, 45.
29. Parsons, D.G. and Wingfield, J.N.: 1976a, Inorg. Chim. Acta 17, 225.
30. Pedersen, C.J.: 1967, J. Amer. Chem. Soc. 89, 7017.
31. Pedersen, C.J.: 1970, J. Amer. Chem. Soc. 92, 386.
32. Phillips, S.E.V.: 1976, personal communication.
33. Phillips, S.E.V. and Truter, M.R.: 1975a, J. Chem. Soc. (Dalton), 1071.
34a. Phillips, S.E.V., and Truter, M.R.:1975b, J. Chem. Soc. (Dalton), 1066.
34b. Poonia, N.S. and Truter, M.R.: 1973, J. Chem. Soc. (Dalton), 1062.
35. Rees, D.A. and Smith, P.J.C.: 1975, J. Chem. Soc. (Perkin II), 830.
36. Seiler, P., Dobler, M., and Dunitz, J.D.: 1974, Acta Cryst. B30, 2744-48.
37. Truter, M.R.: 1973, Structure and Bonding 16, 111.

DISCUSSION

SIMON: Do you have indications for enantiomer selectivity of the chiral crown compounds you studied?

TRUTER: We have tried to separate the enantiomers by reaction with potassium D-mandelate. This was not succesful. We have not made a determined attack by other methods of resolution or by asymmetric synthesis.

A. PULLMAN: It is probably interesting to mention that a theoretical ab initio computation of the capture of an ion by a model crown compound has been performed in our laboratory (A. Pullman, C. Giessner Prettre, and Yu. V. Kruglyak, Chem. Phys. Letters 35, 156, 1975). Due to limitations in the size of our program, we choose 12 crown-4 and the smallest alkali ion Li$^+$. We performed first an STO 3G SCF study of the most sta-

maxidentate   alternate

ble conformation of two forms of the empty crown (assuming an axis of
symmetry): the *maxidentate* where all the oxygens point on the same side
of the mean plane of the carbons, and the *alternate*, where two successive
oxygens of the ring alternatively above and below the plane (see figure).
We found that the most stable form of the maxidentate is obtained for
a conformation where all methylene couples are *staggered*; opening further
the structure decreases the stability; closing it until all methylene
groups are eclipsed reduces the stability by 24 kcal/mole. Similarly
the most stable structure of the alternate form was obtained in a
nearly staggered conformation (this time 10° away from it).

In a second step we approached a Li^{+} cation axially towards the
oxygens of the maxidentate in its most stable form and allowed the
crown to open-up or fold down. The table gives the essential results
obtained: it shows that complexing of Li^{+} with the most stable form of
the empty maxidentate is relatively unfavorable: in order to achieve
better binding the polyether closes its structure; although this involves
a loss in stability, this loss is overcompensated by the gain in binding
energy to the cation achieved when each COC group comes closer to a
coplanar arrangement with the ion. The best compromise is achieved for
$\tau = 30^{\circ}$. Beyond this value the $Li^{+}O$ distance becomes too small and the
exchange repulsion dominates. As a result of this situation the cation
takes up an external position, out of the plane of the oxygen atoms,
about 0.5 Å from it.

A similar study has shown that when the cation is put in the center
of the alternate comformer of the crown, the cage closes gradually on
it until a balance is reached between the loss in internal conformational
energy and the gain in $Li^{+}O$ binding.

TABLE
Variation of the binding energies, $- \Delta E$ (kcal/mole) of Li^{+} to
the maxidentate form of 12 crown 4, upon variation of the
torsion angle around the CC bonds τ (degrees) and of the
distance d (Å) from the cation to the plane of the oxygen atoms.
$\tau = 60^{\circ}$ is the most stable empty form of the crown

$\tau_{O_4 O_3 O_2 O_1}$	d = 0	d = 0.4	d = 0.8
0	217.9	218.0	209.8
15	217.9	218.7	210.8
30	207.7	220.1	213.4
45	202.2	216.8	212.3
60	181.6	198.4	

CATION SELECTIVITIES SHOWN BY CYCLIC POLYETHERS AND THEIR THIA DERIVATIVES*

R.M. IZATT, L.D. HANSEN, D.J. EATOUGH, J.S. BRADSHAW, and
J.J. CHRISTENSEN
*Depts. of Chemistry and Chemical Engineering, Brigham Young
University, Provo, Utah 84602, U.S.A.*

1. INTRODUCTION

The first multidentate macrocyclic compound with proven ability to form
stable alkali metal complexes was prepared in 1962 by Pedersen (1967,
1977). This compound, 2,3,11,12-dibenzo-1,4,7,10,13,16-hexaoxacyclooc-
tadeca-2,11-diene, was nicknamed dibenzo-18-crown-6 (Figure 1:IIb) by
its discoverer, and was the precursor of several hundred synthetic macrocyclic compounds of the cyclic polyether type which have since been prepared (Christensen et al., 1974; Bradshaw, 1977). The nomenclature here
is taken from that proposed by Pedersen (1967). The original fibrous
crystals of dibenzo-18-crown-6 were relatively insoluble in methanol,
but the solubility was markedly increased in the presence of NaOH, or
any soluble sodium salt. This result was unexpected; but following the
further discovery that the compound was cyclic, it was deduced that
its solubility in methanol was related to its complexation with Na^+. The
ability to complex alkali metal cations was the first of many interesting
properties these compounds have been found to have. Their novel reactions
with cations have led to their use in a steadily increasing number of
interesting ways (Ovchinnikov et al., 1974; Izatt et al., 1973; Pedersen
and Frensdorff, 1972) including the following: ion detection in devices
such as cation selective electrodes; phase transfer catalysis; selective
complexation of given cations in studies of reaction mechanisms; elucidation of biological transport mechanisms; study of solvation and ion
pairing effects; solubilization of salts in solvents of low polarity by
complexation of the cation and formation of an ion-pair leading to new
or improved organic synthesis and increased reaction yields; separation
of different cations from one another; separation of Ca isotopes; development of novel electrochemical cells; development of carrier membrane
systems with possible application in the removal of heavy metal ions
from wastes; and use as drugs to remove cations from or introduce them
into living systems. In addition to these applications involving metallic
cations, Cram and his co-workers (Cram and Cram, 1974; Helgeson et al.,

* Contribution No. 86 from the Center for Thermochemical Studies, Brigham
Young University, Provo, Utah 84602, U.S.A.

*B. Pullman and N. Goldblum (eds.), Metal-Ligand Interactions in Organic
Chemistry and Biochemistry, part 1 , 337-361. All Rights Reserved.*

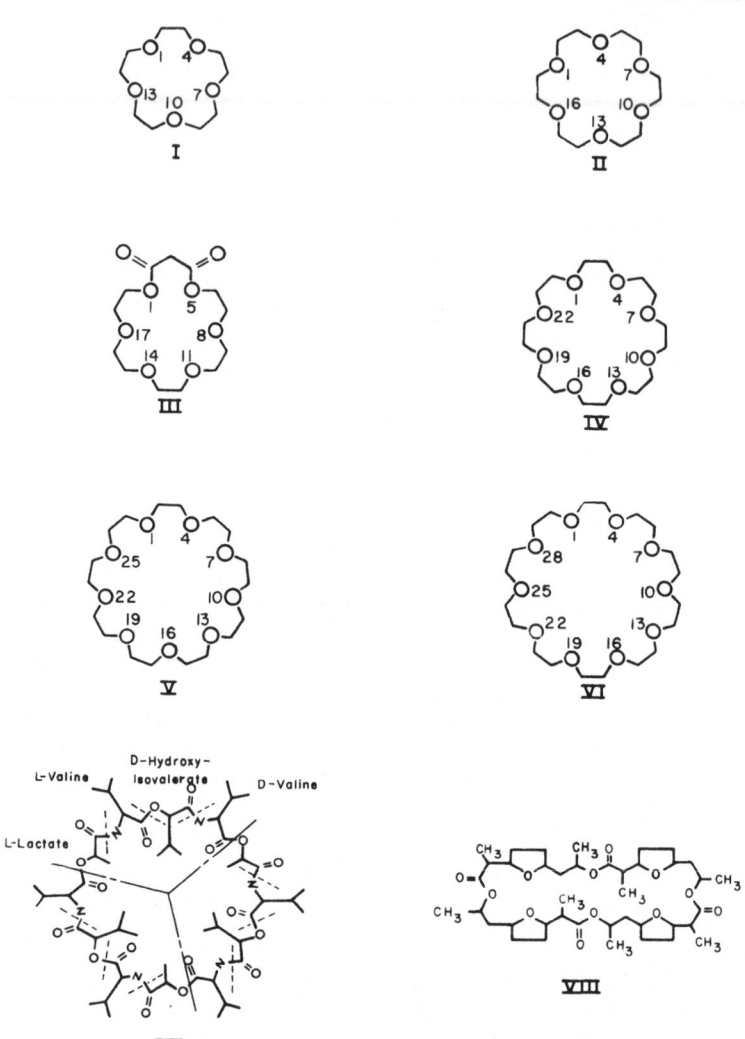

Fig. 1. Representative macrocyclic compounds.
Structure of parent compound only shown. I(a) 15-crown-5,
(b) 2,3-benzo-15-crown-5. II(a) 18-crown-6, (b) 2,3 and 11,12-
dibenzo-18-crown-6, (c) 2,3 and 11,12-dicyclohexo-18-crown-6,
(d) 1,4-dithia-18-crown-6, (e) 1,10-dithia-18-crown-6.
III(a) 19-crown-6-malonate, IV(a) 24-crown-8, (b) 2,3 and 1.'
15-dibenzo-24-crown-8, (c) 2,3 and 14,15-dicyclohexo-24-crown·
(d) 1,4-dithia-24-crown-8, (e) 1,13-dithia-24-crown-8. V(a) 2,'
and 14,15-dibenzo-27-crown-9. VI(a) 2,3 and 17-18-dibenzo-30
crown-10. VII valinomycin. VIII nonactin.

1974) have been instrumental in developing the area of host-guest chemistry which involves complexes of organic cations, and have demonstrated
the feasibility of using macrocyclic compounds for chiral recognition.
It is evident from these examples that differentiation among cations,

organic as well as inorganic, is of interest in many industrial and academic areas.

The design of ligands having cation selective properties has been a goal of many analytical and coordination chemists, but success in this endeavor has been limited. Early designs of complexing agents usually involved polydentate, non-macrocyclic ligands, which generally showed only small differences in log K within a given series of cations (Mellor, 1964). It has been known for some time that biological systems are remarkably effective both in concentrating cations, often against large concentration gradients, and in using cations selectively in specific enzymes and proteins (Williams, 1970). The discovery that cyclic antibiotics of the valinomycin and actin classes selectively complex certain alkali metal cations opened a new horizon of investigation in the field of selective cation complexation. The cyclic depsipeptide valinomycin was first isolated in 1955 from extracts of *Streptomyces fulvissimus*. Interest in valinomycin (Figure 1, VII) and related macrocyclic antibiotics has resulted in a large number of investigations which have led to the elucidation of their structures in several solvents, the kinetics of their metal complexation reactions, and the determination of stability orders for their reactions with metal ions (Ovchinnikov, 1974). The goal of selectivity among closely similar cations was realized in the reaction of valinomycin with K^+ and Na^+, the stability of the K^+ complex being more than 10 000 times that of the Na^+ complex (Grell et al., 1975). These cation selectivity characteristics led to studies of transport across lipid bilayer membranes. The results of these investigations together with efforts to explain cation transport across membranes on a molecular level have been summarized (Ovchinnikov, 1974; Laprade et al., 1975; Eisenman and Krasne, 1975).

The similarities between certain of the cyclic polyethers and the cyclic antibiotics in structure, cation complexing properties, and as mediators of cation transport across membranes have attracted the attention of many investigators. Of particular importance is the observation that one of these, dibenzo-18-crown-6, shows cation selectivity properties toward K^+, Rb^+ and Cs^+ in CH_3OH which are very similar to those of valinomycin (Ovchinnikov, 1974; Grell et al., 1975). Although these two compounds may have similar ring cavity sizes in the complexed form, they differ in several important respects making their similar complexing behavior the more remarkable. First, the possible conformational changes upon complexation are relatively small with an 18-crown-6 ligand, but large with valinomycin. Second, the donor groups involved in the complexation are different, ether oxygens in the case of 18-crown-6 and ester carbonyl oxygens in the case of valinomycin. Third, the cation and ligand donor atoms lie nearly in a plane in the case of 18-crown-6, but have nearly perfect O_h symmetry in the case of valinomycin. Fourth, stabilization of the complex by hydrogen bonding occurs with valinomycin, but not with 18-crown-6. Finally, the log K value for formation of the dibenzo-18-crown-6 complex with Na^+ in CH_3OH is nearly as large as that for the K^+ complex, 4.36 compared to 5.00 (Ovchinnikov, 1974), whereas a large difference is seen in the log K values for formation of the valinomycin complexes with these two cations, 0.67 compared to 4.90 (Ovchinnikov, 1974). The recent synthesis of (19-crown-6) malonate

(Bradshaw et al., 1975) (Figure 1:III) and related compounds should make possible closer comparisons between the cation complexing properties of synthetic macrocyclic compounds and cyclic antibiotics. A large number and variety of macrocyclic compounds of the cyclic polyether type have been synthesized (Christensen et al., 1974; Bradshaw, 1977), and the availability of these compounds has allowed detailed investigations to be made of several of their structures in both the complexed and un-complexed forms as well as the determination of log K values associated with the formation of their complexes.

Attempts have been made to identify the underlying bases of equilib-rium cation selectivity by neutral macrocyclic molecules (Simon et al., 1973; Eisenman et al., 1973). Such treatments are extremely valuable in identifying and quantitatively evaluating the parameters which collec-tively constitute the ΔG, ΔH, and ΔS values for complex formation. Suc-cess in this endeavor should provide valuable insight into cation com-plexation by macrocyclic compounds and could serve as a predictive device for the sythesis of compounds having desired metal complexing properties. The similarity in cation selective complexing behavior between ligands containing the 18-crown-6 cavity and valinomycin suggest that if one could adequately account for the energies involved in cation complexation with this less complicated molecule, a model to study cation selective behavior in progressively larger molecules with greater conformational and donor atom variability could be produced.

In the following sections pertinent thermodynamic and structural data which have given insight into the cation complexing properties of small ring macrocyclic compounds are presented and discussed. Those ligand, cation and solvent parameters which are known to affect the cation selectivities of macrocyclic compounds and the stabilities of their complexes will be discussed with particular emphasis on results recently obtained in our laboratory. We also summarize a relationship whose derivation is given elsewhere (Hansen et al.,to be submitted) betwee ΔG for complex formation and the component ΔG values from which it is derived for an 18-crown-6 type ligand. This relationship predicts the magnitudes of ΔG as a function of cation crystal radius for the reaction of uni- and bivalent cations with an 18-crown-6 type ligand. In addition, the plot of log K vs. cation radius generated from the equation repro-duces several distinctive features of the experimental plot for three 18-crown-6 type ligands.

2. MATERIALS AND METHODS

The macrocyclic compounds were either synthesized by us or purchased from commercial sources (15-crown-5, 18-crown-6). Equilibrium constants for the systems studied here were determined using a calorimetric titra-tion procedure which has proven particularly useful for studies involving neutral ligands since in these systems the concentration of the free ligand is not H^+ dependent (Izatt et al., 1971) thus eliminating several common methods for equilibrium constant determination. In addition, the calorimetric titration procedure is useful for solvents other than water since the experimental quantity measured is a general one and does not re-

quire definition in each solvent used. This procedure is directly applicable to systems where the equilibrium constant, K, is approximately
10 000 or less and ΔH for the calorimeter reaction is measurable different from zero (Christensen et al., 1972). The micro isoperibol and isothermal calorimeters used (Tronac, Inc.) have been described (Hansen et
al., 1974a, 1974b, 1975; Izatt et al., 1974; Christensen et al., 1972;
Eatough et al., 1972b, 1972c, 1975).

3. CYCLIC POLYETHERS

3.1. *Effect on Complex Stability of Cation Diameter: Ligand Diameter Ratio*

A correlation between complex stability and the ratio of metal ion to
cyclic polyether cavity diameters was suggested by Pederson (1967) and
has since been confirmed by extensive equilibrium constant measurements
(Izatt et al., submitted, a, b, c). Estimated ligand cavity and metal
ion diameters are given in Table I

Table I
Cation ionic and polyether cavity diameters

Cation	Ionic diameter, Å [a]	Cation	Ionic diameter, Å [b]
Li^+	1.36	Ag^+	2.52
Na^+	1.94	K^+	2.66
Ca^{2+}	1.98	Ba^{2+}	2.68
Hg^{2+}	2.20	Rb^+	2.94
Sr^{2+}	2.24	Tl^+	2.94
Pb^+	2.40	NH_4^+	2.96
		Cs^+	3.34

Polyether	Cavity diameter, Å [c]	Polyether	Cavity diameter, Å [c]
12-crown-4	1.2 — 1.5	18-crown-6	2.6 — 3.2
15-crown-5	1.7 — 2.2	21-crown-7	3.4 — 4.3

[a] Pauling-type diameters (univalent set) from Ahrens (1952).

[b] From Pauling (1960).

[c] From Pedersen (1977) based on Corey-Pauling-Kolton (1st number) and
Fisher-Hirschfelder-Taylor (2nd number) atomic models.

We have recently completed calorimetric titration studies of the
interaction of Na^+, K^+, Rb^+, and Cs^+ with 15-crown-5, 18-crown-6, and
24-crown-8 ligands in both aqueous (Izatt et al., submitted, b) and
70% CH_3OH solvents (Izatt et al., submitted, c), and of Ca^{2+}, Sr^{2+}, Ba^{2+}
and Pb^{2+} with these ligands in aqueous solution (Izatt et al., submitted,

b) and with 18-crown-6 in 70% CH_3OH (Izatt et al., submitted, c). These data give the log K variation with the ratio of cation size to ligand cavity size in these solvents and the relative contribution of the ΔH and ΔS to its magnitude. It is recognized in the following discussion that we presently do not have enough information to differentiate between bonding and solvent contributions to ΔH, or between solvent and conformational change contributions to ΔS. Nevertheless, the ΔH and ΔS data give meaningful hints, when coupled with other experimental results, concerning the processes occuring during complexation.

In Figures 2 and 3 are plotted log K values for the reactions in aqueous and 70 wt.% CH_3OH solution, respectively, of uni- and bivalent metal ions with ligands of increasing cavity diameter. The ligand with

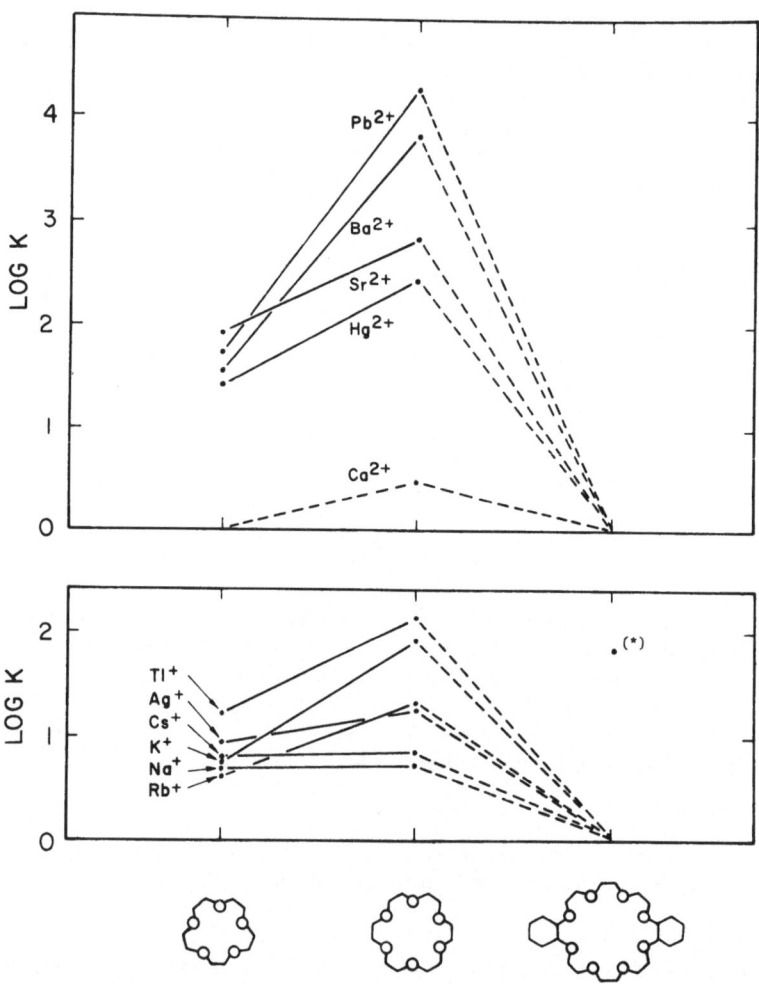

Fig. 2. Comparison of log K values for the reaction of selected cations with ligands of increasing cavity size. T = 25°, solvent = H_2O. *Value from Frensdorff (1971).

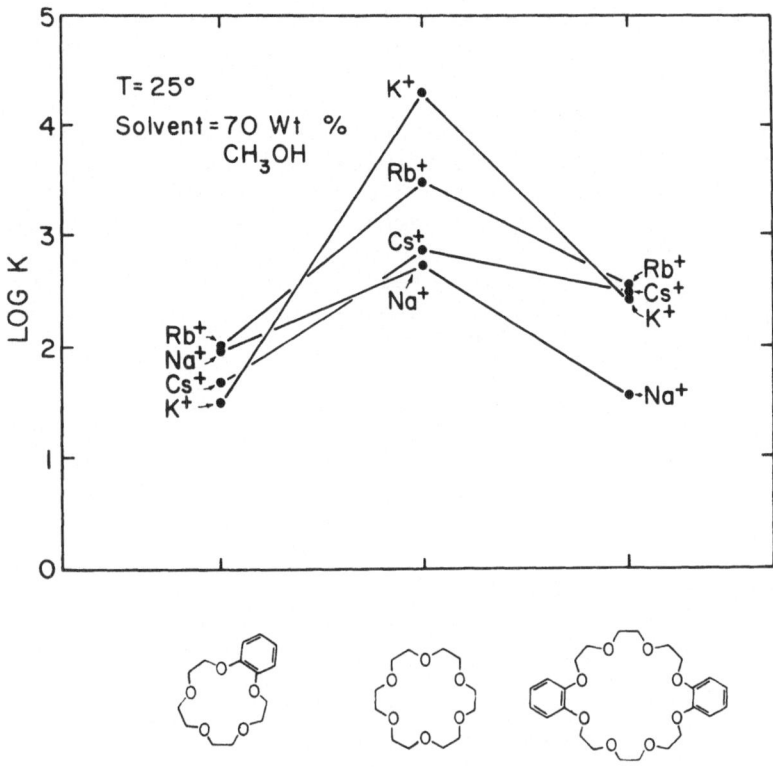

Fig. 3. Comparison of log K values for the reaction of selected cations with ligands of increasing cavity size. T = 25°, solvent = 70 wt.% CH$_3$OH.

the intermediate size cavity, 18-crown-6, is seen to form the most stable complexes in both solvents with all cations studied. Also, the ability to differentiate among cations increases in the order dicyclohexo-24-crown-8 < 15-crown-5 < 18-crown-6 in aqueous solution, but 15-crown-5 < dicyclohexo-24-crown-8 < 18-crown-6 (except for Na$^+$) in 70 wt.% CH$_3$OH. The data in Table I indicate that the cavity in 15-crown-5 should accommodate Na$^+$ but not the larger alkali metal cations. It is probable that the relatively large hydration energy of Na$^+$ is the basis for weak complexation in aqueous solution by 15-crown-5. The absence of complexation in aqueous solution of any of the metal ions with dicyclohexo-24-crown-8 is puzzling although comparison of the log K data in Figures 2 and 3 suggests that little interaction of alkali metal ions with the 24-crown-8 ligand would be expected. On the basis of the interaction in aqueous solution between cations and 18-crown-6 where ΔH values are quite large: K$^+$, -6.21; Cs$^+$, -3.79; and Ba^{2+}, -7.58 kcal mole^{-1} (Izatt et al., submitted, b), we would expect significant ΔH values in the case of a 24-crown-8 ligand particularly for Cs$^+$ which should have a favorable size to fit in the ligand cavity (Table I). Also, ΔH for the reaction of Cs$^+$ with dibenzo-24-crown-8 in 70 wt.% CH$_3$OH is -8.93 kcal mole^{-1} (Izatt et al., submitted, c). The measured heat for the reaction of each cation in Figure 2 and NH_4^+ with dicyclohexo-24-crown-8 was identical within experimental uncertainty to the heat of dilution for each metal ion titrant. From these results, we suspect that no reaction occured with any of the cations studied since it is highly unlikely that all ΔH values would be zero. To test this suspicion, we titrated 18-crown-6 into 1:1 aqueous mixtures of K^+, Cs^+, and Ba^{2+} with dicyclohexo-24-crown-8. There was no difference within experimental uncertainty between the thermograms for these reactions and those for the corresponding reactions of 18-crown-6 with these cations and the calculated log K values were nearly identical. We conclude from the above that the reaction between these cations and 24-crown-8 in aqueous solution is very

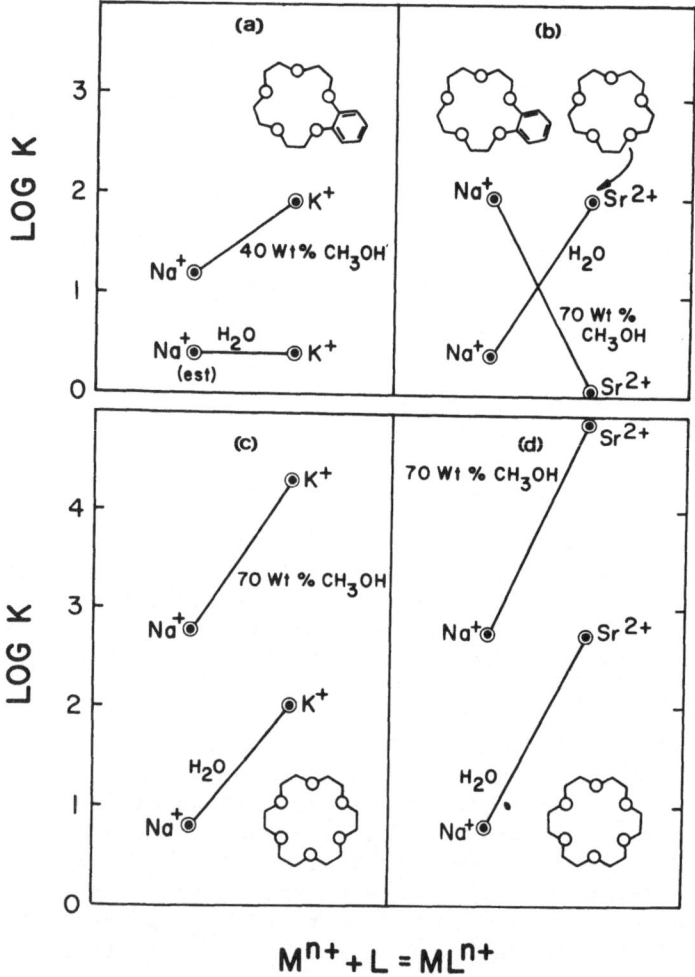

$$M^{n+} + L = ML^{n+}$$

Fig. 4. Comparison of the change in log K values for the reaction of selected cation pairs with the indicated ligands in water and CH_3OH solvents.

small or nonexistent. The only other study of metal ion complexation by this ligand in aqueous solution is that of Frensdorff (1971) who reported a log K value of 1.9 based on selective ion electrode data for its reaction with Cs^+. Except in this instance, agreement between our calorimetric and his selective ion electrode log K values is good to excellent, e.g., Ag^+-18-crown-6, our log K = 1.50 (Izatt et al., submitted, b) that of Frensdorff (1971) = 1.60).

Thermograms show no evidence of a reaction between Ca^{2+}, Sr^{2+}, or Ba^{2+} and benzo-15-crown-5 in 70 wt.% CH_3OH (Izatt et al., submitted, c). This result is interesting since these cations form stable complexes with 18-crown-6 in this solvent (Izatt et al., submitted, c) and alkali metal ions form stable 15-crown-5 complexes in CH_3OH solvents. In Figure 4 are plotted log K values valid in H_2O and H_2O-CH_3OH solvents for several cation pairs with 18-crown-6 and benzo-15-crown-5 (15-crown-5 in the case of Sr^{2+}). A regular increase is seen for either the Na^+, K^+, or Na^+, Sr^{2+} pair in the case of 18-crown-6 (Figure 4c and 4d) but for benzo-15-crown-5 differentiation between either K^+ and Na^+ (Figure 4a) or Sr^{2+} and Na^+ (Figure 4b) is enhanced in the indicated CH_3OH solvent. Thus, a change in solvent results in a marked change in ligand selectivity for these cations. This is especially true of the Na^+/Sr^{2+} case where selectivity of Na^+ over Sr^{2+} (or other alkaline Earth cations) is markedly enhanced by increasing the fraction of the CH_3OH component of the solvent. These results may have relevance for biological systems where cation transport processes involve ion transport between water-like and lipid-like regions.

The ΔH and ΔS values for formation of these complexes (Tables II and III) give added information concerning the bonding. In Table II, log K is seen to increase and reach a maximum at K^+ for the reaction of the alkali metal ions with 18-crown-6. The log K values for these reactions are directly proportional to the corresponding -ΔH values which also rise to a maximum at K^+, and are indirectly proportional to the corresponding TΔS values which are most unfavorable with K^+.

TABLE II
Log K, ΔH, and ΔS values for formation of 1:1 alkali cation-18-crown-6 complexes at 25° in aqueous solution

M^+	Log K	ΔH (kcal/mole)	TΔS (kcal/mole)
Na^+	0.80	−2.25	−1.11
K^+	2.03	−6.21	−3.42
Rb^+	1.52	−3.86	−1.71
Cs^+	0.99	−3.79	−2.43

The data in Table III show the effect on the thermodynamic quantities for Na^+ and K^+ complexation in 70 wt.% CH_3OH as ring size increases. The lowered stability of the crown-8 and crown-9 complexes results from greatly increased unfavorable entropy changes in these cases. Manipulation of space filling models shows that one can fold crown-8 and larger

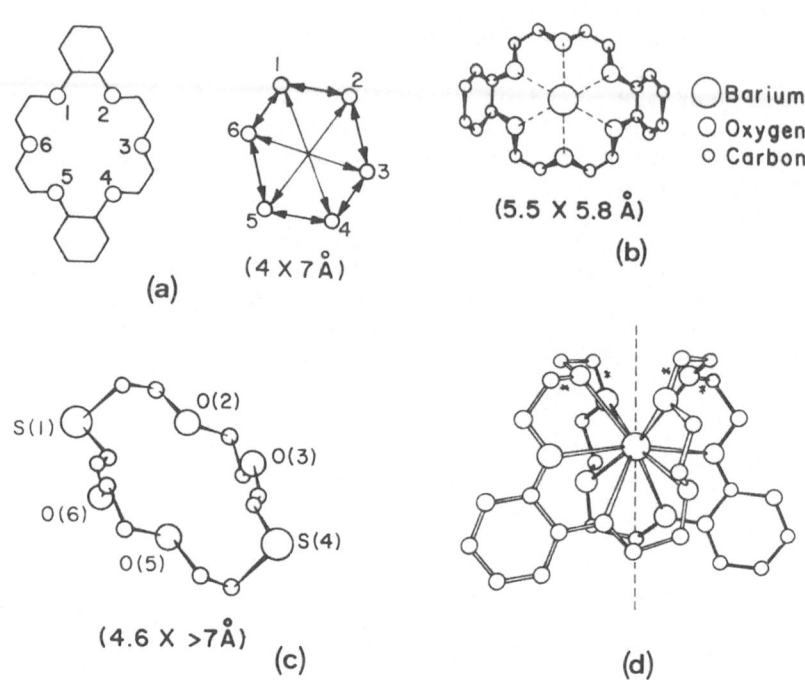

Fig. 5.   Crystal structures of several macrocyclic compounds
and their metal complexes. (a) Uncomplexed cis-syn-cis isomer
of dicyclohexo-18-crown-6, (Dalley et al., 1975a). (b) Ba²⁺
complex of cis-syn-cis isomer of dicyclohexo-18-crown-6,
(Dalley et al., 1972). (c) Uncomplexed 1,10-dithia-18-crown-6,
(Dalley et al., 1975b). (d) The potassium iodide-dibenzo-30-
crown-10 complex, (Truter, 1973). Asterisks mark bonds which
have different torsion angles in the uncomplexed molecules.

rings in a manner similar to that observed with valinomycin while this
type of conformation change is not possible with an 18-crown-6 ring.
Complexation of Ba²⁺ by dicyclohexo-18-crown-6 (Figure 5: a and b) involves an adjustment of the ring from elliptical (4 × 7 Å) (Dalley et al.
1975a) to nearly circular (5.5 × 5.8 Å) (Dalley et al., 1972) while reaction of 30-crown-10 with K⁺ (Figure 5:d) results in the ligand wrapping
around the cation in a folded type structure (Truter, 1973) Kinetic data
(Liesegang et al., in press) confirm a fast conformational rearrangement
of 18-crown-6 immediately prior to its slower complexation with K⁺. It
is probable that the large unfavorable TΔS values associated with the
large ring systems are a result of sizeable ring conformation changes.
It is expected that the smaller Na⁺ would have a greater ability to or-
ganize the ligand than K⁺, and the ΔH and ΔS data for the two larger
ring compounds support this contention. Also, the large change in the
TΔS values for Na⁺ between 18-crown-6 and dibenzo-24-crown-8 is consistent with the differing capacities of these ligands to undergo conformational change. These large negative TΔS values more than offset the
favorable ΔH changes causing the log K values for the two larger ring

TABLE III
Log K, ΔH, and ΔS values for the reaction:
$M^+ + L = ML^+$ in 70wt.% CH_3OH at 25°

M^+	L	Log K	ΔH (kcal/mole)	TΔS (kcal/mole)
Na^+	Benzo-15-crown-5	1.99	-3.82	-1.11
Na^+	18-crown-6	2.76	-4.89	-1.14
Na^+	Dibenzo-24-crown-8	1.54	-7.75	-5.67
Na^+	Dibenzo-27-crown-9	1.50	-11.74	-9.75
K^+	Benzo-15-crown-5	1.50		
K^+	18-crown-6	4.33	-9.68	-3.81
K^+	Dibenzo-24-crown-8	2.42	-8.54	-5.28
K^+	Dibenzo-27-crown-9	2.86	-9.50	-5.64

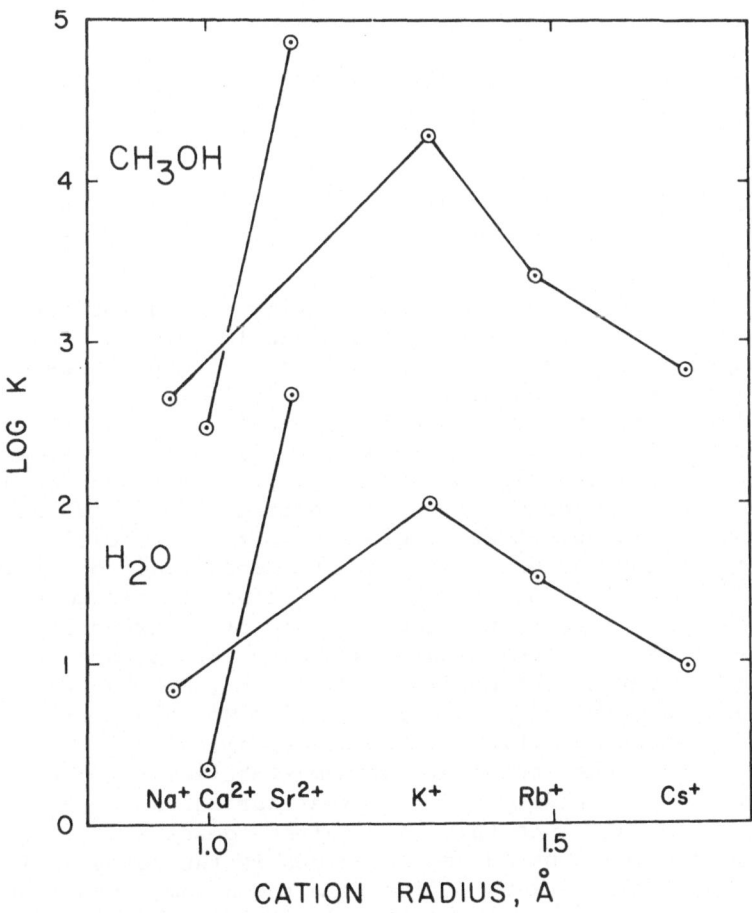

Fig. 6. Plot of log K for the reaction of several univalent
and bivalent metal ions with 18-crown-6 in aqueous and 70 wt.%
CH_3OH solvents vs. cation radius.

compounds to be smaller than that in the case of 18-crown-6.

The effect of solvent on log K values for the reaction of several alkali and alkaline earth metal ions with 18-crown-6 is given in Figure 6. Lowering the dielectric constant of the solvent results in log K increasing by a nearly constant amount for each metal ion studied. As seen in Table IV, for Na^+, K^+, and Sr^{2+}, the stability increase between H_2O and 70 wt.% CH_3OH is entirely a ΔH effect while for Rb^+ and Cs^+ a compensation is found with ΔH being dominant in each case.

TABLE IV
Log K, ΔH, and ΔS values for the reaction $M^+ + L = ML^+$ (L=18-crown-6) in H_2O and 70 wt.% CH_3OH. T=25°

M^+	Log K		ΔH(kcal/mol)		$T\Delta S$(kcal/mol)	
	H_2O	70%CH_3OH	H_2O	70%CH_3OH	H_2O	70%CH_3OH
Na^+	0.80	2.76	−2.25	−4.89	−1.11	−1.14
K^+	2.03	4.33	−6.21	−9.68	−3.42	−3.86
Rb^+	1.52	3.46	−3.86	−9.27	−1.71	−4.59
Cs^+	0.99	2.84	−3.79	−8.09	−2.43	−4.23
Ca^{2+}	<0.5	2.51		−4.27		−0.84
Sr^{2+}	2.72	5.0	−3.61	−7.49	+0.09	−0.75

Log K values for 1:1 reactions of several uni-and bivalent metal ions with 18-crown-6 and two isomers of dicyclohexo-18-crown-6 are plotted vs. cation radius in Figure 7. Tl^+ and Ag^+ fall off the lines connecting the alkali metal cations, and more will be said of this in the following section. The log K values for formation of metal complexes of the cis-syn-cis isomer are larger than those for the cis-anti-cis isomer for all cations studied. Possible origins of this difference have been discussed (Izatt, 1971). Two features of this plot pertinent to the discussion in section 5 should be noted. First, regions of optimum stability for uni- and bivalent metal ions occur at or near the ionic radii of K^+ and Pb^{2+}, respectively. Second, for cations near the region of optimum stability, complexes of bivalent metal ions have significantly higher log K values than do complexes of univalent metal ions while for cations of small ionic radii, the reverse is true (i.e., $Ba^{2+} > K^+$, but $Na^+ > Ca^{2+}$). This reversal has also been noted in the case of 1:1 M^{2+}-dibenzo-18-crown-6 interactions in aqueous solution by Shchori et al. (1975). The data in Figure 7 suggest that the region of optimum stability for bivalent cations occurs at smaller ionic radii than does that for univalent cations. However, we are not certain that this is a correct observation. As discussed in the next section, the ΔH and ΔS values in the cases of Tl^+ and Ag^+ indicate that these cations differ from the alkali metal ions in their binding properties with these ligands. Similar behavior might be expected with Hg^{2+} and Pb^{2+} with Pb^{2+} lying above the curve connecting the alkaline Earth cations, and the curve maximum lying between Sr^{2+} and Ba^{2+}. In the case of the alkali metal ions, the curve maximum may lie to

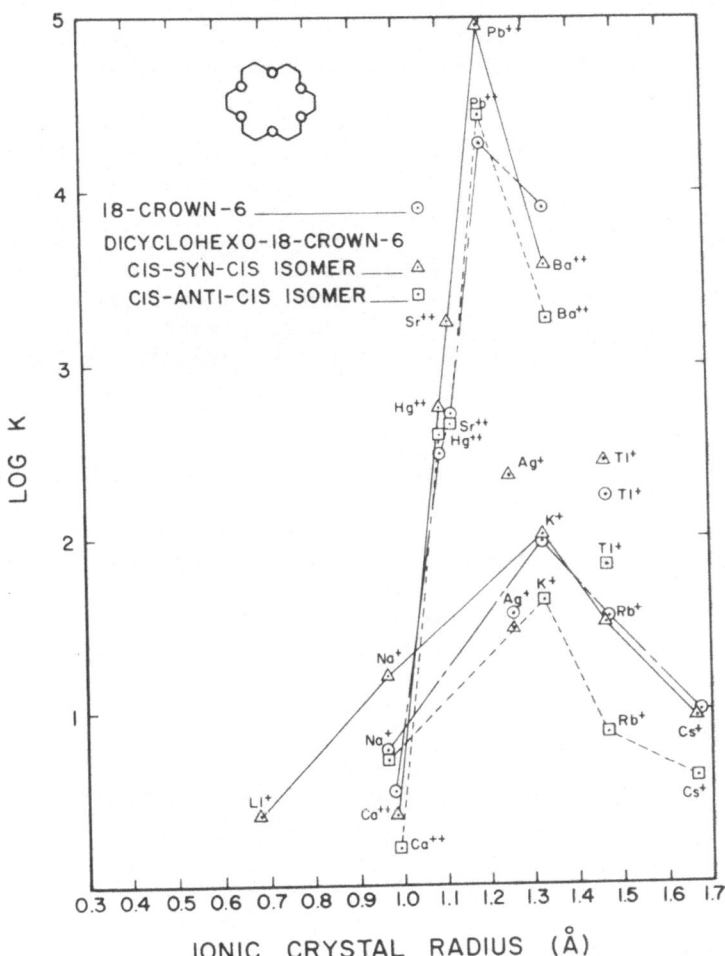

Fig. 7. Plot of log K for the aqueous solution reaction $M^{z+}+L=ML^{z+}$ vs. cation crystal radius. T = 25°.

the left of K^{+}. These ideas cannot be checked experimentally since cations of appropriate size are not available; However, we have been successful in reproducing the essential features of the plot in Figure 7 using an equation based on simple electrostatic theory (see Section 5). This equation predicts that the regions of optimum stability for uni- and bivalent cations are at the same cation radius.

3.2. *Effect of Cation Type on Complex Stability*

Included in Figure 7 are log K values for the reaction of several non-alkali and non-alkaline Earth cations with the 18-crown-6 ligands. The post transition series bivalent cations, Hg^{2+} and Pb^{2+} appear to fit nicely with the alkaline Earth cations, Ca^{2+}, Sr^{2+}, and Ba^{2+}; however, Ag^+ and Tl^+ show significant deviations from the curves for the alkali metal ions.

The log K values for each of the three Tl^+-crown-6 complexes are approximately 0.9 log K unit larger than would be expected based on the curves for the alkali metal ions. Comparison of the log K, ΔH, and ΔS data (Table V) for Tl^+ with those for Rb^+ (same ionic radius as Tl^+, Table I) shows that both the ΔH and ΔS values contribute to the greater log K values in the case of the reaction of Tl^+ with each crown-6 ligand. The greater stabilities of the Tl^+-crown-6 complexes probably reflect the greater polarizability of Tl^+ and, thus, greater covalent contribution to the bonding. The same result might be expected for Ag^+ inter-

TABLE V
Comparison of log K, ΔH, and ΔS data for the 1:1 reaction of 18-crown-6 with Tl^+ and Rb^+ T = 25, Solvent = H_2O

M^+	Log K	ΔH(kcal/mole)	$T\Delta S$(kcal/mole)
		18-crown-6	
Rb^+	1.56	−3.82	−1.71
Tl^+	2.27	−4.44	−1.35
	(A) cis-syn-cis isomer, dicyclohexo-18-crown-6		
Rb^+	1.52	−3.33	−1.26
Tl^+	2.44	−3.62	−0.30
	(B) cis-anti-cis isomer, dicyclohexol-18-crown-6		
Rb^+	0.87	−3.97	−2.79
Tl^+	1.83	−4.29	−1.80

action with these ligands; however, the data in Figure 7 show that only in the case of the cis-syn-cis isomer of dicyclohexo-18-crown-6 is log K appreciably larger than that expected from the curves for the alkali metal ions; and, in this case, the larger log K value is primarily a result of a favorable ΔS value, ΔH being nearly zero (Table VI). Comparison of the log K, ΔH and ΔS values for the reactions of K^+ and Ag^+ with the three crown-6-ligands in Table VI show ΔH to be less favorable and ΔS more favorable in all cases for the reaction of these ligands with Ag^+. The lower ΔH values could be a reflection of the inability of Ag^+ to achieve a favorable bonding arrangement with these ligands because of the peculiar stereochemical requirements of Ag^+.

Shchori et al. (1975) have determined log K values by a spectrophotometric procedure for the reaction in aqueous solution of univalent and bivalent cations with dibenzo-18-crown-6. They find essentially the same relationship for this ligand between the log K values for Tl^+ and Ag^+, and those for the alkali metal ions that we have reported for 18-crown-6 and the two isomers of dicyclohexo-18-crown-6.

It is of interest to compare the results with Tl^+ and Ag^+ with those reported by Eisenman and Krasne (1975). These authors conclude, based on ΔG values for cation interaction with nonactin, valinomycin,

TABLE VI

Comparison of log K, ΔH and ΔS data for the 1:1 reaction of 18-crown-6 ligands with Ag^+ and K^+. T = 25°, Solvent: H_2O

M^+	Log K	ΔH(kcal/mole)	$T\Delta S$(kcal/mole)
		18-crown-6	
K^+	2.03	-6.21	-3.42
Ag^+	1.50	-2.17	-0.12
	(A) cis-syn-cis isomer, dicyclohexol-18-crown-6		
K^+	2.02	-3.88	-1.04
Ag^+	2.36	+0.07	+3.30
	(B) cis-anti-cis isomer, dicyclohexol-18-crown-6		
K^+	1.63	-5.07	-2.88
Ag^+	1.59	-2.09	+0.09

hexadecavalinomycin, and dicyclohexo-18-crown-6, that it is the general rule for Tl^+ and Ag^+ to show excess binding energies with macrocyclic compounds having carbonyl and ether ligands. They point out that these cations have energies of hydration larger than those of alkali metal ions of comparable radii and that this is also the case with the energy of solvation of these ions by the carbonyl group of the solvent propylene carbonate. They attribute these effects to the greater polarizability of these cations and the subsequently greater covalent character of the bonds they form with the donor oxygen atoms. It would be expected that these effects would result in larger negative ΔH values and this is consistent with our results for Tl^+ where ΔH values are always larger than those for Rb^+, but not for Ag^+ where the ΔH values are much less than those for K^+, and the ΔS terms make large contributions to the stabilities of the complexes.

One of the more interesting and unusual systems investigated is that involving the reactions of NH_4^+ with 15-crown-5 and 18-crown-6. The ionic radius of NH_4^+ is nearly identical to that of Rb^+ (Table I) and its salts are isomorphous with those of Rb^+. Log K values for the reaction of NH_4^+ with 15-crown-5 are larger, while those for reaction with 18-crown-6 are lower than the corresponding values for Rb^+ (Table VII).

TABLE VII

Log K, ΔH, and ΔS values for the 1:1 reaction of Rb^+ and NH_4^+ with 15-crown-5 and 18-crown-6

T = 25°, Solvent: H_2O

Cation	Log K	ΔH(kcal/mol)	$T\Delta S$(kcal/mol)
		15-crown-5	
Rb^+	0.62	-1.90	-1.05
NH_4^+	1.71	-0.24	+2.10
		18-crown-6	
Rb^+	1.56	-3.82	-1.71
NH_4^+	1.23	-2.34	-0.66

This order is opposite to that seen for the reaction of Rb^+ and other inorganic cations with these ligands (Figure 2). Comparable differences of 1.66 and 1.48 kcal/mol separate the ΔH values of Rb^+ and NH_4^+ for 15-crown-5 and 18-crown-6, respectively. Thus, the reversal in the relative magnitudes of the log K values for these ligands is an entropy effect, the difference in ΔS values for the two cations being three times greater for the 15-crown-5. Eisenman and Krasne (1975) have attributed the higher affinity of NH_4^+ for nonactin than for valinomycin to the existence of a tetrahedral arrangement of oxygen atoms in the former compared to an octahedral arrangement in the latter. It is possible that the smaller 15-crown-5 offers better hydrogen bonding possibilities than 18-crown-6. Goldberg (1976) reports hydrogen bonding to be important in a tert-butyl ammonium ion-18-crown-5 crystal.

3.3. *Effect on Complex Stoichiometry of Ligand Cavity and Metal Ion Diameter Ratio*

Complexes formed between metal ions and cyclic polyethers sometimes have other than 1:1 stoichiometry (Pedersen, 1976; Christensen et al., 1974; Ovchinnikov, 1974). One would expect this behavior especially when the cation is larger than the ring. However, the solvent can play an important role in determining whether such complexes form. An example is found in the reaction of K^+ with benzo-15-crown-5. Log β_i values for this reaction valid in several solvents are given in Table VIII together with related data valid in 70 wt.% CH_3OH for the reaction of K^+ with dibenzo-24-crown-8 and dibenzo-27-crown-9. Complexes of stoichiometry 1:1 are formed between K^+ and benzo-15-crown-5 in aqueous solution and in 20 and 40% CH_3OH. However, as the water content of the solvent de-

TABLE VIII
Log β_i values (i=1, 2) for the reaction $K^+ + iL = KL_i^+$ in CH_3OH-H_2O solvents (given as wt.% CH_3OH). T = 25°

Solvent	i	Benzo-15-crown-5	Dibenzo-24-crown-8[a]	Dibenzo-27-crown-9[a]
0	1	0.38		
20	1	1.20		
40	1	1.92		
70	1	1.5	2.42	2.86
	2	4.1		
80	1	2.2		
	2	4.8		

[a] Experiments were performed only at 70 wt.% CH_3OH.

creases, it is necessary to assume the additional formation of 1:2 complexes in order to fit the calorimetric titration curves. Thus, β_1 and β_2 values are reported for the reaction in 70 and 80% CH_3OH with $K_2 > K_1$ in these cases. The crystal structure of the 1:2 potassium iodide-benzo-15-

crown-5 complex is known to be a sandwich with the K^+ located between the two ligands. Only 1:1 complexes are found with the larger cyclic polyethers in the 70% CH_3OH solvent suggesting that in these cases, the K^+ is positioned in the polyether cavity.

4. DITHIA CYCLIC POLYETHERS

4.1. *Effect on Complex Stability of Substitution of Sulfur for Oxygen*

Partial replacement of oxygen by sulfur in 18-crown-6 and 24-crown-8 (Figure 1: IId, IIe, IVd, IVe) results in a division of the cations studied into three groups (Izatt et al., submitted, a): the alkali and alkaline Earth ions which have decreased affinity, Tl^+ and Pb^{2+} which have relatively unchanged affinity, and Ag^+ and Hg^{2+} which show greatly increased affinity for each thia substituted ligand compared to the parent cyclic polyether. The lack of affinity by the thia substituted ligands for alkali and alkaline Earth cations can be rationalized in the light of the X-ray crystallographic data summarized in Figure 5. The 1,10-dithia-18-crown-6 ligand (Figure 5: c) resembles 18-crown-6 in that the ring is elliptical with the donor atoms pointing away from the cavity (Dalley et al., 1975 b). Our log K results suggest that K^+, Ba^{2+}, and other alkali and alkaline Earth cations lack the affinity for sulfur necessary to maintain the dithia ligands in the conformation required for complexation. On the other hand, Ag^+ and Hg^{2+} apparently have this ability although this conclusion requires X-ray data which have not yet been determined. The reaction of Ag^+ and Hg^{2+} with the thia substituted ligands results in extremely large ΔH values. The favorable enthalpy change in each case is partially offset by a large unfavorable entropy change as seen in Table IX. The large unfavorable ΔS values support the explanation given above for the absence of reactions between these ligands and the alkali and alkaline Earth metal ions. The ΔH values for reaction of these cations with these ligands would be expected to be small. Thus, the ΔS term would predominate and no complexation would occur.

TABLE IX
Log K, ΔH, and $T\Delta S$ values for the reaction, $Ag^+ + L = AgL^+$, in H_2O at 25°

Ligand	Log K	ΔH(kcal/mole)	$T\Delta S$(kcal/mole)
18-crown-6	1.50 (1.60)[a]	−2.17	−0.12
1,4-dithia-18-crown-6	3.02	−15.70	−11.6
1,10-dithia-18-crown-6	4.34	−16.74	−11.0

[a] Frensdorff (1971)

4.2. *Effect on Stoichiometry of Substitution of Sulfur for Oxygen*

Two sets of thia derivatives were studied, those with sulfur atoms
adjacent or 1,4 (Figure 1, IId and IVd), and those with sulfur atoms
opposite (Figure 1, IIe and IVe). Complexation of Ag^+ and Hg^{2+} with
these ligands produced unexpected results. Complexes of only 1:1 M^{z+}
-ligand stoichiometry were found for both the 1,10-dithia-18-crown-6
and 1,13-dithia-24-crown-8 ligands; but both 1:1 and 1:2 complexes were
found when the sulfur atoms were in the 1,4 positions. A typical calorimetric titration curve illustrating this effect in the case of Ag^+
is given in Figure 8. The curve in Figure 8a has a slope change at the
1:2 Ag^+:L ratio corresponding to the formation of AgL_2^+. The rounded
region of the curve as the 1:1 Ag^+:L ratio is approached is a result of
equilibrium between the AgL_2^+ and AgL^+ species. In curve 8b, a stoichiometric end point is found corresponding to formation of the 1:1 Ag^+:L
complex. Only 1:1 complexes were found with all four ligands when Pb^{2+}
or Tl^+ were the cations. Elucidation of these effects must await further

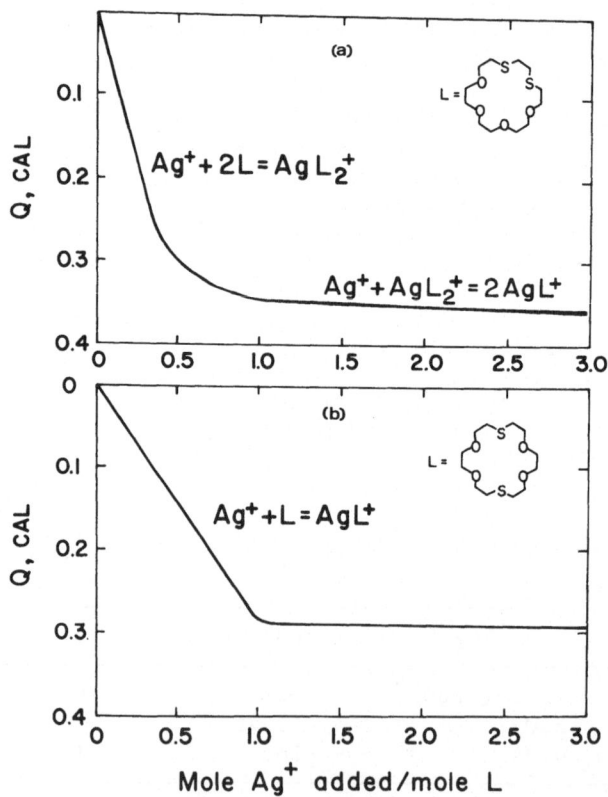

Fig. 8. Calorimetric titration of L with $AgNO_3$ in aqueous
solution at 25°. Formation of (a) both AgL_2^+ and AgL^+ by 1,4-
dithia-18-crown-6, and (b) only AgL^+ by 1,10-dithia-18-crown-6
are illustrated.

study, particularly structure determinations of the complexes involved.

5. MATHEMATICAL RELATIONSHIP BETWEEN ΔG FOR COMPLEX FORMATION AND ITS COMPONENT ENERGY TERMS

We have attempted to identify and estimate values for those quantities which collectively determine the magnitude of the ΔG values for the reactions of macrocyclic compounds with cations (Hansen et al., submitted). We have constructed a simple thermodynamic cycle (Figure 9) for the reaction of a macrocyclic ligand, L, with a metal ion, M^{z+}. From this cycle, Equation (1) can be written:

$$\Delta G_o = \Delta G_1 + \Delta G_2 + \Delta G_3 + \Delta G_4 \tag{1}$$

The terms comprising ΔG_1, ΔG_2, ΔG_3, and ΔG_4 consist of the following quantities: metal ion charge, electron charge, metal ion radius, solvent dielectric constant, solvent molecule diameter, free energy changes for vaporization of the solid ligand and for dissolution of the solid ligand in the solvent, number of ligand donor atoms, dipole moment of the donor atom group in the ligand, cation-ligand dipole distance, ligand cavity radius, radius of the metal-ligand complex, angle between the dipole and a line joining the center of the dipole and the center of the cation, Van der Waals radius for the oxygen atom, angles between the dipoles of the complexed and uncomplexed ligand, and the saturated value of the bulk dielectric constant. Our use of these quantities to calculate the several ΔG values involved analyzing each process in Figure 9,

Fig. 9. Thermodynamic cycle for the reaction of a metal ion, M, of charge z+ with a macrocyclic ligand, L.

determining what molecular parameters would be involved in the process, and devising mathematical relationships which accounted for the involvement. For example, attempts were made to account for the non-planarity of the ligand donor atoms and cation, ligand conformational changes upon complexation, and the solvation of the complex. The procedure involved a certain amount of trial and error as we attempted to produce a plot of log K vs. cation radius which would reproduce the essential features of

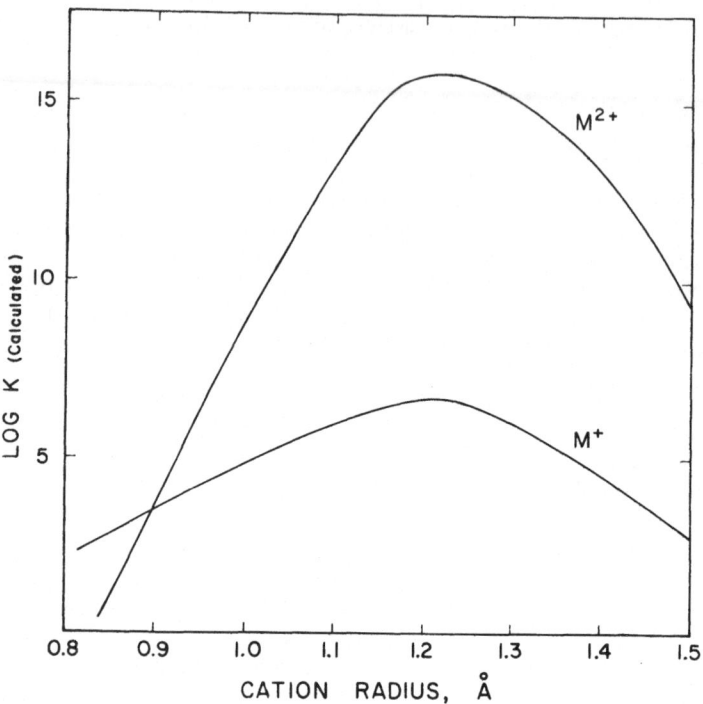

Fig. 10.   Plot of log K vs. cation radius generated using
Equation (1)

the experimental plot as depicted in Figure 7 and described in section
3.1 without forcing the equations.
     The plot generated from equation (1), assuming univalent and biva-
lent cation interaction with 18-crown-6, is given in Figure 10. Although
the numerical values for log K are two to three times higher than those
found experimentally, several features of the experimental curve are
reproduced in Figure 10. The predicted stabilities of the complexes
formed with both +1 and +2 cations show a maximum with the bivalent
cations forming more stable complexes in the region of the maximum. Also, the stabilities of complexes formed by small bivalent cations are
predicted to be lower than those formed by small univalent cations, and
the cation radius at which the maximum occurs is predicted to be the
same for univalent and bivalent cations.

6. CONCLUSIONS

Cyclic polyethers with the 18-crown-6 ring form more stable complexes
with metal ions in aqueous and water-methanol solvents than do ligands
with either 15-crown-5 or 24-crown-8 rings. Reasons for the stability
of the 18-crown-6 complexes may include good cation-ligand fit and relatively little ligand conformation change prior to complexation. Complex-

ation of metal ions by 18-crown-6 involves a compensation between ΔH and ΔS with ΔH usually being the dominant term. The increased stability of metal complexes of 18-crown-6 in methanol solvents compared to water is due to a more favorable ΔH term, the change in ΔS being small except in the case of larger cations, l.e., Cs^+.

Cation type is important in determining complex stability. Complexes between Tl^+ and three 18-crown-6 ligands are more stable than might be expected based on the similar size cation, Rb^+. Both ΔH and ΔS contribute to the increased stability which is probably related to the greater polarizability of Tl^+ compared to Rb^+. Similar behavior might be expected in the case of Ag^+; however, a complex more stable than the corresponding K^+ complex was found with only one of three 18-crown-6 ligands. In all cases, ΔH was less favorable and ΔS much more favorable than in the corresponding K^+ complexes. NH_4^+ is unusual among all cations studied in forming a more stable complex with 15-crown-5 than with 18-crown-6. This finding may be significant in that the cation parameter of hydrogen bonding is likely important in the case of NH_4^+ but is absent in the case of metal ions. The smaller 15-crown-5 may offer a more favorable environment for hydrogen bonding.

Formation of cyclic polyether complexes with 1:2 metal:ligand stoichiometry is a function of both the ratio of metal ion and ring cavity diameters and solvent. The complex between K^+ and 15-crown-5 is 1:1 in water and solvents of low CH_3OH content (0–40 wt.%), but 1:2 in 70 and 80 wt.% CH_3OH. However, the complex between K^+ and either dibenzo-24-crown-8 or dibenzo-27-crown-9 is 1:1 in 70 wt.% CH_3OH.

Substitution of two sulfur atoms for oxygen atoms in either neighboring or opposite ring donor atom positions of 18-crown-6 and 24-crown-8 result in a loss of reactivity in aqueous solution with alkali and alkaline Earth metal ions, little change for Tl^+ and Pb^{2+}, and greatly enhanced stabilities (largely a ΔH effect) for Ag^+ and Hg^{2+}. Complexes of 1:2 as well as 1:1 stoichiometry are formed by Ag^+ and Hg^+ when the sulfur atoms are adjacent (1,4) while only 1:1 complexes are formed when they are opposite. Only 1:1 complexes are formed with Tl^+ and Pb^{2+} in either case with both ligands.

Calculation of the terms in a simple thermodynamic cycle has enabled us to reproduce quite closely the important features of a plot of log K vs. cation radius for the reaction of uni-and bivalent metal ions with an 18-crown-6 ligand as well as to estimate the log K values for the re actions. The features reproduced correctly are: (1) peak selectivities with the +2 cations being predicted to be much more stable than the +1 cations in the region of maximum stability and (2) a reversal of this order of stabilities at much smaller cation radii where the +1 cations are predicted to form more stable complexes than do the +2 cations.

ACKNOWLEDGEMENT

The authors express their appreciation to Dr N. Kent Dalley for many helpful discussions and to the many graduate and undergraduate students whose efforts have made these results possible. This work was supported in part by U.S. Public Health Service NIH Grant GM 18811 and National Science Foundation Grant GP-33536X.

REFERENCES

Ahrens, L.J.: 1952, The Use of Ionization Potentials. Part I. Ionic Radii of the Elements, Geochim. et Cosmochim. Acta 2, 155—169.

Bradshaw, J.S., Hansen, L.D., Nielsen, S.F., Thompson, M.D., Reeder, R.A., Izatt, R.M., and Christensen, J.J.: 1975, A New Class of Macrocyclic Ether-Ester Ligands, J. Chem. Soc. Chem. Commun., 874-875.

Bradshaw, J.S.: 1977, Synthesis of Multidentate Compounds, in R.M. Izatt and J.J. Christensen (eds.), Synthetic Multidentate Macrocyclic Compounds, Academic Press, New York, in preparation.

Christensen, J.J., Eatough, D.J., and Izatt, R.M.: 1974, The Synthesis and Ion Binding of Synthetic Multidentate Macrocyclic Compounds, Chem. Rev. 74, 351-384.

Christensen, J.J., Eatough, D.J., Ruckman, J., and Izatt, R.M.: 1972, Introduction to Titration Calorimetry. Part I, Thermochim. Acta 3, 203.

Cram, D.J. and Cram, J.M.: 1974, Host-Guest Chemistry, Science 183, 803-809.

Dalley, N.K., Smith, D.E., Izatt, R.M., and Christensen, J.J.: 1972, X-Ray Crystal Structure of the Barium Thiocyanate Complex of the Cyclic Polyether Dicyclohexyl-18-Crown-6 (Isomer A), Chem. Commun., 90-91.

Dalley, N.K., Smith, J.S., Larson, S.B., Christensen, J.J., and Izatt, R.M.: 1975a, Crystal Structures of Two Isomers of Dicyclohexyl-18-Crown-6, J. Chem. Soc. Chem. Commun., 43-44.

Dalley, N.K., Smith, J.S., Larson, S.B., Matheson, K.L., Christensen, J.J., and Izatt, R.M.: 1975b, X-Ray Crystal Structures of Three Cyclic Thioethers, J. Chem. Soc. Chem. Commun., 84-85.

Eatough, D.J., Christensen, J.J., and Izatt, R.M.: 1972b, Determination of Equilibrium Constants by Titration Calorimetry. Part II. Data Reduction and Calculation Techniques, Thermochimica Acta 3, 219-242.

Eatough, D.J., Izatt, R.M., and Christensen, J.J.: 1972c, Determination of Equilibrium Constants by Titration Calorimetry. Part III. Application of Method to Several Chemical Systems, Thermochimica Acta 3, 233-246.

Eatough, D.J., Christensen, J.J., and Izatt, R.M.: 1975, Determination of the Enthalpy of Solution of Tris-(hydroxymethyl) Aminomethane in 0.1 M HCl Solution and the Enthalpy of Neutralization of $HClO_4$ with NaOH at low Ionic Strengths by use of an Improved Isothermal Titration Calorimeter, J. Chem. Thermodyn. 7, 417-422.

Eisenman, G., Szabo, G., Ciani, S., McLaughlin, S., and Krasne, S.: 1973, Ion Binding and Ion Transport Produced by Neutral Lipid-Soluble Molecules, in J.F. Danielli et al. (eds.), Progress in Surface and Membrane Science, vol. 6, Academic Press, New York, pp. 139—241.

Eisenman, G. and Krasne, S.J.: 1975, The Ion Selectivity of Carrier Molecules, Membranes and Enzymes, in C.F. Fox (ed.) MTP International Review of Science, Biochemistry Series, vol. 2, Butterworths, London.

Frensdorff, H.K.: 1971, Stability Constants of Cyclic Polyether Complexes with Univalent Cations, J. Amer. Chem. Soc. 93, 600-606.

Goldberg, I.: 1976, Structure and Binding in Molecular Complexes of Cyclic Polyethers. II. Hydrogen-Bonding and Ion Pairing in a Complex of a Macrocyclic Polydentate Ligand with tert-Butylamine at 120 K, Acta Crystallographica, B31, 2592-2600.

Grell, E., Funck, T., and Eggers, F.: 1975, Structure and Dynamic Properties of Ion-Specific Antibiotics, in G. Eisenman (ed.), Membranes,
vol. 3, Marcel Dekker, New York, pp. 1-126.

Hansen, L.D., Izatt, R.M., and Christensen, J.J.: 1974a, Application of
Thermometric Titrimetry to Analytical Chemistry, in J. Jordan (ed.),
New Developments in Titrimetry, Marcel Dekker, New York.

Hansen, L.D., Izatt, R.M., Eatough, D.J., Jensen, T.E., and Christensen,
J.J.: 1974b, Recent Advances in Titration Calorimetry, Analytical
Calorimetry, vol. 3, Plenum Press, New York, pp. 7-16.

Hansen, L.D., Jensen, T.E., Mayne, S., Eatough, D.J., Izatt, R.M., and
Christensen, J.J.: 1975, Heat-loss Corrections for Small Isoperibol-
Calorimeter Reaction Vessels, J. Chem. Thermodyn. 7, 919-926.

Hansen, L.D., Izatt, R.M., Terry, R.E., and Christensen, J.J.: A Theoretical Basis for Predicting Cation Selectivity Behavior by Cyclic-
polyethers, to be submitted; cf. R. Terry, Ph.D. Dissertation, Brigham
Young University, Provo, Utah, 1976.

Helgeson, R.C., Timko, J.M., Moreau, P., Peacock, S.C., Mayer, J.M., and
Cram, D.J.: 1974, Models for Chiral Recognition in Molecular Complexation, J. Amer. Chem. Soc. 96, 6762-6763.

Izatt, R.M., Eatough, D.J., and Christensen, J.J.: 1973, Thermodynamics
of Cation-Macrocyclic Compound Interaction, Structure Bonding 16,
161-189.

Izatt, R.M., Hansen, L.D., Eatough, D.J., Jensen, T.E., and Christensen,
J.J.: 1974, Recent Analytical Applications of Titration Calorimetry,
Analytical Calorimetry, vol. 3, Plenum Press, New York, pp. 237-248.

Izatt, R.M., Nelson, D.P., Rytting, J.H., Haymore, B.L., and Christensen,
J.J.: 1971, A Calorimetric Study of the Interaction in Aqueous Solution
of Several Uni- and Bivalent Metal Ions with the Cyclic Polyether
Dicyclohexyl-18-Crown-6 at 10, 25, and 40°, J. Amer. Chem. Soc. 93,
1619-1623.

Izatt, R.M., Terry, R.E., Hansen, L.D., Avondet, A.G., Bradshaw, J.S.,
Dalley, N.K., Jensen, T.E., Haymore, B.L., and Christensen, J.J.: (a)
A Calorimetric Titration Study of Uni- and Bivalent Metal Ion Inter-
action with Several Oxathiapentadecanes, Dicyclohexo-24-Crown-8, and
Several Thia Derivatives of 9-Crown-3, 12-Crown-4, 15-Crown-5, 18-
Crown-6, and 24-Crown-8 in Water and Water-Methanol Solvents at 25°,
J. Amer. Chem. Soc., submitted.

Izatt, R.M., Terry, R.E., Hansen, L.D., Dalley, N.K., Avondet, A.G.,
Haymore, B.L., and Christensen, J.J.: (b) A Calorimetric Titration
Study of the Interaction of Several Uni- and Bivalent Cations with
15-Crown-5, 18-Crown-6, and Two Isomers of Dicyclohexo-18-Crown-6 in
Aqueous Solution at 25°, J. Amer. Chem. Soc., submitted.

Izatt, R.M., Terry, R.E., Nelson, D.P., Chan, Y., Eatough, D.J., Bradshaw,
J.S., Hansen, L.D., and Christensen, J.J.: (c), A Calorimetric Titration
Study of the Interaction of Some Uni- and Bivalent Cations with Benzo-
15-Crown-5, 18-Crown-6, Dibenzo-24-Crown-8 and Dibenzo-27-Crown-9 in
Methanol-Water Solvents at 25°, J. Amer. Chem. Soc., submitted.

Laprade, R., Ciani, S., Eisenman, G., and Szabo, G: 1975, The Kinetics
of Carrier-Mediated Ion Permeation in Lipid Bilayers and its Theoretical
Interpretation, in G. Eisenman (ed.), Membranes, vol. 3, Marcel Dekker,
New York, pp. 127-214.

Liesegang, G.W., Farrow, M.W., Purdie, N., and Eyring, E.M.: Ultrasonic
Absorption Kinetic Studies of the Complexation of Aqueous Potassium

and Cesium Ions by 18-Crown-6, J. Amer. Chem. Soc., in press.
Mellor, D.P.: 1964, Historical Background and Fundamental Concepts, in
    F.P. Dwyer and D.P. Mellor (eds.), Chelating Agents and Metal Chelates,
    Academic Press, New York.
Ovchinnikov, Yu., Ivanov, V.T., and Shkrob, A.M.: 1974, Membrane Active
    Complexones, Elsevier, New York.
Pauling, L.: 1960, The Nature of the Chemical Bond, Cornell University
    Press, Ithaca, New York.
Pedersen, C.J.: 1967, Cyclic Polyethers and Their Complexes with Metal
    Salts, J. Amer. Chem. Soc. 89, 7017-7036.
Pedersen, C.J.: 1976, Historical Introduction, Synthetic Multidentate
    Macrocyclic Compounds, Academic Press, New York, in preparation.
Pederson, C.J. and Frensdorff, H.K.: 1972, Macrocyclic Polyethers and their
    Complexes, Angewandte Chemie 11, 16-25.
Shchori, E., Nae, N., and Jagur-Grodzinski, J.: 1975, Stability Constants
    of a Series of Metal Cations with 6,7,9,10,17,18,20,21-Octahydrodi-
    benzo [b, k] [1,4,7,10,13,16] hexa-oxa-cyclo-octadecin(Dibenzo-18-
    Crown-6) in Aqueous Solutions, J. Chem. Soc. Dalton, 2381-2386.
Simon, W., Morf, W.E., and Meier, P.Ch.: 1973, Specificity for Alkali
    and Alkaline Earth Cations of Synthetic and Natural Organic Complexing
    Agents in Membranes, Structure Bonding 16, 113-160.
Truter, M.R.: 1973, Structures of Organic Complexes with Alkali Metal
    Ions, Structure Bonding 16, 71-111.
Williams, R.J.P.: 1970, The Biochemistry of Sodium, Potassium, Magnesium
    and Calcium, Quart. Rev., 311-365.

DISCUSSION

A. *Pullman:* My comment concerns the binding of NH_4^+.
    When we studied the hydration of NH_4^+ theoretically we started by
looking for the most stable position of one water molecule around the
cation, by probing the potential surface moving the water molecule around
NH_4^+. We found, in so doing, that the best location of the water oxygen
is definitely in the direction of the NH bonds (a) rather than in a
bissecting position (b). This is confirmed by the observations made in
measuring in the gas phase the enthalpies of stepwise addition of water
molecules to NH_4^+: a break in the curve $\Delta H_{n,n-1}$ in function of the number n of water molecules added occurs for $n=4$, indicating the completion
of the first hydration shell and the beginning of a second shell. Thus,
it appears that the directionnal preference for ligand binding in the
case of NH_4^+ is in favor of a tetrahedral arrangement of 4 ligands, a
situation which does not occur with the much more spherical alkali ions.
This appears as a good reason for observing that NH_4^+ behaves in a different way than Rb^+, in spite of the similarity of their radii. (The
theoretical studies and the references to the experimental work of
Kebarle et al. may be found in A. Pullman and A.M. Armbruster, Int. J.
of Quant. Chem. S8, 169, 1974; ibid. Chem. Phys. Letters, 36, 558, 1975).
    Izatt: This comment ties in very well with the observations of
Eisenman and Krasne (1975) discussed in connection with Table VII.
    Hanlon: I find these ΔH changes in the reactions you describe sur-

prising high. Can you comment upon the reasons for the magnitude of these values?

Izatt: The ΔH values reflect the exchange of solvent molecules for ligand donor atoms. The reactions in 70 wt.% CH_3OH have more favorable ΔH values associated with them than do the corresponding reactions in water reflecting differences in solvation energies of these solvents. We do not have enough data to account for the magnitudes of the ΔH values, even in aqueous solvents, but can use these values as a basis for understanding the behavior of some cations (e.g., Rb^+, Tl^+, see Table V).

X-RAY STRUCTURE OF A SYNTHETIC, NON-CYCLIC, CHIRAL POLYETHER COMPLEX
AS ANALOG OF NIGERICIN ANTIBIOTICS: TETRAETHYLENEGLYCOL-BIS(8- OXYQUINO-
LINE)-ETHER·RUBIDIUM IODIDE COMPLEX

WOLFRAM SAENGER*, HERBERT BRAND, FRITZ VÖGTLE** and EDWIN
WEBER**
*Abteilung Chemie, Max-Planck-Institut für experimentelle
Medizin, Hermann-Rein-Strasse 3, 3400 Göttingen, West-Germany*

1. INTRODUCTION

1.1. *The ionophore antibiotics*

The family of ionophore (cation-carrier) antibiotics can be subdivided
into several groups with different chemical constitution [1]. Amino acids
are building blocks of the gramicidins, antamanides, alamethicin and some
cyclopeptides from mitochondrial membranes while valinomycin, the ennia-
tins and monamycins consist of depsipeptide units. Nonactins are formed
by hydroxy acid residues with attached tetrahydrofuran rings and the
nigericin antibiotics nigericin, grisorixin, dianemycin, monensin, X-206
and X-537A contain tetrahydrofuran and -pyran groups fused together by
spiro and single C–C linkages (Figure 1). In general, the ionophore mol
ecules have cyclic structures with the exception of some gramicidins
(these are probably helical [2]) and of the nigericin antibiotics. Although diverse from a chemical point of view, the ionophore antibiotics
display rather similar biological activities. They show a broad antimicrobial spectrum against Gram-positive hacteria, yeasts and fungi and
have found application in the prevention and treatment of several diseases
such as coccidiosis in chicken [3].
    The ionophores are of great interest owing to their ability to com-
plex metal ions. In these complexes, the antibiotic molecules assume bracelet shaped structures. The cavities are lined by hetero atoms which serve
as ligands for metal ions entrapped in the center of the cavity. At the
periphery of the bracelet, hydrophobic groups are arranged which allow
the complexes to dissolve in organic solvents and to act as cation carriers through biological and artificial membranes. Owing to the different
sizes of the cavities within these molecules and to different substituents
at the molecule periphery, a considerable ion selectivity is observed
which is well pronounced in the cyclic antibiotics but less developed in
those belonging to the noncyclic nigericin family [1].

---

 * to whom correspondence should be adressed.
** Institut für Organische Chemie und Biochemie der Universität Bonn.

*B. Pullman and N. Goldblum (eds.), Metal-Ligand Interactions in Organic
Chemistry and Biochemistry, part 1, 363-374. All Rights Reserved.
Copyright © 1977 by D. Reidel Publishing Company, Dordrecht-Holland.*

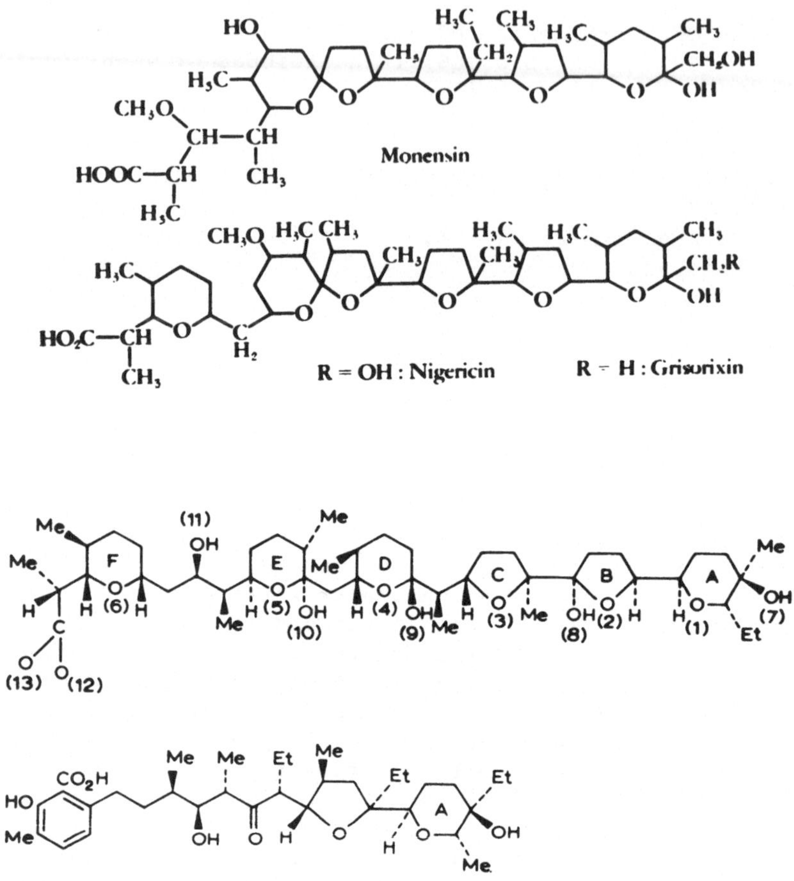

Fig. 1. Structural formulae of nigericin antibiotics. From top: monensin, nigericin and grisorixin, X-206 and X-537 A. Note the carboxylate and hydroxyl functions on both ends of these molecules.

1.2. *The group of nigericin antibiotics*

Several metal complexes of nigericin antibiotics have been investigated by X-ray crystallographic methods, Figure 2 [1]. It was observed in all of these structures that the molecules are wrapped around the metal ion such that one or two head-to tail hydrogen bonds can form between the terminal carboxylate and hydroxyl groups, i.e. if the hydrogen bond is considered as a bond, these complexes may be regarded as macrocyclic structures. In contrast to other ionophore antibiotics, the complexes between nigericin type molecules and monovalent cations are electroneutral because the positive charge of the metal ion is balanced by the negative charge of the carboxylate group. Although less ion-selective than the cyclic antibiotics, the nigericin family has retained preference for some monovalent cations. Thus nigericin binds potassium more strongly than

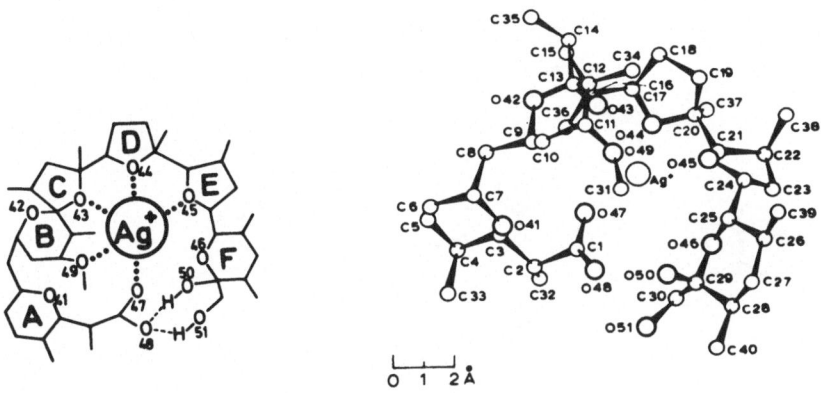

Fig. 2.   Schematic and X-ray crystal structure of the silver salt of nigericin [26, 27].

other cations, monensin prefers sodium but dianemycin is less particular. It is remarkable in this respect that the nigericin·potassium complex is more stable than any other potassium complex of the cyclic antibiotics [4].

1.3. *Polyethers as Analogs for Ionophore Antibiotics*

The astonishing complex formation characteristics of the ionophore antibiotics has promted chemists to design molecules with similar metal-complexing abilities. The first analogs were the cyclic crown ethers and thioethers wich could be synthesized with different ring diameters and substituted with benzene and cyclohexane residues in order to improve their hydrophobic periphery [5, 6]. X-ray studies revealed that, similar to the antibiotics, the crown ethers complex the cations with their ether oxygen atoms as ligands ([7, 8], Figures 3, 4) and thermodynamic measurements indicated that they display considerable ion-selectivity [5]. Experiments with natural and artificial membranes demonstrated that cyclic polyethers have similar ion carrier properties as the ionophore antibiotics [9—12].

Later, macrotricyclic diaminopolyethers were designed in which both amino- and oxygen atoms participate in the chelation of the cations [13]. These cryptates comprise three dimensional cavities which bind the metal ions even more tightly than is known from the crown ethers [14, 15, Figures 5, 6]. The rates of complex formation are reduced, however, because the cavities are more difficult to approach.

Crown ethers and cryptates are such potent complexing agents that even insoluble salts like $BaSO_4$ can be solubilized in water. Since both the cation complex and the anion dissolve without hydration shell in organic solvents, new chemical properties of the naked ions have been observed and have found application in synthetic and mechanistic organic chemistry [5, 16].

In a further step towards a more ideal model for cyclic antibiotics,

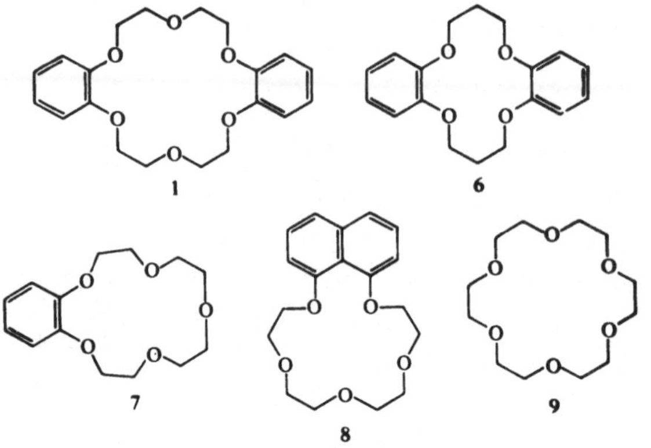

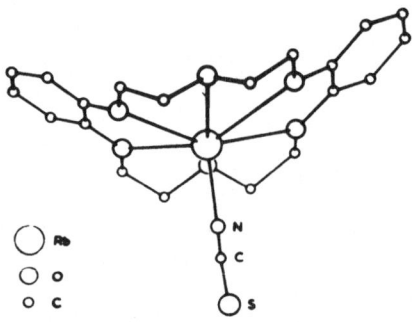

Fig. 3. A few example of crown ethers. Other structures are feasible and have been synthesized, many of them with S or N as hetero atoms.

Fig. 4. The crystal structure of the Rb·NCS complex of dibenzo 18-crown-6, molecule 1 in Figure 3. [25].

chirality has been introduced into crown ether molecules through implementation of binaphthyl residues ([17, 18], Figure 7) and other optically active moieties [19, 20]. Because the cavities in these chiral molecules have sizes which allow not only metal ions but also organic cations to be complexed, racemic mixtures of, e.g., the α-phenylethylammonium ion and of amino acids could be separated (chiral recognition) [17, 18].

1.4. *Tetraethyleneglycol-bis(8-Oxyquinoline)-Ether·RbI, an Analog of Nigericin Antibiotics*

More recently, polyether analogs of the non-cyclic nigericin family antibiotics were synthesized. They display cation carrier abilities reminiscent of those known from cyclic crown ethers and their ion selectivity depends on the groups attached to both ends of the polyether chain [21, 22]. In an ionophore related to these molecules, the polyether chain is augmented on both ends by two attached 8-oxyquinoline residues ([23],

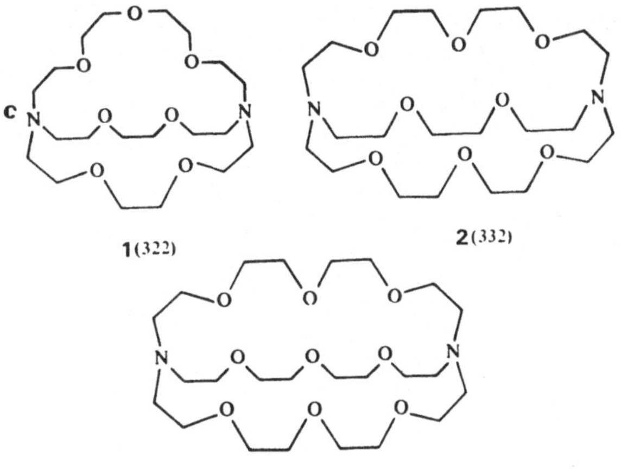

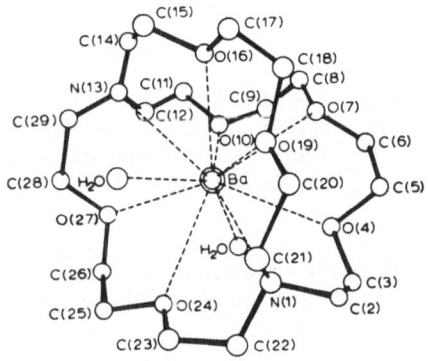

Fig. 5. A few examples of kryptate structures. Many other molecules have been synthesized, also with S as heteroatoms.

Fig. 6. The crystal structure of (322) . Ba(NCS)$_2$ · H$_2$O, a complex with molecule 1 of Figure 5 [28].

Figure 8) to yield tetraethyleneglycol-bis(8-oxyquinoline)-ether (TGOE). While the uncomplexed molecule TGOE is an oily fluid, complexes with cations can be obtained as crystalline solids. Thus far, 1:1 complexes with K$^+$, NH$_4^+$, Rb$^+$, and Ba^{2+} have been prepared while Pr^{3+} and Nd^{3+} yield 2:3 complexes (2(TGOE) : 3 cations). The stoichiometry of Ag$^+$ and Co^{2+} complexes could not be clarified [23].

NMR and IR spectroscopic data suggest that the quinoline systems are involved in complex formation [23]. Since no details about the structure of the complexes could be derived from these studies, the RbI complex of TGOE was subjected to an X-ray crystal analysis which is the

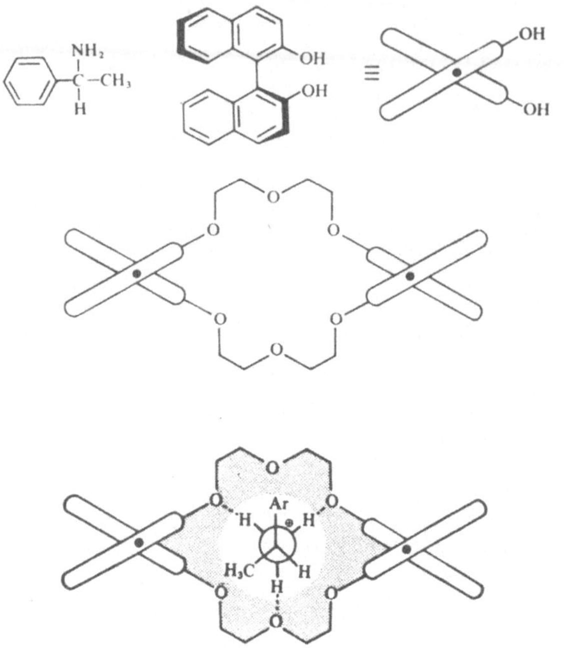

Fig. 7. Schematic structure of a chiral crown ether containing
binaphthyl moieties (top) and complex of this molecule with
α-phenyl-ethyl-ammonium ion (bottom). A recemic mixture of the
latter was separated by means of this complex [17, 18].

subject of this contribution. Crystallographic data are presented in
Table I and details of the structure are depicted in Figures 9 to 11.

TABLE I
Crystallographic data

Chemical formula = $C_{26}H_{28}N_2O_5 \cdot RbI$.
Needle shaped, yellowish crystals, needle axis along [010].

Space group monoclinic, $P2_{1/c}$

Cell constants a =   9.649(2) Å
                b = 17.666(3) Å
                c = 15.868(3) Å
                β = 100.15(2)°

calculated density = 1.622 g cm^{-3}
Number of molecules/unit cell Z = 4
Number of reflection data = 4715
R-factor $(\Sigma ||F_0|-|F_c||/\Sigma|F_0|)$ = 0.80

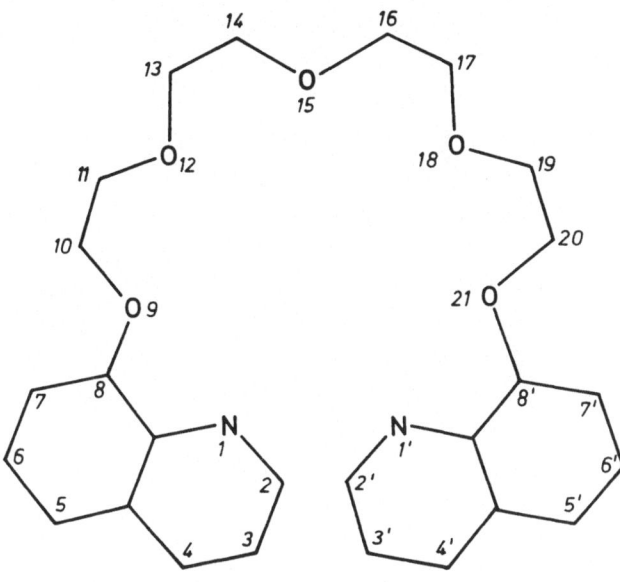

Fig. 8.   The chemical structure and numbering scheme of tetra-
ethyleneglycol-bis(8-oxyquinoline)-ether (TGOE).

The iodide ions is omitted from the figures because it has no direct
contact with Rb^+ or any heteroatom but is located between molecules and
surrounded by hydrogen atoms.
      The Rb^+-ion is linganded by all the seven heteroatoms (Figure 9).
From the sum of atomic and ionic radii, 1.40 Å and 1.50 Å for oxygen and
nitrogen resp. but 1.47 Å for Rb^+ [24], the $Rb^+\cdots N$ distance is expected
to be 2.97 Å and $Rb^+\cdots O$ should be 2.87 Å; the dibenzo-18-crown-6·RbNCS
structure exhibits six $Rb^+\cdots O$ distances in the range 2.861 Å to 2.939 Å
[25]. In the TGOE·RbI complex the $Rb^+\cdots N$ distances, 2.91 Å and 2.96 Å,
are close to the theoretical value but the $Rb^+\cdots O$ distances deviate
markedly and display a certain degree of symmetry with respect to the
line passing through Rb^+ and O(15), Figure 9. The $Rb^+\cdots O(15)$ distance, 2.88 Å, approaches the theoretical 2.87 Å, but distances
$Rb^+\cdots O(12)$ and $Rb^+\cdots O(18)$, 2.99 Å and 2.94 Å, are significantly longer.
In the $Rb^+\cdots O(9)$ and $Rb^+\cdots O(21)$ distances of 3.07 Å and 3.09 Å, the
ideal value is exceeded even by about 0.2 Å. This distance distribution
could be due to differences in electronegativity of the oxygen atoms
bonded to the aliphatic chains and to aromatic system (partial positive
charge from mesomeric effects). Steric hindrance appears improbable because the polyether chain of TGOE is rather flexible.
      The angles with Rb^+ as vertex belong to three different categories.
The $O\cdots Rb^+\cdots O$ angles of 54° to 59° are determined by the geometry of
the $O-CH_2-CH_2-O$ fragments which all show the gauche conformation while
the $N\cdots Rb^+\cdots O$ angles, 52° and 53°, are given by the dimensions of the
quinoline systems. The $N\cdots Rb^+\cdots N$ angle, 77°, results from the noncyclic
structure: if the polyether chain were longer by one ethyleneglycol

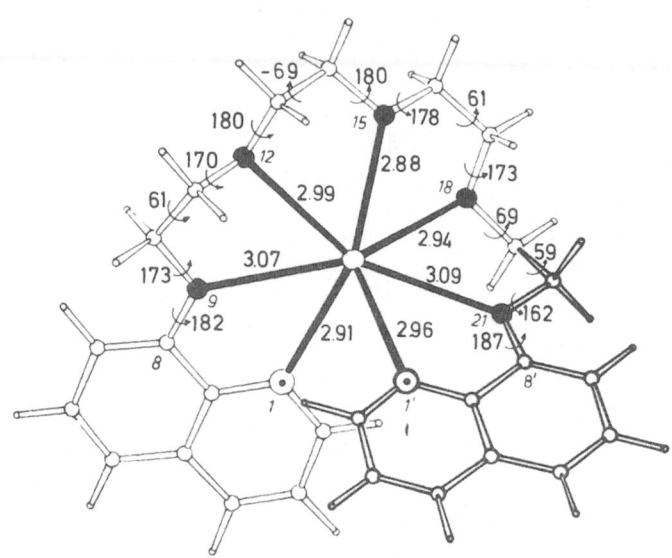

Fig. 9. A view of the TGOE·Rb⁺ complex perpendicular to the
plane comprising atoms O(9), O(15) and O(21). The iodide ion
is not drawn. o = Hydrogen, O = carbon, ● = oxygen, ◉ = nitrogen. The quinoline ring closer to the viewer is drawn with
heavy lines, Rb···O and Rb···N coordination bonds are marked
by solid lines. Small italics indicate atom numbering (see
Figure 8), other numbers refer to bond distances and dihedral
angles A—B—C—D resp. The latter is defined as zero when looking
along the central bond B—C, the bonds A—B and C—D are co-planar
Counting is positive when the far bond is rotated clockwise
with respect to the near bond.

group*, this angle should be reduced by about 58°.
      Commonly in cyclic crown ether complexes, the oxygen atoms liganded
to the metal ion are comprised in one plane and the ion lies within or
near this plane, depending on its radius and on the size of the cavity
[1, 7, 25]. Further, the conformational angles along the polyether chain
are gauche for C—C bonds but trans about the C—O bonds. In TGOE, the
chain formed by the oxygen and nitrogen ligands it too long to achieve
proper chelation of the Rb⁺-ion if all the hetero atoms are in one plane.
Therefore, only atoms N(1), O(9), O(12), O(15), and O(18) are located
in a well defined plane, and all the conformational angles are gauche
for C—C and trans for C—O bonds. However, in order to avoid collision
between the two quinoline systems, the dihedral angle C(17)—O(18)—C(19)
—C(20) is gauche instead of trans, lifting the heteroatoms O(21) and N(1'
and with them the attached quinoline system well above the plane through
the other hetero atoms. This peculiar structural feature gives the TGOE·R

* Work corrently being in progress in our laboratories.

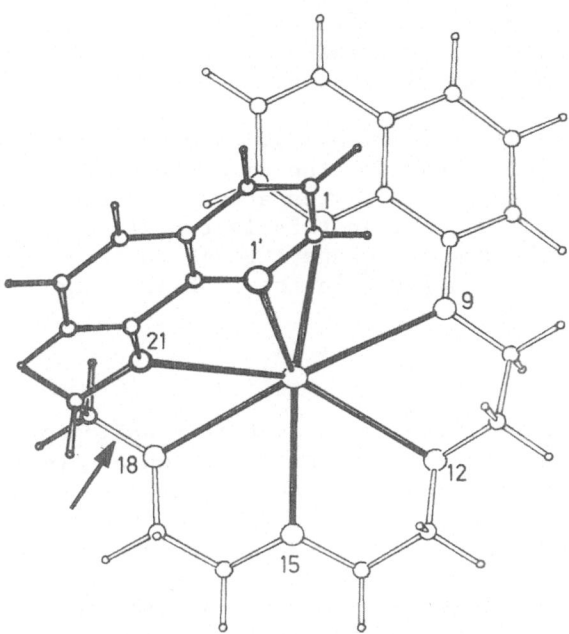

Fig. 10.   The structure of the TGOE · Rb$^+$ comples viewed approximately perpendicular to the plane formed by atoms N(1), O(9),
O(12), O(15), O(18). At atom C(19), the molecule performs a
sharp turn and moves towards the viewer. The atoms located
nearer to the viewer are drawn with heavier lines. The dihedral
angle with gauche conformation causing the helical structure
of the molecule is indicated by an arrow. Circles in increasing
size represent H, C, O, N.

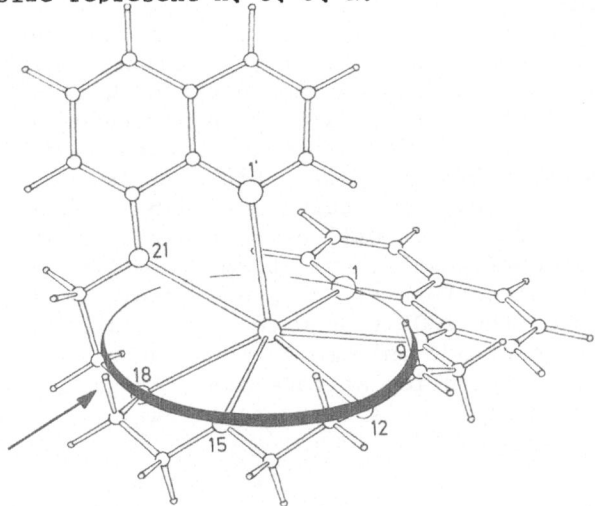

Fig. 11.   Similar to Figure 10 but the molecule is rotated by
nearly 90°. The large circle represents the plane comprising
the heteroatoms N(1), O(9), O(12), O(15), O(18).

complex the appearance of a one-turn helix. Helices are always chiral
and since this crystal structure is centrosymmetric (space group $P2_1/c$),
both left and right handed helices are present in the lattice. The former is depicted in the figures.

The coplanar heteroatms N(1), O(9), O(12), O(15) and O(18) are at
five corners of a rather regular hexagon, the dimensions of which resemble those of the hexagon formed by the oxygen atoms in the di-benzo-18-
crown-6·Rb NCS complex [25], Figure 4. In this latter structure, the
Rb^+-ion is removed by 0.942 Å from the plane through the liganding oxygen
atoms, a value comparable to the 0.748 Å in the TGOE·RbI complex.

2. CONCLUSIONS

At the outset, TGOE·RbI was introduced as a synthetic analog of the
nigericin family antibiotics. These antibiotics all contain terminal
carboxylate and hydroxyl groups which close the bracelet structures by
formation of hydrogen bonds and neutralize the charge of the liganded
(monovalent) cation. In the TGOE·RbI complex, the bracelet has degener-
ated into an open helix, a structure which was proposed for the non-
cyclic gramicidins [2]. Having these different structures in mind, the
TGOE·ligand can be considered as a general model for non-cyclic antibiotics. With one ethyleneglycol group less, its complexes with metal
ions should approach more closely the cyclic nigericin antibiotics but
with more ethyleneglycol groups it could build up helical complexes
with more than one turn, corresponding to the proposed helical structure
of gramicidins. In that case, trans-membrane channels could be constructed in which cations are carried from one side of a membrane to the other
not by diffusion of the whole complex but by movement of the cation
through the stationary polyether helix — in analogy to the model proposed
for the helical gramicidin molecules [2].

Thus far, the metal complex formation ability of TGOE has been studied qualitatively but its ion-selectivity is yet unknown. Owing to their
open structures, TGOE and similar non-cyclic molecules will display properties resembling those of crown ethers with probably reduced ion-selectivity. It will be of interest to find out what structures molecules like
TGOE but with increased polyethyleneglycol chain will assume.

The chirality of the helical structures formed by TGOE opens interesting applications provided that is will be possible to synthesize or
to separate the two individual enantiomeric forms. Similar to the cyclic
polyethers containing chiral centers [17—20], one should be able to design non-cyclic polymers of the TGOE type with cavity sizes sufficiently
wide to allow the inclusion of not only metal ions but also of organic
cations. In that case, the separation of racemats of organic cations or
of amino acids should be feasible on a large scale basis.

REFERENCES

1. Ovchinnikov, Yu.A., Ivanov, V.T., and Shkrob, A.M.: Membraneactive
   complexones, B.B.A. Library, vol. 12, Elsevier Scientific Publ. Co.,
   Amsterdam, 1974.

2. Urry, D.W.: in 'Conformation of Biological Molecules and Polymers', The Jerusalem Symposia on Quantum Chemistry and Biochemistry, vol. 5, (ed. by E.D. Bergmann and B. Pullman), Academic Press, New York, 1973.
3. Shunnard, R.F. and Callender, M.E.: Antimicrob. Agents Chemother. 369 (1967).
4. Feinstein, M.B. and Felsenfeld, H.: Proc. Natl. Acad. U.S. 68, 2037 (1971).
5. Pedersen, C.J. and Frensdorff, H.K.: Angew. Chem. 84, 16 (1972).
6. Christensen, J.J., Eatough, D.J., and Izatt, R.M.: Chem. Rev. 74, 351 (1974).
7. Truter, M.R.: Struct. Bond. 16, 71 (1973).
8. Truter, M.R. and Pedersen, C.J.: Endeavour, 142 (1971).
9. Lardy, H.: Fed. Proc. 27, 1278 (1968).
10. Eisenmann, G., Ciani, S.M., and Szabo, G.: Fed. Proc. 27, 1289 (1968).
11. Pressman, B.C.: Fed. Proc. 27, 1283 (1968).
12. Pedersen, C.J.: Fed. Proc. 27, 1305 (1968).
13. Dietrich, B., Lehn, J.M., Sauvage, J.P., and Blanzat, J.: Tetrahedron 29, 1629 (1973).
14. Moras, D., Metz, B., and Weiss, R.: Acta Crystallogr. B29, 388 (1973).
15. Lehn, J.M.: Struc. Bond. 16, 1 (1973).
16. Vögtle, F. and Neumann, P.: Chemiker-Ztg. 97, 600 (1973).
17. Kyba, E.B., Koga, K., Sousa, L.R., Siegel, M.G., and Cram, D.J.: J. Amer. Chem. Soc. 95, 2692 (1973).
18. Helgeson, R.G., Timko, J.M., Moreau, P., Peacock, S.C., Mayer, J.M., and Cram, D.J.: J. Amer. Chem. Soc. 96, 6762 (1974).
19. Layton, A.J., Mallinson, P.R., Parsons, D.G., and Truter, M.: J.C.S. Chem. Commun. 694 (1973).
20. Parsons, D.G.: J.C.S. Perkin I245 (1975).
21. Pretsch, E., Ammann, D., and Simon, W.: Research/Development 25, 20 (1974).
22. Ammann, D., Bissig, R., Güggi, M., Pretsch, E., Simon, W., Borowitz, I.J., and Weiss, L.: Helv. Chim. Acta 58, 1535 (1975).
23. Weber, E. and Vögtle, F.: Tetrahedron Lett. 2415 (1975).
24. Handbook of Chemistry and Physics ed. and publ. by The Chemical Rubber Co., Cleveland, Ohio (U.S.A.), 1964.
25. Bright, D. and Truter, M.R.: J. Chem. Soc. (b), 1544 (1970).
26. Steinrauf, L.K., Pinkerton, M., and Chamberlin, J.W.: Biochem. Biophys. Res. Commun. 33, 29 (1968).
27. Shiro, M. and Koyama, H.: J. Chem. Soc. (B) 243 (1970).
28. Metz, B., Moras, D., and Weiss, R.: J. Amer. Chem. Soc. 93, 1806 (1971).

DISCUSSION

Jagur-Grodzinski: In the complex formed by the linear polyether investigated by you, differences in the M^+—O bond distances are observed for oxygens which differ in their basicity (adjacent to aromatic ring and CH_2-groups respectively). Such differences were not observed for the dibenzo-18-crown-6-complexes studied by Dr Truter and her coworkers.

Accordingly, low complexing ability of the dibenzo derivative as

compared to the aliphatic crown compound, was attributed to conformational factors only. Your result seems to indicate that in the macrocyclic system the M—O bond distances are made equal by the specific geometric requirement of the system. Differences in the respective oxygens
basicity may play after all an important role in determining complex
stability!

Truter: In some cyclic polyether complexes we have found apparently
significant differences between M^+—O(aliphatic) and M^+—O(aromatic distances but these are not consistent or persistent from one structure to
another. If the M^+-O(aliphatic) and M^+-O(aromatic) distance
reported for the open chain molecule is found in several structures of related molecules and/or with different cations then the effect
is real and a chemical explanation can be sought. The idea that the distance to the more basic, i.e. aliphatic, oxygen atoms is shorter is most
attractive.

Werber: Does Mn^{++} also form a complex with your non-cyclic ether,
since it has many similarities with Ca^{++} that does bind to it?

Saenger: To my knowledge Mn^{++}-ions were not investigated in connection with this ligand.

A. Pullman: Coming back to the basicity question, I would like to
indicate that it is not necessarily parallel to the affinity for cations,
at least of group IA and IIA cations. For instance we have just termi-
nated the theoretical study of the binding of sodium to the bases of the
nucleic acids and the results for cytosine show that the intrinsic preference of Na^+ is for the oxygen of the carbonyl group whereas the preference for protonation is towards the nitrogen atom.

Duax: In your complex, the helical nature of the ligand appears to
be dependent upon the presence of the complex. In contrast, Gramacidin
A is proposed to have a helical structure in the absence of cations. In
your efforts to synthesize a long chain polyether that will be helical,
are you incorporating features that will stabilize a helix in the absence
of cations? And I would also like to ask if you have crystallyzed the
uncomplexed form of the ligand?

Saenger: Let me answer your second question first. The uncomplexed
tetraethyleneglycol-bis-(8-oxyquinoline)-ether molecule is an oily substance and, thus far, could not be crystallized. Therefore, its structure
is unknown but we assume that it will be different from the regular helical structure found in the complex. To answer your first question: We are
attempting to synthesize longer ligands in order to obtain metal complexes in which the ligand can build up helices with more than one turn. As
we are still at the beginning of the experiments we have no idea whether
the helical structure has to be predetermined in the uncomplexed ligand.
If so, groups with suitable stereochemical constraints could be incorporated in the polyether chain of the ligand.

Izatt: What is the evidence for metal complexation by this ligand
and what stoichiometry is observed?

Saenger: Complex formation of this ligand in solution can be followed by NMR techniques. The protons of the pyridine-rings of the quinoline
moieties are shifted upfield by $\delta = 0.52$ ppm when KSCN is complexed.
Further, IR spectra of the KSCN complex (in KBr) display an additional
band at 2060 cm^{-1} which can be attributed to the $SCN^{\ominus}$ ion. The observed
stoichiometries range from 1:1 to 2(ligands):3(cations) in case of Pr^{3+}
and Nd^{3+} cations.

METAL ION – LIGAND INTERACTIONS FOR THE TETRACYCLINE DERIVATIVE
ANTIBIOTICS: A STRUCTURAL APPROACH

JOHN J. STEZOWSKI
*Institut für Organische Chemie Biochemie und Isotopenforschung
der Universität Stuttgart, Pfaffenwaldring 55, 7000 Stuttgart
80, Federal Republic of Germany*

The tetracycline derivatives (I) display broad spectrum antimicrobial
activity, which may in part result from their ability to form metal
ion complexes (Franklin and Snow, 1973). Identification of the ligand
site(s) utilized in complex formation has been the goal of numerous
investigations employing various physical-chemical techniques to examine
the metal ion-ligand interactions. Complexes have been described in which
the nature of the ligand ranges from the zwitterionic free base, MH_2L,

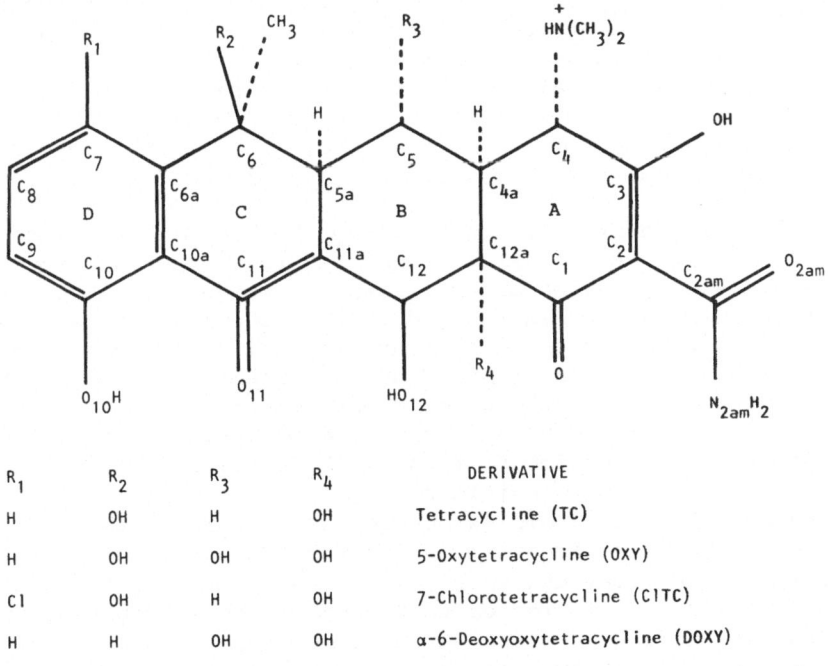

R_1	R_2	R_3	R_4	DERIVATIVE
H	OH	H	OH	Tetracycline (TC)
H	OH	OH	OH	5-Oxytetracycline (OXY)
Cl	OH	H	OH	7-Chlorotetracycline (ClTC)
H	H	OH	OH	α-6-Deoxyoxytetracycline (DOXY)
H	OH	$OCOCH_3$	$OCOCH_3$	5,12a-Diacetyloxytetracycyline (DAOXY)

I

(Jogun and Stezowski, 1976), a monovalent anion, MHL, and a divalent anion, ML, (Silva and Dias, 1972). Examples have also been reported in which there are two ligands per metal ion, ML_2, (Albert and Rees, 1956; Doluisio and Martin, 1963) and in which more than one metal ion interacts with one ligand (Mitscher et al., 1969 and 1972; Jogun and Stezowski, 1976).

There are two chromophoric regions in the tetracycline molecule, the BCD chromophore and the A-ring chromophore. Functional groups of both chromophores have been proposed to be the principal coordination site for metal ion complex formation. The bathochromic shift in the long wavelength UV absorption spectrum resulting from complex formation of the tetracycline derivatives and model compounds led Conover (1956) to propose that complex formation occurs through oxygen atom coordination to the BCD chromophore. This conclusion is supported by Mitscher et al. (1969, 1972) as a result of the analysis of extensive CD spectral measurements for a series of tetracycline derivatives and metal ions. Baker and Brown (1966) have proposed A-ring coordination through oxygen atoms of the tricarbonylmethane system as the likely site. The latter proposal is based on their analysis of reflectance IR spectra and has recently been supported by analysis of extensive pmr measurements of a series of paramagnetic and diamagnetic metal ion complexes with tetracycline in dimethylsulfoxide solutions (Willamson and Everett, 1975) and from ^{13}C nmr measurements in the same solvent (Gulbis and Everett, 1975). Chelation involving both the BCD and A-ring chromophores has also been proposed (Colaizzi et al., 1965; Caswell and Hutchison, 1971). Recent crystallographic analyses in conjunction with renewed examination of the long wavelength UV absorption spectra of oxytetracycline in the presence of various metal ions extend this proposal (Jogun and Stezowski, 1976).

Much of the difficulty in defining the coordination site(s) of the tetracycline derivatives arises from the presence of numerous functional groups in the composition of the ligand. Consequently, the tetracycline derivatives display considerably different chemical properties (and chemical structures) under different chemical conditions which complicates the interpretation of most physical-chemical data. In the course of our investigation of the chemical-structural properties of the tetracycline antibiotics, we have gained some insight into the chemical structure and conformation of these multifunctional organic molecules under various chemical conditions.

One aspect of the complex chemical structure of the tetracycline antibiotics is their amphoteric character. Rigler et al. (1965) have demonstrated that four pKa values may be determined for the tetracycline derivatives and have indicated that they may be attributed, in order of decreasing acidity, to the A-ring tricarbonylmethane moiety, a hydroxyl group of the BCD chromophore, the protonated dimethylamine group, and the second hydroxyl group of the BCD chromophore. The ionization state of the molecule has a strong influence on the chemical structure and character of the chromophoric groups.

Abnormal bond distances in the amide substituent observed in the crystal structure analysis of ClTC-HCl led Donohue et al. (1963) to propose an unusual chemical structure for this moiety in which the amide oxygen atom is protonated. A recent crystal structure determination for

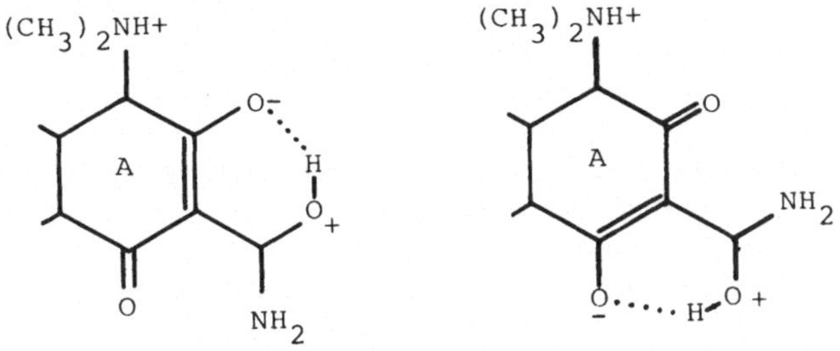

II

DOXY-HCl, in which all hydrogen atoms were located by difference Fourier techniques and confirmed by least-squares refinement, has established the correctness of the proposed chemical structure and furthermore demonstrated the existance of two tautomeric forms of the A-ring chromophore in the fully protonated zwitterionic molecule, II (Stezowski, 1976a). Thus the fully protonated tetracycline derivatives may be characterized as zwitterionic cations.

The dissociation of the first proton in the overall deprotonation scheme of the tetracycline derivatives is generally accepted to be associated with the tricarbonylmethane system of the A-ring, thus giving rise to a zwitterionic structure for the free base. Mitscher et al. (1972) have demonstrated that, in aqueous solution, the free base displays a conformation very similar to that of the fully protonated molecule and have indicated that the conformation is similar to that reported for the crystalline hydrohalide salts of the biologically active derivatives (Donohue et al., 1963; Cid-Dresdner, 1965; Palenik and Mathews, 1972; and Stezowski, 1976a) Crystal structure analyses for TC and OXY free bases crystallized from aqueous-organic solvents have confirmed the similarities in the conformations of the zwitterionic free bases and fully protonated salts and have further demonstrated the existence of tautomeric forms for the A-ring chromophore, III (Stezowski, 1976b).

A second conformation, differing greatly from that of the zwitterionic free base, has been proposed by Schach von Wittenau and Blackwood (1968) for oxytetracycline derivatives and subsequently observed in the crystal structure analysis of DAOXY free base (Von Dreele

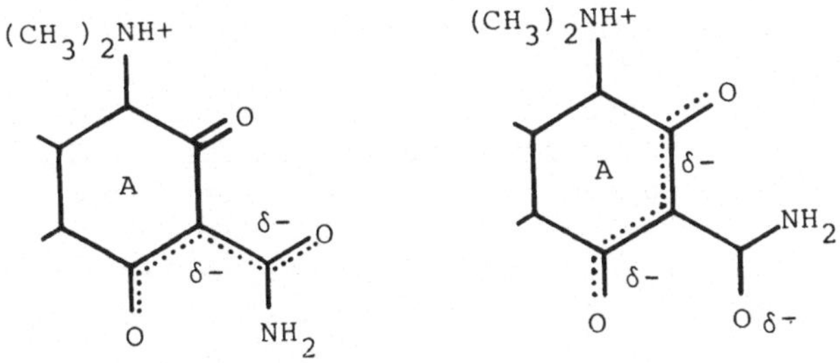

III

and Hughes, 1971). Extensive crystallization experiments in our laboratory with OXY free base employing several aqueous-organic solvent
systems have consistently yielded the crystalline zwitterionic OXY
dihydrate. In contrast, when OXY free base is crystallized from dry
toluene or benzene an anhydrous material is obtained. The crystal
structures of two modifications* of anhydrous OXY have been determined
(Stezowski, 1976b, Prewo and Stezowski, 1976) in which the free base
diplays a nonionized molecular structure in the conformation proposed
by Schach von Wittenau and Blackwood. This conformation has been reported to be limited to tetracycline derivatives possessing the 5-hydroxyl
group** (Mitscher et al., 1972), however, it has been pointed out that
the evidence drawn upon to support this conclusion is not entirely relevant considering the conditions under which this conformation has been
observed (Stezowski, 1976b). It has also been demonstrated, in the latter
report, that for steric reasons this conformation is not accessible to
the 4-*epi*-tetracycline derivatives, a series of compounds which display
only minimal biological activity.

Whether or not the tetracycline derivatives not possessing the 5-
hydroxyl group adopt a similar conformation to that observed for
nonionized OXY, under similar conditions they almost certainly adopt
the chemical structure displayed by the A-ring chromophore; that is a
chemical structure displaying a nonzwitterionic A-ring is adopted by
the free base derivates in nearly anhydrous solutions***. In this structural modification the free base displays a clearly enolic structure in
which the hydroxyl group at C(3) is very strongly hydrogen bonded to
the oxygen atom of the amide carbonyl, IV. While no examples of
tautomerism analogous to I have been reported from crystal structure
analyses of tetracycline derivatives, there is nmr evidence that model

---

*
The two crystal structure modifications observed for nonionized
oxytetracycline free base have been obtained under remarkably similar
conditions by slow evaporation of a warm toluene solution. In each case,
the toluene solution was contained in a small beaker placed on a warm
(*ca.* 70°C) hot plate. The conformation and chemical structure of the
nonionized molecules are nearly identical but there are differences in
the intra- and intermolecular hydrogen bonds.

**
The observation of the Schach von Wittenau and Balckwood conformation
for crystalline DAOXY free base by Von Dreele and Hughes was attributed
to steric effects by Mitscher et al. in the discussion of their analysis
of the conditions under which this conformation is observed. A contrary
effect would be expected to result from steric effects considering the
close nonbonded contact between the 5- and 12a- oxygen atoms: 2.86 A.

***
As reported in Stezowski (1976b), the tetracycline derivatives TC
and ClTC appear to display different crystal symmetry when the free
base is recrystallized from anhydrous toluene. Miyazaki et al, (1975)
also report different modifications for TC, OXY and ClTC free bases
upon recrystallization from aqueous solutions or anhydrous methanol.

IV

compounds for the A-ring chromophore display similar tautomerism* (Dudek
and Volpp, 1965).

The second macroscopic dissociation constant in the overall deprotonation scheme for the tetracycline derivatives is generally accepted
to correspond largely, though not exclusively, to loss of a hydroxyl
proton of the BCD chromophore. Colaizzi and Klink (1969) have reported
the relative concentrations in aqueous solution of TC in various
ionization states as a function of pH. The relative concentration of
the monovalent anion, characterized as zwitterionic, becomes appreciable
in the pH region 6.0 to 7.0. The position of the long wavelength UV
absorption maximum of OXY in 0.01N aqueous KCl begins to display a
systematic bathochromic shift in the same pH region, Figure 1.Thus, the
initial bathochromic shift can be assigned to the formation of the
monovalent anion in which the BCD chromophore is ionized.

The maximum relative concentration of the monovalent anion for TC
occurs at approximately pH 8.7. Colaizzi and Klink report the following
distribution of ionized species at pH 8.5: -0+ (14.4%), --+ (72.1%),
-00 (7.2%) and --0 (6.3%), where each symbol represents the ionization
state of the A-ring chromophore, the BCD chromophore and the dimethylamine
group, respectively (e.g. the symbol -0+ represents the charge distribution of the zwitterionic free base). From the reported distribution
one may conclude that the BCD chromophore has lost a proton and is
thus negatively charged in approximately 78% of the molecular species
at pH 8.5. A similar distribution is to be expected for OXY, which is
a somewhat stronger acid than TC (Stephens et al., 1956), at ca. pH
8.2. As can be seen from an examination of Figure 1, the spectral shift
curve displaying the position of the long wavelength UV absorption
maximum as a function of pH rises steeply between pH 7 and 8. Thus the
long wavelength UV absorption, which has been attributed to the BCD

---

* Dudek and Volpp have discussed their nmr spectra for 2-carbamoyldimedone
in relation to the chemical structure proposed for ClTC by Donohue et al.
(1963). The lack of the dimethylamine function in this model compound,
a substituent which the authors indicate results in greater acidity of
the A-ring chromophore, suggests that interpretation of the nmr data
from the model compound is more relevant to the tautomeric character
of the nonionized free base in which the dimethylamine group is neither
protonated nor hydrogen bonded.

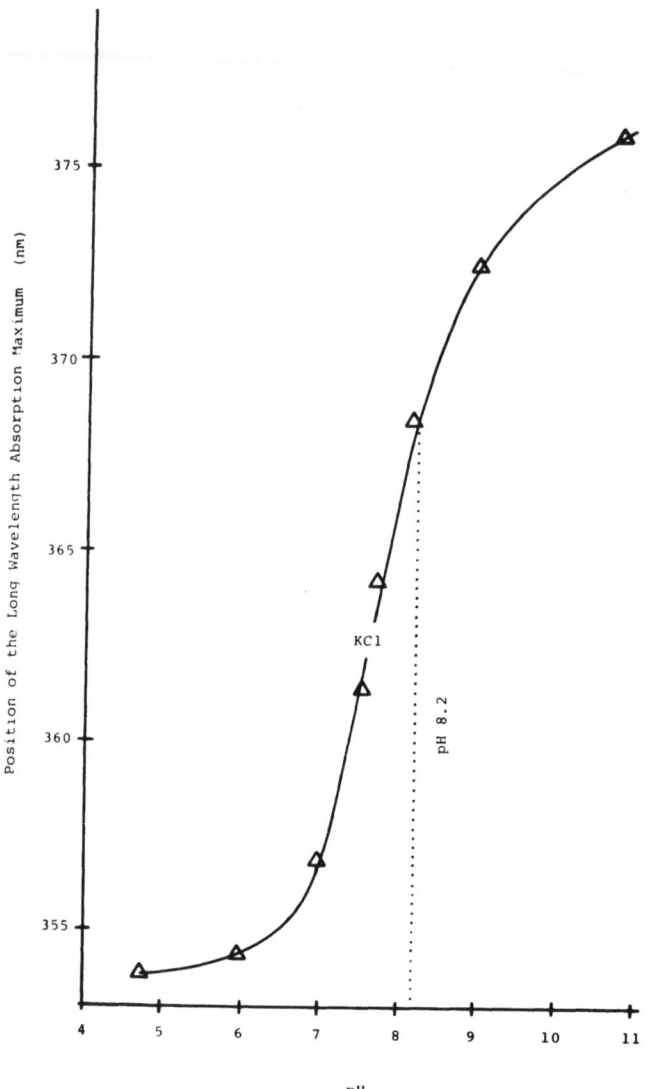

Fig. 1. The position of the long wavelength UV absorption maximum for OXY as a function of pH. All spectra were measured with aqueous 0.1M KCl as an electrolytic solvent. For further details concerning the conditions under which the spectra were measured see Jogun and Stezowski (1976).

chromophore (Conover, 1956) is significantly altered by an increase in the relative concentration of species in which the BCD chromophore is ionized.

Little is known about the conformation of the monovalent anion, either in the simple ionic form, -00, or in the zwitterionic form --+.

Mitscher et al. (1968) have reported that the 262 nm band in the tetracycline CD spectra, which is assigned to the A-ring chromophore, undergoes a striking change as the pH rises above 7. They have attributed the spectral change to the adoption of a conformation in which a hydrogen bond is formed between the 12a-hydroxyl group and the deprotonated dimethylamine group. If, as the authors propose, the dimethylamine group is deprotonated, the species giving rise to the change in the CD spectra is most likely the nonzwitterionic monovalent anion (-00) or at higher pH the divalent anion (--0). Thus virtually no definitive information is presently available concerning the conformation of the zwitterionic monovalent anion (--+), the form which accounts for more than 70% of the TC species present in aqueous solution at pH 8.5. Unfortunately, to date, attempts in our laboratory to crystallize a monovalent anion salt, in either the --+ or -00 form, have not been successful.

The third pKa value for the tetracycline derivatives has been attributed to the dissociation of the proton of the dimethylamine group to give rise to a divalent anion (--0). It appears to have been generally accepted that this is the only form adopted by the divalent anion (Colaizzi and Klink, 1969) and that the dissociation of a second proton from the BCD chromophore may be neglected when considering the acid-base character of the tetracycline derivatives (Rigler, et al., 1965). The pH dependent bathochromic shift in the long wavelength absorption maxima of OXY in 0.1 N aqueous KCl presents evidence contradicting this simplified treatment of the structure of the divalent anion. As can be seen upon examination of Figure 1, the absorption maximum continues, though somewhat less dramatically, to shift to longer wavelengths as the pH is raised above 8.2. It does not seem probable that conversion of the molecular moieties to species with a singly ionized BCD chromophore can account for the magnitude of the spectral shift. It is more consistent with the observed bathochromic shift to assume that a significant fraction of the concentration of the divalent anion is made up of zwitterionic species in which the second hydroxyl group of the BCD chromophore is deprotonated. Such a species may be formulated as -=+, that is a zwitterionic divalent anion in which the BCD chromophore carries a double negative charge and the dimethylamine groups remains protonated. Based on the assumptions that the relative concentrations of the two proposed divalent anionic species resembles that between the monovalent anionic species (Colaizzi and Klink, 1969) in the respective pH regions and that the fourth pKa value determined from the methiodide (Rigler et al., 1965) is not grossly different from that of the normal tetracycline derivatives, one can roughly estimate the relative concentrations of the species in which the BCD chromophore carries a double negative charge, Figure 2. While the relative concentrations involving the -=+ and the +=0 species presented in Figure 2 are no more than casual estimates, the importance of these species in the description of the acid-base character of the tetracycline derivatives is clearly supported by the observed bathochromic shift in the position of the long wavelength UV absorption maximum at higher pH. A more quantative analysis of the UV spectra may help demonstrate the existence of the proposed species.

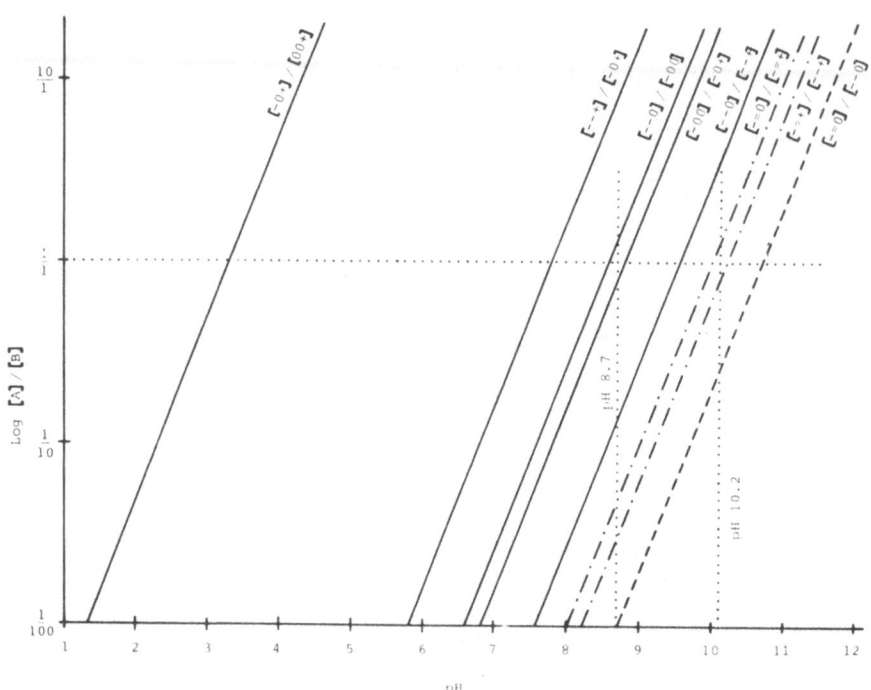

Fig. 2. The relative concentrations of various ionized species
for tetracycline as a function of pH. The (————) lines are
plotted from data presented by Colaizzi and Klink (1969) for
tetracycline. The (- - -) line is constructed with the fourth
pKa value for the overall deprotonation of tetracycline as
derived by Rigler et al. (1965) from tetracycline methiodide.
The (-·-·-) lines represent hypothetical relative concentrations constructed to be analogous to the distribution for
ions with one unit less of negative charge.

The predominant species for the divalent anion, --0, displays a conformation in the crystalline dipotassium salt (Jogun and Stezowski, 1976)
that differs greatly from the conformations observed for the tetracycline
derivatives under less basic conditions, Figure 3. The observed confor-

---

→

Fig. 3. Stereoscopic projections (Johnson, 1965) displaying
the three distinctly different conformations presently demonstrated to exist for OXY under different chemical conditions:
(a) nonionized OXY free base, 000; (b) zwitterionic OXY free
base, -0+; this conformation is nearly identical with that
displayed by the fully protonated derivative and (c) the OXY
divalent anion, --0, drawn without hydrogen atoms. The three
conformations presented are drawn from crystallographic coordinates determined for the respective molecular species. For
additional characterization of these conformations see Stezowski
(1976b) and Jogun and Stezowski (1976).

(a)

(b)

(c)

Fig. 3.

mation was not among those considered by Mitscher et al. (1972) in their
analysis of the CD spectra obtained for OXY in aquous solutions at
higher pH but the similarities with and the differences from* their
proposed conformation for the non-5-hydroxylated tetracycline derivatives under similar conditions are all consistent with the conformational
integrity displayed by the tetracycline derivatives between the solution
and solid state under less basic conditions. Thus it seems probable that
the conformation of the --O divalent anion observed in the potassium
salt is also that adopted in solution. In addition to adopting a different conformation the crystalline --O divalent anion displays a different tautomeric form of the BCD chromophore than that adopted when
the chromophore is not ionized, V. As indicated the tautomeric form

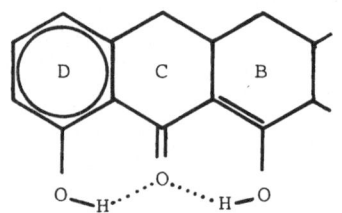 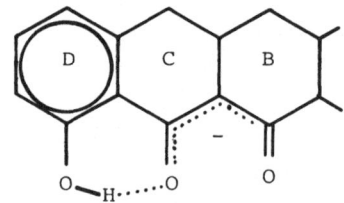

$$\underline{\text{V}}$$

displayed by the nonionized BCD chromophore is stabilized by strong
intramolecular hydrogen bonds. The tautomeric form of the singly ionized
BCD chromophore is particularly interesting in terms of the ligand
character of the tetracycline antibiotics and will be discussed further
in this context.

　　To date the structural aspects of the zwitterionic divalent anion,
-=+, and the trivalent anion, -=O, have not been characterized either
from solution studies or by crystal structure analyses.

　　The rather extensive discussion of the effects of pH on the conformation, ionization state and tautomerism of the tetracycline derivatives is presented to provide a basis for a better understanding of the
interaction of these complex organic molecules with metal ions.

　　Our interpretation of and conclusions concerning metal ion-tetra-
cycline derivative interactions draw heavily on the correlation between
the acid-base character of the ligand in the presence and absence of
metal chelating agents and on the results of crystal structure analyses.

---

*
　Mitscher et al. (1972) indicate that their proposed conformation for
the non-5-hydroxylated tetracycline derivatives is unfavorable for the
5-hydroxylated derivatives because of *peri*-interactions between the
latter group and the 6-methyl group. These relationships between the
observed structure of the OXY divalent anion and the proposed structure
for the non-5-hydroxylated derivatives are discussed in some detail by
Jogun and Stezowski (1976).

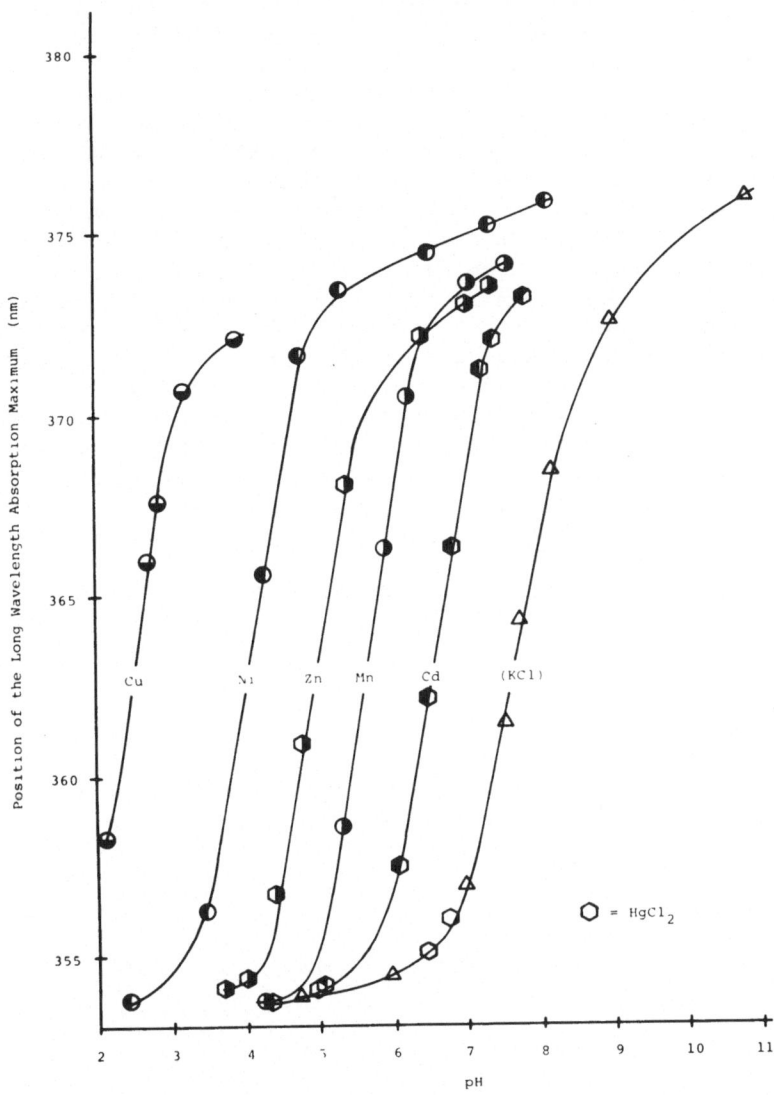

Fig. 4. The position of the long wavelength UV absorption maximum for OXY as a function of pH in the presence and absence of the indicated transition metal divalent cations.

Central to this investigation has been a reexamination of the effect of metal ion chelating agents on the pH dependence of the bathochromic shift in the position of the long wavelength UV absorption maximum initially reported by Conover (1956). The experimental details concerning the determination of the spectra are reported elsewhere (Jogun and Stezowski, 1976).

The effect of transition metal ions on the position of the long

wavelength absorption maximum of OXY is presented graphically in Figure
4. The curves are generally similar in form to that of OXY in the 0.1 N
KCl solution (the KCl solution was used as the electrolytic solvent
for all measurements), however, they are displaced to lower pH. The
displacement has been shown to be correlated (Jogun and Stezowski, 1976)
to the stability constants reported for the MHL and/or ML complexes
(Silva and Dias, 1972). We believe that this correlation clearly demon-
strates that there is a significant metal ion interaction with the BCD
chromophore in the MHL and ML complexes. Collaborative evidence that
such an interaction arises from direct coordination to the BCD chromophore has been developed from a comparison of the pH dependence of
the long wavelength absorption spectra of OXY in the presence of $HgCl_2$
and the crystal structure analysis of the complex of $HgCl_2$ with
zwitterionic oxytetracycline free base, a MH_2L complex (Jogun and
Stezowski, 1976). The crystalline complex displays the $HgCl_2$ moiety
coordinated through the mercury atom and oxygen atoms O(1) and O(2am)
of the OXY zwitterion; thus the complex is formed by coordination to
the A-ring tricarbonylmethane moiety, Figure 5. The long wavelength
absorption spectra of OXY in the presence of $HgCl_2$ has been found to
be atypical of those observed for the other transition metals investigated. No displacement of the curve relative to that for OXY in the KCl

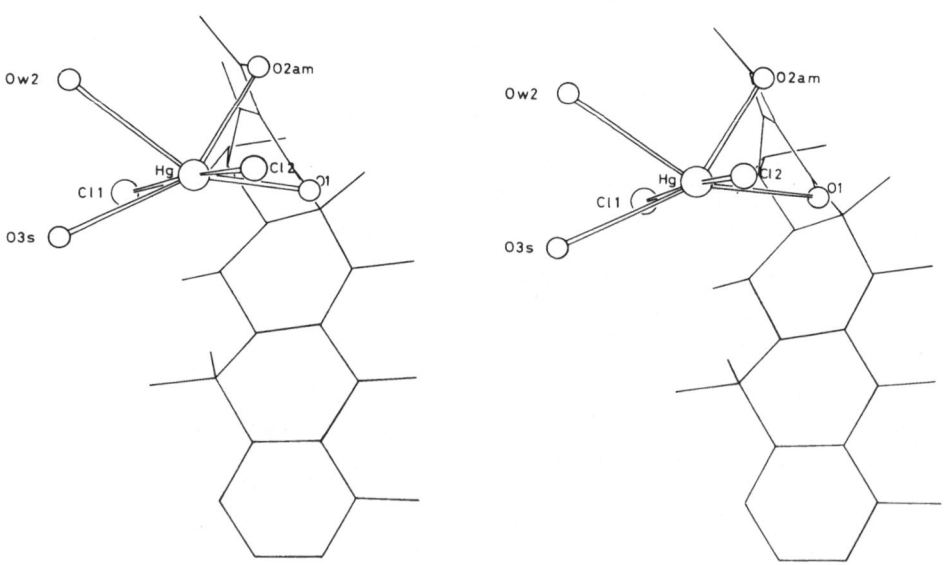

Fig. 5.   A stereoscopic projection of the $HgCl_2$ complex with
zwitterionic OXY. Hydrogen atomic coordinates were not determined in the crystal structure analysis upon which the drawing
is based (Jogun and Stezowski, 1976).

electrolyte was observed in the spectra measured for solutions up to pH 7, Figure 4, (Above pH 7, the spectra became time dependent). Thus it does not appear that the spectral shift observed for the ligand in the presence of metal ions can be accounted for by A-ring coordination alone.

A series of crystalline complexes with OXY and several transition metals, Mn(II), Co(II), Ni(II) and Zn(II), have been obtained from solutions in which the ratio of OXY to metal ion concentration is 2 or greater. Apparently isomorphous crystals, which have space-group symmetry $P6_3$, have been obtained for the above metal ions. The crystals display an unusually large unit cell volume and appear to contain 18 formula units*, $M(OXY)_2$, and an unknown amount of solvent per unit cell. Considering the crystallographic symmetry, the crystals contain three symmetry independent formula units of the 2:1, ligand to metal ratio, complex. The crystalline Co and Ni complexes were initially obtained by dissolving the components (anhydrous OXY free base, $Co(NO_3)_2 \cdot 6H_2O$ or $Ni(NO_3)_2 \cdot 6H_2O$) in methanol and adding dilute aqueous NaOH dropwise until a precipitate formed. The reaction mixture was subsequently filtered and the precipitate dissolved in 50% methanol-dimethylsulfoxide after which the solution was adjusted to 75% methanol content. After some time (several hours to 3 or 4 days) large hexagonal bipyramidal crystals were obtained. The crystalline product was successfully re-crystallized from methanol, acetonitrile or dimethylsulfoxide. The crystalline cobalt complex has also been obtained without the use of dimethylsulfoxide. A redbrown precipitate is obtained upon addition of aqueous cobalt (II) nitrate to a methanol solution of OXY free base. The precipitate can be redissolved by addition of acetone to give a deep red colored solution which yields large hexagonal crystals after several days at reduced temperature (ca. 8°C). The isomorphous Zn and Mn complexes were also obtained from solvent systems consisting of water, methanol and acetone.

A second crystalline modification, which displays space-group symmetry $P2_13$, has been obtained for the Co-OXY complex. Once again there are three fromula units of what we believe to be the $M(OXY)_2$ complex per crystallographic asymmetric unit. This cubic modification was obtained when dioxane was used to redissolve the precipitate obtained upon addition of aqueous cobalt(II) nitrate to OXY dissolved in a 9:1 acetone-methanol solvent system. Apparently isomorphous crystals** have been obtained for the Ni complex under similar conditions.

In an effort to obtain crystals more suitable for a crystal structure analysis, we have also carried out crystallization experiments

---

*   The symbol (OXY) used in the formula of this and other complexes is presented without indication of the valence state of the ligand.

** The crystals of the Ni complex display similar morphology to those obtained for the Co complex and remain black in all orientations when observed with crossed polarizers.

TABLE I
Lattice parameters and volume of the crystallographic
assymmetric unit for Co(OXY) complexes

Space-Group	$P6_3$	$P2_13$	$P2_12_12_1$
a	25.92(1)A	35.77(1)A	38.00(2)A
b	=a	=a	40.31(2)
c	45.27(1)	=a	31.78(2)
V_a	4392 A^3	3809 A^3	12170 A^3

with 4-*epi*-OXY, α-DOXY and ClTC. We have obtained large single crystals
of a Co-4-*epi*-OXY complex which display space-group symmetry $P2_12_12_1$.
Unfortunately the volume of the asymmetric unit is approximately
3 times that observed for the two modifications thus far crystallized
in which the ligand has the normal configuration. The crystalline complexes obtained to date with α-DOXY and ClTC appear to be isomopophous
with the hexagonal modification of $M(OXY)_2$. The isomorphous character
is indicated by the crystal morphology and optical extinction properties
displayed when the crystals are examined with polarized light. The
lattice parameter data for the Co example of the crystals obtained thus
far are presented in Table I. Each crystalline derivative has a rather
high solvent content some of which appears to be only loosely held in
the crystal lattice. The crystals lose solvent rapidly when exposed to
the air and have also been found to interact quickly with most solvent
mixtures in a manner that makes density determination by the usual
flotation technique highly impractical. The crystal quality in terms of
their x-ray diffraction character appears to be reasonably good as long
as the crystals are maintained in an environment in which some mother
liquor is present. A single crystal structure analysis has not yet been
carried out for any of these derivatives though such an analysis is
believed to be feasible for either the hexagonal or cubic modifications.
     The lack of a successful analysis of the crystal and molecular
structure of a typical transition metal complex of the tetracycline
derivatives precludes a definitive assignment of the coordination site(s)
utilized by the ligand or an explanation of the proclivity of the complexes to adopt a packing unit upon crystallization consisting of what
appears to be three formula units of a $M(OXY)_2$ complex. However, it is
possible to drawn some inferences concerning the probable coordination
sites from an examination of the environments displayed by the potassium
ions in the crystalline dipotassium salt (Jogun and Stezowski, 1976).
As a result of crystallographic symmetry, the crystal structure adopted
by the dipotassium salt, while containing one M_2L formula unit per
asymmetric unit, provides an interesting model for the ML_2 complexes.
Two potassium ions lie on a crystallographic two-fold symmetry axis
which allows each of them to interact with two symmetry related divalent
anions. A third potassium ion occupies a general crystallographic position and interacts with three symmetry related divalent anions. The
interactions of the three potassium ions are presented in part in ste-

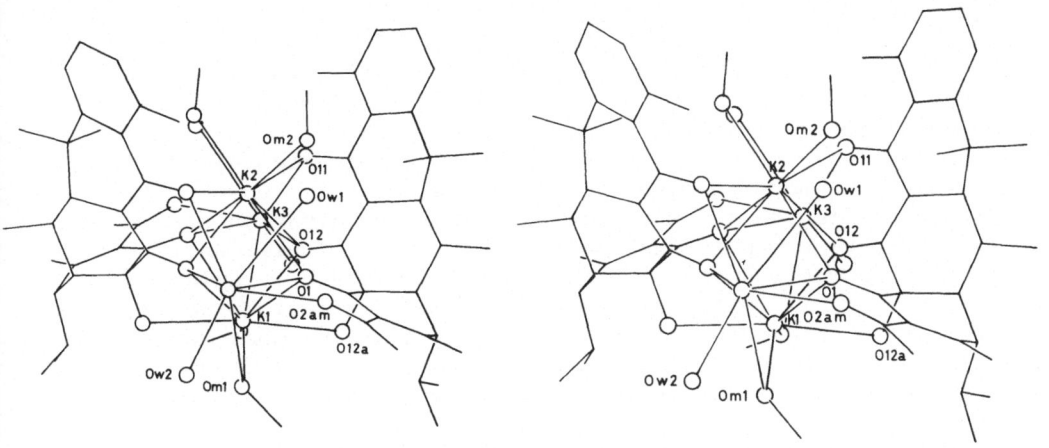

Fig. 6. The potassium ion environments for a crystalline dipotassium salt of OXY (Jogun and Stezowski, 1976) displayed in stereoscopic projection. Ions K(2) and K(3) lie on a crystallographic two-fold symmetry axis relating the two divalent anions shown.

reoscopic projection (Johnson, 1965) in Figure 6.

The two potassium ions on the crystallographic two-fold symmetry axis are very interesting in terms of a model for the coordination in the transition metal complexes. They present intramolecular metal ion-ligand interactions which are consistent with the effects of complex formation on the long wavelength UV absorption spectra of the ligand and also display an interaction with the A-ring chromophore. The intramolecular interaction of both of these cations with ketonate groups of the two chromophores is made possible by the tautomeric form displayed by the ionized BCD chromophore described above. The more interesting metal ion-ligand interactions are displayed by ion K(2) which may be described as being coordinated to atoms O(1), O(12) and O(11). The interaction of ion K(2) with these three oxygen atoms extends the interaction of the metal ion and the BCD chromophore in a manner analogous to that in transition metal β-diketonate complexes. The intramolecular interaction of ion K(3) with the two chromophores occurs through atoms O(1), O(12a) and O(12). This coordination site seems to be less favorable for complex formation due to the interaction of the metal ion with a nonionized hydroxyl group through atom O(12a), a group that is a less suitable electron donor than the formally ionized hydroxyl group at C(11). Potassium ion K(1), the ion in a general crystallographic position, also displays interactions with both chromophoric groups, however, the interactions are of an intermolecular nature. Ion K(1) is coordinated to the A-ring chromophore through atoms O(1) and O(2am) of one molecular anion and to atoms O(11) and O(12) of a symmetry related anion; this cation also interacts with atom O(3) of a third symmetry related molecular anion. Thus all three symmetry independent potassium ions interact with both chromophoric regions of the divalent anion. This makes selection of a specific cation as representative of the metal ion complexes of the tetracycline derivatives premature, however, the interactions displayed by ion K(2) do appear to be the most suitable for the maximum ligand metal ion interaction.

The lack characterization of the zwitterionic monovalent anion, --+, severely hinders the meaningful extension of the observations from the divalent anion to the MHL complexes. The coordination site utilizing the 12a-hydroxyl group is that most likely to be significantly affected since the hydrogen bond between this group and the dimethylamine group, if it exists at all in the zwitterionic monovalent anion, would clearly not utilize the hydrogen atom of the hydroxyl group. Consequently, a different orientation of the hydroxyl group hydrogen atom would result, perhaps making the latter atom available for intermolecular hydrogen bonding.

The complex nature displayed by all the crystalline examples of complexes we have obtained with the first row transition metal ions may indicate that the asymmetric unit consists of complexes in which the ligand utilizes different coordination sites or it may also result from a complex network of intermolecular hydrogen bonds. The latter possibility is supported to some extent by the apparent high, though loosely held, solvent content of the crystals.

We have also investigated the interaction of OXY with alkaline earth cations. To date only very small crystals have been obtained for what is believed to be $Ca(OXY)_2$ and $Ba(OXY)_2$ complexes. While no space-group determinations have been possible by X-ray diffraction techniques, the optical properties and morphology of the crystals are similar to those of the hexagonal modification of the transition metal complexes described above. Thus the $M(OXY)_2$ complexes of the alkaline earth and transition metals may prove to be isomorphous.

The spectral shift curves displaying the position of the long wavelength UV absorption maximum as a function of pH for OXY in the presence of excess alkaline earth ions, Figure 7, indicate that a divergence in the similarity between the alkaline earth complexes and the transition metal complexes formed in the presence of excess metal ion occurs in the pH regions in which the divalent anion concentration becomes important. As can be readily seen from a comparison of Figures 4 and 7, the shift in the absorption maximum is very similar until the pH at which the concentration of the divalent anion becomes significant. Unlike the curves determined in the presence of the transition metal ions, there are two distictly different forms displayed by the spectral shift curves for OXY in the presence of alkaline earth ions. The extreme cases are presented by the spectra measured in the presence of excess Ca and Mg ions. We interpret these differences as arising from complex formation between a second cation and the OXY divalent anion in which the second coordination site differs for Ca and Mg ions and furthermore that the relative distribution between the zwitterionic divalents anion, -=+, and the nonzwitterionic form, --0, is strongly effected by the formation of the respective M_2L complexes. The differences in the coordination behavior have been formulated schematically (Jogun and Stezowski, 1976) and are presented in Scheme I.

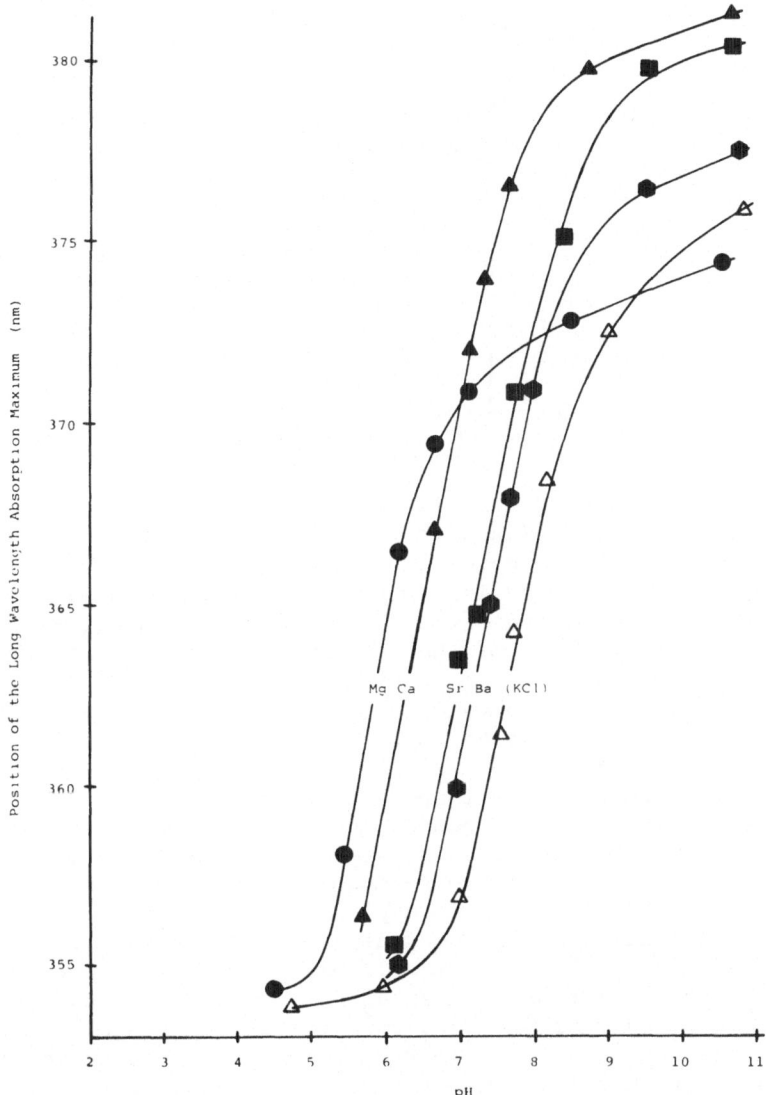

Fig. 7. The pH dependence of the position of the long wavelength UV absorption maximum for OXY in the presence and
absence of excess alkaline earth divalent cations. The Mg and
Ca ion curves cross at pH 7; the pH at which the same wavelength (370.5 nm) is observed as the absorption maximum for
uncomplexed OXY (KCl curve) is 8.5.

The continued rise in the spectral shift curve, at a nearly constant
slope, for OXY in the presence of excess Ca or Sr ions indicates a
continued change in the electronic structure of the BCD chromophore

Scheme I

$$(--+)M^{2+} \quad + \quad M^{2+} \quad \underset{K_O}{\overset{K_N}{\rightleftarrows}} \quad \begin{matrix} (--0)(M^{2+})_2 \quad + \quad H^+ \\ \\ (-=+)(M^{2+})_2 \quad + \quad H^+ \end{matrix}$$

$$K_N = \frac{\left[(--0)(M^{2+})_2\right]\left[H^+\right]}{\left[(--+)M^{2+}\right]\left[M^{2+}\right]}$$

$$K_O = \frac{\left[(-=+)(M^{2+})_2\right]\left[H^+\right]}{\left[(--+)M^{2+}\right]\left[M^{2+}\right]}$$

beyond that observed in the KCl electrolyte. We believe this indicates formation of a $M_2(-=+)$ complex with the second metal ion probably co-ordinated to the BCD chromophore. In contrast, the much more gradual change in the position of the long wavelength absorption maximum in the presence of excess Mg ion indicates the formation of a $Mg_2(--0)$ complex, perhaps through coordination of the second Mg ion to the dimethylamine group. Formation of the latter complex would be expected to favor forma-tion of the nonzwitterionic divalent anion. Thus in the case of excess Ca, Sr and to a lesser extent Ba ions $K_O > K_N$, while in the presence of excess Mg ion $K_N > K_O$.

The observations of Caswell and Hutchison (1971) concerning the conformational differences between Ca and Mg complexes with the tetra-cycline antibiotics in nonaqueous solvents (90% methanol) may also be explainable in terms of the above formulation. The authors have described the complexes investigated as 1:1 complexes, however, the published data indicate that at least some of the measurements were made in the presence of excess metal ion and at sufficiently high pH (7.4) to correspond to the conditions for formation of the $Ca_2(-=+)$ complex in aqueous solution. A consideration of our observations concerning the change in the chemical structure of the free base from a zwitterion, -0+, to a nonionized form, 000, when the solvent system is changed from an aqueous rich medium to a nearly anhydrous one raises the possibility that an analogous mechanism is responsible for the reported conformational change for the Ca complex. The parallel is achieved if the ligand undergoes a solvent sensitive chemical structure change from a zwitterionic form, -=+, to a divalent anion with a nonionized A-ring, 0=0. The latter form of the divalent anion can adopt an enolic structure for the A-ring chromophore similar to that of the nonionized free base. The lack of a similar conformational change for Mg ion is consistent with our proposal that the coordination of a second Mg ion to the ligand favors formation of the --0 divalent anion. The apparent lack of formation of a 0=0 divalent anion in nearly anhydrous solvents when excess Mg ion is present, in combination with our observations in aqueous solutions, may indicate that the second Mg ion coordinates to the dimethylamine group and to an oxygen atom of the

A-ring chromophore. The chemical structure of the A-ring is such that this Mg ion may adopt a coordination site analogous to one of the glycinate rings of the ethylenediaminetetraacetatoaquomagnesate(II) ion (Stezowski et al., 1973). The formation of such a complex, through the coordination of a magnesium ion to the dimethylamine group and the carbonylate group at C(3), would be expected to hinder the formation of the enolate structure of the A-ring chromophore postulated to explain the conformational change observed for the Ca complex*.

The role of metal ion complexation in the antimicrobial activity of the tetracycline antibiotics is not well understood. Complex formation with Mg ion is considered to play a role of some consequence in the mechanism of binding the antibiotic to the ribosome (Franklin and Snow, 1973). From the position of the spectral shift curve of OXY in the presence of excess Mg ion, the formation of an MHL complex seems most likely at physiological pH levels. It is highly probable that the metal ion is bonded to the BCD chromophore and also likely that it is bonded simultaneously to the A-ring chromophore. The ligand coordination site occupied by ion K(2) in the dipotassium salt provides a highly suitable model for such coordination. Since the formation of the hydrogen bond between the hydroxyl group at C(12a) and the dimethylamine group is less likely to occur in the monovalent anion, the effect of complex formation may be viewed as a reduction in the positive charge near the coordination site by one unit and if as suspected, the ligand is zwitterionic, --+, a transfer of one unit of positive charge from the site of the Mg ion to that of the dimethylamine group. The change in charge density and charge distribution resulting from the formation of such a complex provide a plausible explanation for the disruption of protein synthesis caused by the tetracycline antibiotics.

ACKNOWLEDGEMENT

The author wishes to thank the Institut für Organische Chemie, Biochemie und Isotopenforschung der Universität Stuttgart for making this research possible. Special thanks are due K.H. Jogun, R. Prewo, F.E. Leohard and M. Chohan for their many contributions to this effort.

---

* It should be noted that the occurrence of a structural change similar to that observed for the noncomplexed tetracycline free base derivatives and proposed for the Ca_2(OXY) complex is also possible for complexes with the monovalent anion, MHL. The structural change may be formulated as conversion from a M(--+) complex to a M(0-0) complex. Docktor and Magnuson (1973) report a marked increase in the relative fluorescence of ClTC in the presence of Mg ion (in the presence of excess ClTC ligand) when the spectra are measured for a 50% methanol solution. Only a modest increase is noted in comparison with noncomplexed ClTC when the spectra are measured in aqueous solution. These changes in the spectra are in accord with the postulated structural changes for complexes in which a zwitterionic A-ring structure is available to the tetracycline ligand.

REFERENCES

Albert, A. and Rees, C.W.: 1956, Nature 177, 433-434.
Baker Jr., W.A. and Brown, P.M.: 1966, J. Amer. Chem. Soc. 88, 1314-1317.
Caswell, A.H. and Hutchison, J.D.: 1971, Biochem. Biophys. Res. Commum. 43, 625-630.
Cid-Dresdner, H.: 1965, Z. Kristallogr. Kristallgeometrie Kristallphys. Kristallchem. 121, 170-189.
Colaizzi, J.L. and Klink, P.R.: 1969, J. Pharm. Sci. 58, 1184-1189.
Colaizzi, J.L., Knevel, A.M., and Martin, A.N.: 1965, J. Pharm. Sci. 54, 1425-1436.
Conover, L.H.: 1956, Symposium on Antibiotics and Mould Metabolites, Special Publication No 5, the Chemical Society, London, pp. 48-81.
Docktor, M.E. and Magnuson, J.A.: 1973, Biochem. Biophys. Res. Commun. 54, 790-795.
Doluisio, J.T. and Martin, A.N.: 1963, J. Med. Chem. 6, 16-20.
Donohue, J., Dunitz, J.D., Trueblood, K.N., and Webster, M.S.: 1963, J. Amer. Chem. Soc. 85, 851-856.
Dudek, G.O. and Volpp, G.P.: 1965, J. Org. Chem. 30, 50-54.
Franklin, T.J. and Snow, G.A.: 1973, Biochemie Antimikrobieller Wirkstoff, Springer-Verlag, Berlin, pp. 109-112, transl. by Goebel, W. from Biochemistry of Antimicrobial Action, 1971, Chapmann and Hall Ltd., London.
Gulbis, J. and Everett Jr., G.W.: 1975, J. Amer. Chem. Soc. 97, 6248-6249
Jogun, K.H. and Stezowski, J.J.: 1976, J. Amer. Chem. Soc., accepted for publication.
Johnson, C.K.: 1965, OR-TEP, A Fortran Thermal Ellipsoid Plot Program for Crystal Structure Illustrations, ORNL-TM-3794, Oak Ridge National Laboratory, Oak Ridge, Tenn.
Mitscher, L.A., Bonacci, A.C., Slater-Eng, B., Hacker, A.K., and Sokoloski, T.D.: 1969, Antimicrob. Ag. Chemother. 111-115.
Mitscher, L.A., Bonacci, A.C., and Sokoloski, T.D.: 1968, Tetrahedron Lett., 5361-5364.
Mitscher, L.A., Slater-Eng, B., and Sokoloski, T.D.: 1972, Antimicrob. Ag. Chemother. 2, 66-72.
Miyazaki, S., Nakano, M., and Arita, T.: 1975, Chem. Pharm. Bull. 23, 552-558.
Palenik, G.J. and Mathew, M.: 1972, Acta Crystallogr. A28, S47..
Prewo, R. and Stezowski, J.J.: 1976, J. Amer. Chem. Soc. Accepted for publication.
Rigler, N.E., Bag, S.P., Leyden, D.E., Sudmeier, J.L., and Reilley, C.N.: 1965, Anal. Chem., 37, 872-875.
Schach von Wittenau, M., and Blackwood, R.K.: 1966, J. Org. Chem. 31, 613-615.
Silva, J.J.R.F., and Dias, H.M.H.: 1972, Rev. Port. Quim. 14, 159-169.
Stephens, C.R., Murai, K., Brunings, K.J., and Woodward, R.B.: 1956, J. Amer. Chem. Soc. 78, 4155-4158.
Stezowski, J.J.: 1976a, J. Amer. Chem. Soc. Accepted for publication.
Stezowski, J.J.: 1976b, J. Amer. Chem. Soc., accepted for publication.
Stezowski, J.J., Countryman, R., and Hoard, J.L.: 1973, Inorg. Chem. 12, 1749-1754.
Von Dreele, R.B. and Hughes, R.E., 1971, J. Amer. Chem. Soc. 93, 7290-7296.

Williamson, D.E. and Everett Jr., G.W.: 1975, J. Amer. Chem. Soc., <u>97</u>,
   2397-2405.

DISCUSSION

EISENMAN: I am fascinated by your spectral shift curves for the alkaline earth cations because, assuming that the spectral shift reflects
strength of binding, your data show a beautiful set of selectivity
sequences:

VII	Mg > Ca > Sr > Ba
VI	Ca > Mg > Sr > Ba
V	Ca > Sr > Mg > Ba
IV	Ca > Sr > Ba > Mg

which are exactly analogues to the wellknown sequences for monovalent
cations (Eisenman, 1961, 1962) and also directly predicted from a simple
electrostatic model (Eisenman, 1965; Sherry, 1969(?)). Perhaps they could
therefore be interpretable as due to changes in field strength of the
divalent complexing site as a function of H^+ dissociation, although
superficially the effects seem backwards.
   The full set of divalent sequences are as follows (where it will
be seen that Li^+ is analogous to Be^{++}, Na^+ is analogous to Mg^{++}, etc.)

Selectivity order	Divalent pattern	Monovalent pattern
XI	Be > Mg > Ca > Sr > Ba	Li > Na > K > Br > Cs
X	Mg > Be > Ca > Sr > Ba	Na > Li > K > Rb > Cs
IX	Mg > Ca > Be > Sr > Ba	Na > k > Li > Rb > Cs
VIII	Mg > Ca > Sr > Be > Ba	Na > K > Rb > Li > Cs
VII	Mg > Ca > Sr > Ba > Be	Na > K > Rb > Cs > Li
VI	Ca > Mg > Sr > Ba > Be	K > Na > Rb > Cs > Li
V	Ca > Sr > Mg > Ba > Be	K > Rb > Ba > Cs > Li
IV	Ca > Sr > Ba > Mg > Be	K > Rb > Cs > Na > Li
III	Sr > Ca > Ba > Mg > Be	Rb > K > Cs > Na > Li
II	Sr > Ba > Ca > Mg > Be	Rb > Cs > K > Na > Li
I	Ba > Sr > Ca > Mg > Be	Cs > Rb > K > Na > Li

STEZOWSKI: With respect to formation of the MHL complex (1 : 1 complex
with the monovalent anion) there is a correlation between the stability
constants and the pH at which the maximum of the long wavelength absorption spectrum is observed at the appropriate wavelength. Thus selectivity
sequence VII is clearly indicated. In extending the description to
sequences VI, V and IV respectively one must keep in mind that the
spectral shift is associated with the BCD chromophore and that the
spectra were measured for solutions with a large excess of cation. It
seems probable that the changes in relative affinity as a function of
pH are relevant to the complex formation with the second cation to the
site involving the BCD chromophore. Thus the ground state of the
ligand may differ significantly for each of the second cation complexes.

This may contribute to the apparent reversal of the trend (VI – V – IV). It should also be remembered that Mg^{2+} ion is most likely coordinating to a second site not involving the BCD chromophore.

IZATT: (1) Are microspecies involved in proton ionization with this system. (2) Could you give and comment on possible reasons that in the systems you describe Hg prefers O to N and Mg prefers N to O.

STEZOWSKI: (1) There is a complex series of microspecies in the deprotonation scheme for these molecules. The scheme describing the nature of the microspecies has been discussed in part by Colaizzi and Klink. That discussion did not consider the 4th dissociation constant which we feel plays an important role in the acid base behavior of these antibiotics. We also believe that the microspecies associated with the doubly ionized BCD chromophore are important in the metal complexes of the ML (divalent anion) type. In schematic representation we believe the following species are important in aqueous solution: (a) Cation (+O+), (b) free base (-O+), (c) monovalent anion (--+) > (-OO), (d) divalent anion (--O) > (-=+), (e) trivalent anion (-=O). In non aqueous or nearly anhydrous solutions microspecies with an unionized A-ring become important (e.g. OOO for the free base).

(2) The $HgCl_2$ complex is formed with the zwitterionic form of the free base in which the dimethylamine group is protonated. Above pH 7 the UV absorption spectra become time dependent indicating that Hg interacts with the functional groups of the BCD chromophore, probably in an oxidation reduction reaction.

The proposed complex formation of the second Mg^{2+} ion to the dimethylamine group and oxygen atom O_3 of the A-ring chromophore may be favored by ion size. As indicated the ligand is analogous to a glycinate arm of the EDTA ligand. In the oxytetracycline case the conformation of the ligand may be fixed by the formation of the complex with the first Mg ion in a manner favorable to interaction with Mg^{2+}